Analecta Husserliana

The Yearbook of Phenomenological Research

Volume CXXII

Published under the auspices of
the World Phenomenology Institute
Anna-Teresa Tymieniecka, Founder

More information about this series at http://www.springer.com/series/5621

Calley A. Hornbuckle
Jadwiga S. Smith • William S. Smith
Editors

Phenomenology of the Object and Human Positioning

Human, Non-Human and Posthuman

Springer

Editors
Calley A. Hornbuckle
English Department
Dalton State College
Dalton, GA, USA

Jadwiga S. Smith
Co-President, American Division
World Phenomenology Institute
Bridgewater, MA, USA

William S. Smith
Executive President, World Phenomenology Institute
Bridgewater, MA, USA

ISSN 0167-7276 ISSN 2542-8330 (electronic)
Analecta Husserliana
ISBN 978-3-030-66436-7 ISBN 978-3-030-66437-4 (eBook)
https://doi.org/10.1007/978-3-030-66437-4

This Springer imprint is published by the registered company Springer Nature Switzerland AG
The registered company address is: Gewerbestrasse 11, 6330 Cham, Switzerland

Contents

Contributors

Abdul Rahim Afaki University of Karachi, Karachi, Pakistan

Tony E. Afejuku University of Benin, Benin City, Nigeria

John Nelson Balsavich Bridgewater State University, Bridgewater, Massachusetts, USA

Lucian Delescu St. Francis College, Brooklyn, New York, USA

Michel Dion University of Sherbrooke, Sherbrooke, Quebec, Canada

Antonio Domínguez Rey National University of Distance Education, Madrid, Spain

Natasha Kiran Bahauddin Zakariya University, Multan, Pakistan

Natalia Kućma Jagiellonian University, Krakow, Poland

Sergio Labate University of Macerata, Macerata, Italy

Anna Małecka AGH University of Science and Technology, Krakow, Poland

Piotr Mróz Jagiellonian University, Krakow, Poland

Stefano Polenta University of Macerata, Macerata, Italy

Martina Properzi Pontifical Lateran University, Rome, Italy

Victor G. Rivas López Meritorious University of Puebla, Puebla, Mexico

Bruce Ross Independent Scholar, Bangor, Maine, USA

Marie Antonios Sassine Dominican University College, Ottawa, Ontario, Canada

Christopher S. Schreiner University of Guam, Guam, USA

Ion Soteropoulos Apeiron Centre, Paris, France

Katarzyna Stark AGH University of Science and Technology, Krakow, Poland

Marcella Tarozzi Goldsmith Independent Scholar, New York, New York, USA

Aydan Turanli Istanbul Technical University, Istanbul, Turkey

Daniela Verducci University of Macerata, Macerata, Italy

About the Editors

Calley A. Hornbuckle is an Associate Professor of English in the School of Liberal Arts at Dalton State College, Georgia. Her scholarship focuses on British women writers and the environmental tradition.

She has published on Anna Letitia Barbauld's ecological sensibility and presented several papers on ecological intelligence in the works of Charlotte Smith, Ann Radcliffe, Dorothy Wordsworth, and Beatrix Potter. Currently, she is investigating Mary Robinson's investigations of ethics, ethology, and aesthetics in light of recent developments in neuroscience and embedded cognition. She also serves as an executive editor for *The Explicator*.

Dr. Hornbuckle has presented at the World Phenomenology Institute on numerous occasions.

Jadwiga S. Smith is emerita Professor of English at Bridgewater State University where she has worked for 32 years. She earned two master's degrees in European and Slavic literatures from Jagiellonian University in Krakow, Poland, where she developed an early interest in phenomenology. Graduate studies took Dr. Smith to Duquesne University, where she earned a Ph.D. in English literature; she wrote her dissertation on the phenomenology of Roman Ingarden and its application to the stage play. She has continued to publish countless scholarly articles on phenomenology, literary theory, and drama in the *Analecta Husserliana*, the *Phenomenological Inquiry*, and other journals.

In the early 1980s, Dr. Smith began working with Dr. Anna-Teresa Tymieniecka and became a major collaborator in the World Phenomenology Institute. After Dr. Tymieniecka's death, Dr. Smith became the WPI President for the American Division. She continues this important work today, organizing and administering conferences, conducting and leading conferences, and editing volumes of the *Analecta Husserliana: The Yearbook of Phenomenological Research*. Dr. Smith remains a vital, enthusiastic, inspiring link in the World Phenomenology Institute where she encourages younger scholars to pursue phenomenological studies.

William S. Smith is emeritus Professor of English and Dean of the College of Graduate Studies at Bridgewater State University in Bridgewater, Massachusetts, where he worked for 30 plus years. Graduating from Duquesne University with a Ph.D. in English literature, he has spent 40 years in higher education. Dr. Smith's scholarly interests and areas of specialization include nineteenth-century British literature, particularly the poetry of William Wordsworth and the Gerard Manley Hopkins. Of particular interest are the epistemological traditions extending across the poetic landscape of the nineteenth century. His scholarly articles have appeared in the *Phenomenological Inquiry* and the *Analecta Husserliana*.

Dr. Smith's relationship with the World Phenomenology Institute began in the mid-1980s and continues today as the Executive President of the WPI. These administrative responsibilities are many, including managing the WPI staff, organizing conferences, administering funding sources, and editing the *Analecta Husserliana: The Yearbook of Phenomenological Research*.

Part I
Homage to Anna-Teresa Tymieniecka

The Subject/Object Relationship According to the Phenomenology of Life of Anna-Teresa Tymieniecka: Discovering the Metamorphic Logos of the Ontopoiesis of Life

Daniela Verducci

1 Introduction

Anna-Teresa Tymieniecka did not limit herself to extending the field of phenomenological inquiry to the phenomenology of life. Her phenomenology of life opens a new horizon of meaning for facing the serious problems of our times, connected with the dis-communication between subject and object, such as the questions related to environmental sustainability and post-humanism. Practicing the phenomenology of life, A.-T. Tymieniecka discovered both, life as the objective interweaving between human condition and nature, and the auto-individualizing logos of life as the formative and propulsive force, which is intrinsic to becoming in all its forms. This logos is creative and po(i)etic, according to the terminology of Aristotle's *Poetics* (1996), or better, it is onto-poietic, because it gives rise to the creation-of-being, activating the dynamics of self-individualization of being itself at every level. In the perspective of life, the subject-object dualism appears recomposed, as we can phenomenologically show that there is a unique auto-individualizing logos/force, which leads the becoming in every its forms. The logos of life as sentient and metamorphic is different from the modern rationalistic logos, which, on the contrary, is fixed and extrinsic to life, limited to imposing reduced frameworks to cage the infinitely faceted fluidity of living. However, the logos of life is equally capable to weave that web of meaning that "saves" the new phenomena that appear before our eyes from the dispersion and senselessness to which they are condemned by postmodern deconstructionism and nihilism. Effectively, Tymieniecka, focusing the ontopoietic logos of life, was able to exhibit a real—not only logic—poietic continuity between the constructivism of natural life and the creative evolution of human life. But, despite having achieved such a metaphysical maturity of the logos, a fur-

D. Verducci (✉)
University of Macerata, Macerata, Italy
e-mail: daniela.verducci@unimc.it

C. A. Hornbuckle et al. (eds.), *Phenomenology of the Object and Human Positioning*, Analecta Husserliana 122, https://doi.org/10.1007/978-3-030-66437-4_1

ther dualism between nature and super-nature challenged the phenomenology of life. At this point, the metamorphic quality of the logos of life played a decisive role, because it allowed the unique logos to vary its own form according to the vectors that support its enaction, be they the laws of physics, the animal instinct or man's creative imagination and will. It was precisely by assuming the human will as its new vehicle that the ontopoietic logos of life was able to advance from the vital/ontopoietic round of significance into two new dominions of sense: that of the creative/spiritual and that of the sacred. Herefrom, the ontopoiesis of life opened to the transformative advance of the Great Metamorphosis, that completes life's meaning in a transition from temporal life to hyper-temporality, combining every residual dualism. The path of a New Enlightenment opens out before us and the metaphysical enterprise is once again within our grasp.

2 The Dualistic Phase of the Subject-Object Relationship

The subject-object relationship, that reached in the twentieth century, profoundly suffered the dualisms of philosophical Modernity, principally that of Descartes, which divided the ontological field between *res cogitans* and *res extensa*, and that of Kant, which separated phenomenic appearance and noumenal thingness. Over the centuries and into our times, further dichotomies and polarizations have been added, among them those between intuition and concept, faith and knowledge, sciences of nature and sciences of the spirit, theory and practice, nature and culture, base and superstructure, the conscious and the unconscious, consciousness and world, mind and brain, the élite management/owners and categories kept in a state of inferiority/workers, and so on, in an infinite series of oppositions which, according to the French sociologist Alain Touraine, are nothing other than exemplifications of the fundamental, radical and paralyzing fracture between spirit and life that has been produced through the modern metaphysical conception (2006: 66).

Precisely in this sense, A.-T. Tymieniecka addressed the problem of the relationship between subjectivity and objectivity, working to resolve the disconnection that afflicts it. In fact, she entitled "Logos and Life" the editorial series that includes her first set of four works, published with Kluwer of Dordrecht, beginning in 1988 with books 1 and 2, respectively: *Creative experience and the critique of reason* (*Analecta Husserliana* 24) (Tymieniecka, 1988) and *The Three movement of the soul* (*Analecta Husserliana* 25); then with book 3 in 1990, *The passions of the soul and the elements in ontopoiesis of culture* (*Analecta Husserliana* 28), and closing with the description of the meta-ontopoiesis of life in the monumental book 4 of 2000, *Impetus and Equipoise in the Life-Strategies of Reason* (*Analecta Husserliana* 70) (Tymieniecka, 2000).

Tymieniecka lucidly perceived that the pulsing heart of every correlation appears inaccessible to us: we feel hermeneutically confined and immobilized by the bronze chains of a metaphysics that does not want to deal with "the abundance and variety that our present state of human experience reveals, to say nothing of the expanding perspectives on our horizons" (Tymieniecka, 2009: xxvi) as would be needed to face the problems related to the environment, development, and post-humanism. Even though Max Scheler amply demonstrated how the phenomenon of *Wechsel* (= exchange/interrelation) pervades not only the sphere of the humans and the living being, but also the entire cosmos (Scheler, 1954: 475–476), A.-T. Tymieniecka observes that "the most concretely felt concern emergent at the present, and this is universally so, is with 'communication'" (Tymieniecka, 2005: xxv).

It seemed that Edmund Husserl's phenomenological reflection with the dual devices of *epochè* and intuition of essences overcame the dualistic format of Modern philosophy (Landgrebe, 1961–1962: 133–177) and, with the discovery of the consciousness' constituting intentionality, reached a unified vision of the subject-object relationship, such that in his 32nd lecture in the winter semester of 1923–1924, now in *Erste Philosophie II* (Husserl, 1959: 40), he could use as the starting point for the attainment of the suitable and apodictic evidence of knowledge the statement: "I am—the world is" (Tymieniecka, 2005: xix).

Throughout his life of thought, Husserl always strove to reach the goal of the absolute beginning of philosophizing, but never attained its completion. He understood where the various historical attempts had run aground while seeking an ultimate philosophical clarification of the foundations of our knowledge and experience: from Socrates and Plato onwards, philosophers ended up in skepticism and subjectivism because they misunderstood the nature of cognitive subjectivity, reducing it to the psychological, naturalistic or historicist sense. According to Husserl, only phenomenological reduction can open us to the authentic transcendental subjectivity in which objective reason is constituted.

The Italian phenomenologist Vincenzo Costa delineates with great expressive lucidity the features of this transcendental subjectivity where objective reason is constituted. It is the place of manifestness in general, the place where things manifest themselves, where reality itself becomes a phenomenon and announces itself (Costa, 2007: xii), but, as A.-T. Tymieniecka notes, this transcendental subjectivity "doesn't meet its intrinsic postulates and calls for a further step of interrogation" (Tymieniecka, 2005: xxxiii).

In her recent book, *The Sense of the Things*, Angela Ales Bello, longtime friend of A.-T. Tymieniecka, does not seem to be aware of the metaphysical-foundational question that arises precisely from the constituent stabilization of the phenomenological ontology in which subject and object, consciousness and world, spirit and life are so functionally connected and at the same time imprisoned.

It is true that Husserl's passage in the 1920s from the static phenomenological method, focused on the analysis of pure consciousness, to the genetic one (Husserl, 1973a: 336–346; 1973b: 613–633), "through which consciousness reveals the basic internal temporality of the constitutive moment" (Tymieniecka, 2005: xxxii), opens transcendental consciousness to the pre-categorical sphere, on the *hylectic* basis of passive syntheses, to the world-of-life, showing, as Angela Ales Bello concludes, how transcendental idealism can be converted into a kind of transcendental realism (Ales Bello, 2015: 97–106).

However, there remains unanswered the crucial theoretical question which, as formulated by Max Scheler, asks: How must the foundation of the world be in order for the present phenomenon of this reciprocal conversion of idealism and realism to be founded? Without finding an answer to this question, the phenomenon highlighted "hangs in the air" (Tymieniecka, 2005: xxxiii); as it is not supported by any network of apodicticity, therefore it still needs to be saved from fluctuation and dispersion.

And effectively, as Anna-Teresa Tymieniecka observed in the inaugural speech of the 2005 Conference in Oxford dedicated to the *Logos of Phenomenology and Phenomenology of Logos*, precisely at this point, "when Husserl arrived to investigate the incipient phase of the constituent logos, after having traversed the entire circuit of it and approached it from the eidetic summit" (xxxii), he saw the explosion of "the drama" of the constituting intentionality of consciousness, which functioned as "a bastion of the human mind, playing an exclusive dominating role in the constitution of reality" (xxxiii). The intentionality of consciousness, having revolutionized not only philosophy in the twentieth century, but also all domains of scholarship, now shows its limitations. It is incapable of carrying on the conclusive test outlined by Husserl in the early stage of working out his project that would yield the crowning achievement of his quest for the certain and necessary foundation of all knowledge, the "phenomenology of phenomenology" (xxxiii).

Certainly, once the world is understood as *Lebenswelt*, that is "a domain of life," everything seems to change, because then also the human subject that "constitutes" the manifestation of reality shows that it is rooted in life. As A.-T. Tymieniecka notes, Husserl retained that at this point it was possible to reach the hyletic level of the human subject (Husserl, 1973a: 336–346, 2012: 174–178; Ales Bello, 2005), where consciousness in its primal state before the constitutive process enters into play. However, Elisabeth Ströker already underlined that not even in its most elementary state, constituent consciousness can avoid "rendering" objectuality (1972: 245–257). Tymieniecka added that the hyletic dimension itself, even if pre-constitution, cannot be considered outside the constituent powers of consciousness, as something preceding in the absolute sense this active being of consciousness itself. Tymieniecka acknowledges that it is true that Husserl, with the long itinerary of genetic phenomenology, reached the level of sensation and set it in continuity with that of pure intellectual consciousness, but the passage from the generative level of empirics to that of pure abstract forms remains enigmatic because, even though the objective world is constituted by both forms and degrees, when Husserl asked himself about the sources of the pure forms of consciousness he looked exclusively to their transcendental origin, continuing to omit the fact that "the genetic progress of empiria surges from and obviously stems from origins in nature that are physiological," inasmuch as they emerge from within environmental conditions and together with them (Tymieniecka, 2011: 4).

Consequently, once again we see the dualism between subject and object in Husserl and Fink's discussion of the *Sixth Cartesian Meditation*, after their strenuous search for the point of reference from which we may proceed from the lifeworld and accomplish the last founding reduction (Verducci, 2010: 26–29). In the conferences of Prague and Vienna, Husserl lamented the persistent and unsolved dualism expressed by the "paradox of human subjectivity: being a subject for the world and at the same time being an object in the world"—as reported at §53 of *The Crisis of European Sciences and the Transcendental Phenomenology* (1970: 178), which collects the texts of those conferences (Tymieniecka, 2005: xv).

At this point, even the most devoted follower of Husserl might experience some discomfort. It would seem that the curse of Wilhelm Wundt was once again at work here, damning phenomenological research to spin on its own axis in a spiral of sterile self-referentiality (1910: 511–634).

3 Within Ontological Dualism and No Way Out

Husserl could not eliminate the traces of Cartesian dualism. When he came to compare ideas with Descartes in his French homeland, he sought not only to honor him by engaging again in the theoretical form of meditation, but also to radicalize Descartes subjectivistic turn in order to verify whether from the sphere of the thinking substance through phenomenology, he could bridge the transcendental connection with intersubjectivity and extended substance (Cristin, 1994: viii–ix).

Husserl experienced something similar to what Hans Jonas attributes to Descartes. Descartes worked "to drain the spiritual elements off the physical realm" (Jonas, 2001: 13) in order to reinforce the supremacy of the *res cogitans*, by separating it from the *res extensa*. But so doing he also reduced the success chances of the longed-for *mathesis universalis* and deprived himself of the theoretical instrument to restore the lost metaphysical unity between subject and object since, according to Jonas, the just emerged polarity of *res cogitans* and *res extensa* could be overcome only by making it "absorb into a higher unity of existence from which the opposites issue as faces of its being or phases of its becoming" (17). In Descartes, continues Jonas, even if it is interpretable as a transition phase, in the pursuit of a scientifically rigorous ontological reunification, such a procedure obtained the paradoxical effect of marginalizing precisely the living being; this form of being was, instead, crucial for Descartes' and Husserl's objective of the *mathesis universalis*, inasmuch as it is the carrier, at various levels, of the actuality of coexistence and synergy of consciousness and world: it is the living being that "makes methodological *epochè* founder on its rock" (18).

Thus it seemed that the pulsating heart of the life of being itself and its spontaneously communicative generative capability became definitively unattainable. And yet, the Ancients were able to grasp it, as Hans Jonas taught (7). Philippe Hadot (1995) also in *Qu'est-ce que la philosophie antique?* highlighted well how much the whole ancient philosophy "ethically" understood itself as "way of life" and "living action."

But Modernity damaged the Ancient certainty of the living communicative unity among the different levels and forms of being, and since its late Medieval beginning, with the advent of the theory of double truth—which divided truth into two levels, theoretical and practical, that do not communicate with each other (as noticed by Sergio Landucci in *La doppia verità* (2006))—it deprived the communicating life flow of being in human experience of its natural spontaneity. The so-called dis-communicative ontologies of Modernity have been criticized—apart from some significant exceptions such as Leibniz—for their tendency to inhabit the sphere of abstract speculation and thus deem illusionary all experiences that did not possess pre-established transcendental standards. After all, metaphysics—the teaching of which went on until the *Schulphilosophie* of the German Enlightenment (Wundt, 1992) was transferred to Modernity not as living and evermore updating research of the principle of all things, but rather in the "dead" and dis-communicative way of a discipline objectivized in the manuals produced by the Second-Scholastic for a didactical purpose (Courtine, 1990).

Too late was it noted that the surprising scientific and technological progress that began then and continues today, could have developed only because it relied on the mysteriously given but efficient communicative spontaneity of being that Ancient and Medieval traditional philosophy was able to understand and safeguard in its historical flow, in a way that even the rationalistically Modern reductionism could profit from it.

Only in the late Enlightenment, Kant was able to acknowledge in his *Critique of Pure Reason* that the synthetizing process "to unify the manifold of concepts by means of ideas" for a *focus imaginarius* by reason (*Vernunft*) (2007: 533), intrinsically depended on a natural, spontaneous and "irresistible" (*unhaufhaltsam*) force/disposition of reason itself to metaphysics, which he called *metaphysica naturalis* (56).

In this way it became evident that with their own hands Modern people deprived themselves of the enjoyment of spontaneously edifying flow of vital/communicative being which effortlessly kept together subject and object: surrendering to their compulsive enthusiasm for "rebuilding" the whole natural being according to what they presumed to be the true logos, Modern people criticized tradition, but their rationalization of life procured even more sectorialization. In fact, in their analyses of being, they applied only the logic of "instrumental rationality" (*Zweckrationalität*), as Max Weber called it in *Economy and Society* (1978: 24), a strategic or purposive rationality; it is teleological only in a subjective sense (*telos* in an objective sense = *Ziel*) and thus it was voluntaristically and untruthfully used to analyze being. As a result, subject-object dis-communication, growing fragmentation and dis-synergy were produced everywhere: knowledge and actions became increasingly incapable of grasping life, which was perceived as uncontrollable, or, to use a Zygmut Baumann's expression, "liquid" (2005). This led to anthropological "disenchantment" (*Entzauberung*) (Weber, 2004: 13), decadence and nihilism of meaning, as Friedrich Nietzsche first foresaw for the Europeans of the twentieth and twenty-first centuries (1968: Preface § 2) and as Jürgen Habermas recently pointed out and stigmatized in relation to the uncontrolled power of Niklas Luhmann's systemic theory of society and culture (1995: 22).

The anthropological identity had been so invaded by this Modern polarization of subject and object, spirit and life, that the same emergence of human free cultural capabilities was considered a mark of dis-communication. For example, Ernst

Cassirer affirmed that man is an *animal symbolicum*: man is endowed with active freedom to create his own socio-cultural world through relationships, communication, interaction and division of work, thus he is capable of going beyond the boundaries of biological needs and practical interests of animal life. Yet, according to Cassirer's *Das Symbolproblem* (1995), exactly because of this cultural capability, man continuously experiences a struggling separation from "natural life" and a loss of "immanence in reality" (2005: 185–186). Moreover, he fears being defeated by the self-destructive tendencies of spirit, a critical force of self-negation that causes man to constantly fight against himself in opposition to life. According to Cassirer, out of the distancing and emancipating process of symbolization emerges, the "tragic mark of the entire cultural development," influenced by the psychological experience in which the movement of the "I" seems to break "on its own creations" and its original vital flow seems to exhaust itself in the measure in which the extent and power of this creation grow (172).

4 Communicative Virtues of the Phenomenology of Life

4.1 Life as the Objective Interweaving Between Human Condition and Nature

On the contrary, the phenomenology of life found out that the function of logoic creation and ciphering or symbolization does not belong to the spirit of man in spite of his organic dimension and in dis-communication with it; rather, the creative human condition appears at a certain stage of cosmic evolution and crowns the deterministic autopoietic constructivism of life with the ontopoiesis which "generates being" according to the freedom of the *Imaginatio Creatrix* but, at the same time, maintains close cosmic immanence (Tymieniecka, 1988: 28). As Tymieniecka unveils surprisingly:

> the elementary condition of man—the same one through which Husserl and Ingarden attempted in vain to open a breach, extending the expansion of its intentional nexuses and at the same time turning to ante-reduction scientific data—appears to be constituted by the blind element of nature, and yet at the same time this element shows itself to have virtualities for individualization on the vital level and, what is more important, for a specifically human individualization. The latter virtualities we could call 'subliminal spontaneities' (28).

But how could A.-T. Tymieniecka arrive at an intuition of such depth, bringing to light the objective interweaving that holds together the elementary human condition and the blind element of nature?

It was the result of a radical phenomenological reconnaissance, set in motion by Tymieniecka's accepting the challenge, Husserl himself had launched with his "late breakthrough to the plane of nature-life," in order to free himself from "the early theoretic-methodological restrictions that the focus on intentionality [had] imposed on him" (Tymieniecka, 2002: 685b–686a).

According to the typical attitude of feminine thought, oriented primarily to what is alive, as Max Scheler and Edith Stein revealed, Tymieniecka approached phenomenology as an organic phenomenon in vital expansion: it had reached and touched her with its generative/propulsive energy, involving her empathetically in its productive logos. Therefore Tymieniecka spared herself the "effort to interpret phenomenology through its method." On the contrary, taking up the challenge launched to transcendental phenomenology by Ortega y Gasset, who held that not dead ideas but "life was the theme of our times" (Tymieniecka, 1994: 444, 1997b: ix), she set out on the search for the source of the vitality that had animated Husserl's inquiry, soon coming to the realization that Husserl, in his complex and fruitful reflective proceeding, didn't keep to the logic of the "speculative thinker who [extrinsically] seeks to unify his various insights"; rather, to presiding at the succession of phases of the "integral Husserl" (Tymieniecka, 2002: 2b) is the same logos that is at work in the formation of "the planes of human reality" (3a), the living and temporally constructive logos that "carries on the great streaming edifice of life" (Tymieniecka, 2000: 4).

Initiating such novel research, Tymieniecka yielded to the philosophical pressures from, so to speak, the "neophiliac" and progressive atmosphere (Caillé, 1997: 13th Thesis) of Pragmatism and Analytical Philosophy that predominated at the American headquarters of the World Phenomenology Institute, for many years located in Belmont, Cambridge, Massachusetts and later moved to Hanover, New Hampshire. In fact, critical questions about the practical value of classic phenomenology compared to contemporary science had been raised by Stephen C. Pepper and Alfred Tarski in the 1960s when Tymieniecka, newly arrived in the States, was reading Husserl's *Logical Investigations* at the University of California. Thus Tymieniecka herself recalled in a 2008 interview on the occasion of receiving an honorary doctoral degree in philosophy from the University of Bergen in Norway (Torjussen et al., 2008: 25–26).

For this reason, Tymieniecka committed her first studies to achieving the vital-natural rooting for the transcendental consciousness; she overturned the Husserlian assertion of corporeality as the zero-point (*Nullpunkt*) (Husserl, 1989: §41) of the transcendental constitution of the world, focusing on the "own-living-corporeal-conscious" experience (*Leiblich-bewusste*), in which consciousness itself "is corporealized" (*Verleibung*), exhibiting a further and more originary "talent and disposition" (*Uranlage*) (Tymieniecka, 1971: 4–7) that enabled it to enter an intimate relationship with the entire structure of nature (*Naturgefüge*) (2–3). In addition, she relied upon the results of the most recent organic-dynamic psychiatry of Henri Ey (1963) according to which the transcendental consciousness is not closed in its absoluteness, but engages a fruitful relationship with the formative spontaneity (*bildende Spontanität*) of normal and pathological living conscious experience that develops beyond the system of constitution, for example, in the typical elementary formations of the collective imagination or in states of dreaming or mental confusion (Tymieniecka, 1971: 4–7).

Precisely consequent to this openness and contextualization of consciousness and reason in the vaster sphere of investigation that is the "universe of human existence in the unity of all," within "the unity of everything there is alive" (Tymieniecka, 1986: 9, 1988: 6), Tymieniecka found herself immersed in a completely new phenomenic situation: that of "givenness of life," as the "vast sweep of the significant modalities entering into and interplaying in the vertiginous outburst of unfolding forces in the ongoing gigantic play of the manifestation of beingness," with "all the fragments, sequences, segments of complete constructive processes subject to disruption by unforeseeable conditions and influences" (Tymieniecka, 2000: 3). No exposition following the traditional rules of organization may convey this flood-like concatenation of processes that is life! (5). This "streams in all directions and will at any point refract its modalities and their apparatus into innumerable rays that flow concurrently onward" and therefore requires the engagement of "all modes of human functioning, [not only the rational-intentional ones but] all human involvement in the orbit of life" (3).

4.2 *Gaining with the Creative Context of Life Through the Creative Act of the Human Being*

To access the givenness of life, Tymieniecka sought a new "Archimedean point" from where everything is able to find its proper place (Tymieniecka, 1988: 4) and she found it in the creative act of the human being (7).

The phenomenology of life, as realized up to this point, had opened a "vision" unlike the transcendental-structural outlook of classic phenomenology, which begins with the human being as demiurge. Instead, in the phenomenology of life, man is both "caught up in the turmoil of a generating progress" (Tymieniecka, 1986: 10) and at the same time the bearer of the "creative orchestration of human functioning" (Tymieniecka, 1988: 384).

Tymieniecka described the new philosophical position reached in terms of geometry: the plane of the human condition intersects perpendicularly that of the phenomenology of life, since it is by proceeding in the description of the vital flow, understood as evolutive genesis of ever more complex and individualized forms, that one reaches the phase in which the human condition is configured. The human condition is characterized by the presence of original "creative virtualities," that, starting from themselves, that is, horizontally on their own plane, open the vast kingdom constituted by being's propensity to undertake projects (Tymieniecka, 1986: vii–viii).

Therefore, it is the phenomenological description of the natural vital flow that shows that "all the modes of life's forces, forms, energetic complexes [...], synergetic virtualities" reach, in constant auto-individualizing flow, the stage of the human condition, and here are structured according to a new meaningfulness, in the measure to which they pass through the "intergenerative schemas" of that "processor" that is the human person. In other words: the natural vital virtualities, once they have flowed into the level of the individual living human being, continue their development in a sphere where "the [all-embracing] heart of the meaningfulness of life in general" beats (Tymieniecka, 1988: 384); herefrom the natural virtualities are issued as "ciphered" according to the new structure of the meaning of being, always renewing themselves throughout history (Tymieniecka, 2000: 197–205).

Thus, on the one hand, man-source of classic phenomenology, is re-situated within the broader ontological field of inquiry of life and therefore arises from the tangle of vital networks as the "vortex of the universal sense," in which the various orders of those networks meet (Tymieniecka, 1986: 11), according to a metaphysical *topos* common to various personalist thinkers, e.g. Maine de Biran, Renouvier, Blondel, Mounier, and Scheler. On the other hand, the vital virtualities of nature, that reach the human level of life and run through it, are "processed" in such a way that on the flow of "constructive advance" of life or autopoiesis, in the wording of the two Chilean neurophysiologists Umberto Maturana and Francisco Varela (1980), the "composition" of a more elevated "style" imposes itself (Tymieniecka, 1988: 384) and becomes an ontopoietic one, which is a style in conformity to the being as "firstness" of the process of creative existential formation that belongs only to the living human being (Tymieniecka, 1996: 15).

On the basis of these descriptions of phenomenology of life, Tymieniecka can observe that:

> instead of the field of the life-world assumed as the ultimate ground, even by Merleau-Ponty, yet seen as the expansion of the constitutive, objectifying consciousness which is being restricted to the intellectual surface of life, we gain with the "creative context," a full-fledged field of philosophical inquiry into Nature, life, its specifically human meaningfulness and the sense of human orbit, in which human functioning is not cut off at source points but stretches in all directions and into all dimensions—is not dwarfed as some or other modality, rational or sensuous, but comprises them all. (Tymieniecka, 1988: 10)

The subject-object dualism thus appears recomposed, at least as a phenomenon, since Tymieniecka has been able to exhibit a real poietic continuity between the constructivism of natural life and the creative evolution of human life.

It was at this point that the ontopoiesis of life as new philosophical paradigm was announced, precisely in the Round Table at the beginning of the Program of The World Phenomenology Institute at the XXth World Congress of Philosophy, in Boston, in August 1998.

Tymieniecka developed from that speech a long and detailed essay, written as a dialogue with such exponents of the so-called "New Science," as René Thom, Ilya Prigogine, Isabelle Stengers, Alvin Toffler, Stephen Jay Gould, Benoit Mandelbrot (Tymieniecka, 1998: 12–58). She wanted to transmit the idea that "the notion of the ontopoiesis of life, situated within the philosophy of life, its center and its nerve, [...] promises to fulfill the expectation of a new intellectual paradigm"; in fact, it is able to provide the "Prometean worldview that we seek because we are increasingly aware of the possibilities our powers give us for so transforming nature that we will not be subject to its vicissitude" (15).

5 The Unveiling of the Ontopoietic Logos of Life in Its Metaphysical Importance

Tymieniecka did not stop at this phenomenic level of inquiry. An analogous connection between the transcendental dimension and the life-world had been reached by the same genetic phenomenology of Husserl, as has been seen, but it was unsatisfactory from the point of view of ultimate foundation of phenomenology. Husserl himself believed that "'the dream' of the apodictically certain cognitive ground is *ausgeträumt* (=dreamed)" (Tymieniecka, 2005: xxxiv).

In fact, it is not sufficient to inventory facts or speculate about reasons, because "at its core the project of phenomenology is to reach reality in a way that neither subsumes it within general concepts nor reduces it to elements. It is one attempt to make reality foundational and thought immediate, the better to focus and raise sites, to see reality in the round" (Tymieniecka, 2002: 1b).

Taking her lead from the creative phase of phenomenology, which she herself had initiated, and following the creative act of the human being in its constructive thread, Tymieniecka descended to the plane from which the constructive design of self-individualization in beingness takes off. This is the plane of the logos of life, of the constructive logos that carries the entirety of the givenness discovered on the track of impetus and equipoise; it harnesses the universal becoming into the genesis of self-individualizing beingness as it both participates in the universal flux of life within the world, constituting it, and simultaneously makes it present to itself in innumerable perspectives. Here at the ontopoietic level the logos of phenomenological interrogation as logos of life, losing nothing of its postulated cognitive rigor, does not need any further clarification: it reposes in itself as the ultimate that is absolute because in need of no further "reduction," being the yield of the very last reduction (Tymieniecka, 2005: xxxviii).

This logoic-metaphysical maturation of the phenomenology of life was doubtless also due to the constant and deepened exchanges of A.-T. Tymieniecka and of the World Phenomenology Institute with the more than 1000 years philosophical tradition that has flowed into Italian culture, starting with Parmenides of Elea, passing through Saint Thomas Aquinas, to the more recent forms of *Italian Theory*, as it has been termed by Dario Gentili (2012). This contact provided the phenomenological inquiry of the World Phenomenology Institute with a philosophical anchorage in the tradition of Lovejoy's (1964) "The Great Chain of Being" strongly implicated by the "Italian Difference" (Chiesa & Toscano, 2009) that today is rightly re-evaluated in many quarters precisely as "living thought," in the words of Roberto Esposito (2012), as it is the current bearer of the fruit of manifold and complex speculative sedimentations, which elsewhere, instead, were swept away.

But at this point a metaphysical problem arises and Tymieniecka asks herself: "Does this logos stop with the timing of life?" (Tymieniecka, 2007b: 21). Will "the driving force of the logos" that unfolds life in its complete self-individualizing dynamic, be able to conduct it from "the incipient instance of originating life in its self-individualizing process" all the way to "the subsequent striving toward the abyss of the spirit"?

The question is crucial because if the ontopoietic logos of life were qualified to found only the natural dimension and could not bring itself beyond the temporal and chronological constructivism of nature, in the kairic dimension of the freedom of spirit to supernatural eternity, its non-dualistic metaphysical value would be in vain. For this reason Tymieniecka turned her attention once again to the phenomenology of life and the integral ontopoietic dynamic that "the very creative act of the human being itself" reveals in the degree to which it "brings to the sense-giving apparatus of living being the specifically human virtualities that fashion sensorial, emotional, even pre-experiential material into human constitutive-conscious life-significance" (Tymieniecka, 2005: xxxvi).

She thus focuses attention on the fact that the auto-individualizing logos that is intrinsic to life has manifested itself as "a primogenital force striving without end, surging in its impetus and seeking equipoise": it promotes "the constructive prompting" that determines "the progress of life" and "it prepares its own means/organs for its own advance." This advance means the fulfillment of constructive steps toward transformations, that is: "step by step unfolding projects of progressive conversion of constructive forces into new knots of sense." Therefore, "the crucial factum of life" has not appeared without reason, "brought […] out of 'nowhere'"; on the contrary, the "logoic force of life has its purpose"—just like Schelling's living nature, that embodies the "scheme of freedom"—and that purpose reveals itself to be achieved in an ontopoietic way inasmuch as it expresses itself "in preparing scrupulously in a long progression the constructive route of individualizing life so that *Imaginatio Creatrix* emerges as an autonomous modality of force with its own motor, the human will." Crowning its development, "the force of the logos of life," with the will as new modality of force, finds itself able to advance "from the vital/ontopoietic round of significance into two new dominions of sense": that of the creative/spiritual and that of the sacred.

In the terms of traditional ontology, this means that "'substances' undergo a 'transubstantial' change" and also that "the inner modality of the logoic force undergoes an essential transmutation." Therefore, "life, […] as a manifestation of the ontopoietic process […] is far from a wild Heraclitean flux, for it articulates itself." In addition and first of all, "[life] 'times' itself" (Tymieniecka, 2007b: 20), because time reveals itself as "the main artery through which life's pulsating propensities flow, articulating themselves, intergenerating" (Tymieniecka, 1997b: 4).

Therefore the ontopoietic logos of life is characterized in an entirely original way compared to any other previously theorized: in fact, it is metamorphic, inasmuch as, while maintaining constant its being and its essential auto-individualizing function, it varies its own form to remain immanent to the vectors in which it expresses itself each time, for example the laws of physics, animal instinct, or creative imagination. For this reason, the ontopoietic logos of life is endowed with a sentient nature; it is through the sentience of the logos which permeates all the functional moves of the unfolding life and in which the constructive designs are processed, that a continuity is maintained throughout (Tymieniecka, 2007b: 31).

In such a metamorphic sentient capacity that intrinsically qualifies the ontopoietic logos of life, there is the possibility for "the new metaphysical panorama" (Tymieniecka, 2009: xxv), through which we can transcend "the timeless pattern of surrender to nature" and going beyond "the equipoise established through millennia of life between nature and human beings and between the gifts of nature and their use by living beings" (Tymieniecka, 2000: 99) also establishing new nexuses between time as Chronos and Kairos (Tymieniecka, 1997b: 4). The fulcrum of this metamorphosis is that "unique phase of evolutive transmutation" in which the "mature" phase of the platform of life manifests an extraordinary character and gives rise to the Human-Condition within the unity-of-everything-there-is-alive (Tymieniecka, 2007b: 31). The essential differentiation of the human condition amid the unity of life is "a watershed event, essentially a transformation of the significance of life": it brings with itself the "enigmatic" surging of *Imaginatio Creatrix* in the middle of ontopoietic sequence, surging freely as it floats above the inner working of nature (31).

Here we reach—observes Tymieniecka—the most surprising turn of logos of life, because this great shift was being prepared by the logos' constructive steps, starting at the very beginning of self-individualizing of life, but it produces a "countervailing move," that "brings about a complete conversion of its hold on life's individualization and opens the entire horizon of freedom" (32).

Imaginatio Creatrix, rooted within the functioning of Nature-life and yet an autonomous sense giver, introduces three new sense giving factors: the intellective sense, the aesthetic sense, and the moral sense. The moral sense lies at the core of the metamorphosis of the life situation from vital existence into the advent of Human Condition (33): "Indeed"—Tymieniecka exclaims—"through the moral and entirely freely chosen work of the conscience, the self-enclosed ontopoietic course may be undone and remolded in a free redeeming course!" (60).

The logos of life has lead us to a borderline place between the ontopoietic logos of life and logos' sacral turn toward territory that is beyond the reach of the logos of the vital individualization of beingness (60). It is here that the Great Metamorphosis takes place: "Ontopoiesis carries its own necessities and opens to the transformative advance of the Great Metamorphosis that completes life's meaning in a transition from temporal life to a-temporality, or better, hyper-temporality" (67).

All told, concludes A.-T. Tymieniecka, engaging the vast problem of the "evolving God," "what in the cognitive-intellective perspective of human mind appears [...] to be the greatest human "folly," an absurdity and something impossible for sober reason to accept, is a revelation by the logos of life/sacral logos within our now completed human experience of nothing less than the reason of all reasons" (70).

6 Conclusion

Through this *excursus* upon the thought of A.-T. Tymieniecka, in order to highlight the passages that enabled her to re-unite not only the subject-object dualism, but also many of the disconnections that we have inherited from Modernity, we have encountered a "first philosophy" directed to the objective of "theorizing" the overall phenomenon of the new "fullness of the Logos in the key of Life." What has thrown itself wide open before us is a path of theoretical research that we did not believe existed, on which instead we can adventurously embark, renewing the instance of Kant's Enlightenment: "dare to know!" We now catch sight of a unitary logos leading us, that animates the Parmenidean sphere and the same absolute Hegelian Spirit and that, self-individualizing through ontopoiesis, shows it is intrinsically able to connect phenomena emerging bit by bit from the inorganic to the organic, to the human. It weaves a "metapoietic" network of innumerable metamorphic passages of transcendence, that open it in the direction of the divine, newly risen to sight, according to the perspective of *philosophia perennis,* delineated by G. W. Leibniz, when, to rationally understand the truth of the propositions of fact, he introduced the principle of sufficient reason, which, while establishing a foundational dynamics tending toward the infinite, made it possible to construct a solid ladder of truth in order to always better rise to the fullness of the logos, A.-T. Tymieniecka explains in *The Case of God in the New Enlightenment.*

With the phenomenology of life of A.-T. Tymieniecka we are not dealing with a complete and concluded philosophical system, in which the new problems about the environment, development and post-humanism can already find adequate and stable collocation. Do not forget, on the other hand, that the same metaphysical tradition spent many centuries for structuring and stabilizing.

However, Tymieniecka is not unaware that we are in front of an unprecedented expansion of human life, that asks to be faced with an equally new and broadened horizon of meaning. In her talk at the International Conference on "The Development in Question. The Human Forms of Transformation," held in Falconara Marittima, Italy, in November 2006, organized by Professor Francesco Totaro of the Department of Philosophy and Human Sciences of the University of Macerata, A.-T. Tymieniecka asserted that we are in front of an "evolution of the human individual who appears to be already endowed with consciousness greatly enlarged compared to humans of one hundred years ago and appears also to have found the secret of further, seemingly infinite, expansion of human potential" (Tymieniecka, 2007a: 15).

Such expansion—she continued—manifests itself not only within the individual but in the transformative progress of the entire life's and world's network. Each day we are faced with new inventions and turns of mind. We presume an infinite progress ahead. This progress creates new demands upon the individual and society as well as creating new problems which society will have to solve. Ever new sources of force emerge and the human being believes to be able—as master—to foresee and control their effects to lead their course. Yet, like the sorcerer's apprentice, having found the key to release the current of power, human beings do not possess either the key to stop the course nor entirely control its achievements. They remain always subject to the whims of natural, cosmic and human forces. This course involves not only individual natural endowment and inclinations but the entire network of sharing-in-life within the circumambient and further circles. It depends upon ontopoietic fitting directions, ontopoietic rules of the circumambient contexts, on the one hand, and the individual creative genius, on the other.

It is beyond the scope of this paper to address the question: "how to master the routes of the human development within the individual as well as within its interactive world, society, culture while navigating upon the stormy sea between and among conflicting forces without a compass." Our presentation intends merely to take up the ontopoietic groundwork of A.-T. Tymieniecka's phenomenology of life and to draw from it some essential indications. That is by Tymieniecka's words:

in order to control in some way the flux of human development for its existential advantage, human being has to assume a special frame of mind. Keeping in sight the ontopoietic groundwork sketched above, human calculation and balancing out of life's conditions should be handled according to it with measure, proportion and temperance. (Tymieniecka, 2007a: 15–16)

References

Ales Bello, A. (2005). Husserl interprete di Kant. *Dialegesthai. Rivista Telematica, 7*. Retrieved February 5, 2017, from http://mondodomani.org/dialegesthai/aab02.htm.

Ales Bello, A. (2015). *The sense of things: Toward a phenomenological realism* (A. Calcagno, Trans.). In *Analecta Husserliana, 118*, 1–23.

Aristotle. (1996). *Poetics* (M. Heath, Trans.). New York: Penguin.

Baumann, Z. (2005). *Liquid life*. Cambridge, UK: Polity Press.

Caillé, A. (1997). 30 Théses pour une Gauche Nouvelle et Universalisable. *Revue du Mauss, 9*, 297–331.

Cassirer, E. (1995). Das Symbolproblem als Grundproblem der philosophischen Anthropologie. In J. M. Krois (Ed.), *Zur Metaphysik der symbolischen Formen. Nachgelassene Manuskripte und Texte* (Vol. 1, pp. 32–109). Hamburg: Meiner.

Cassirer, E. (2005). 'Geist' und 'Leben' in der Philosophie der Gegenwart. In B. Recki (Ed.), *Gesammelte Werke Hamburger Aufgabe* (Vol. 17, pp. 185–206). Hamburg: Meiner.

Chiesa, L., & Toscano, A. (Eds.). (2009). *The Italian difference: Between nihilism and biopolitics*. Melbourne: Re.press.

Costa, V. (2007). Introduzione: Filosofia Prima e Fenomenologia Trascendentale. In E. Husserl, *Filosofia Prima. Teoria della Riduzione Fenomenologica* (A. Staiti, Trans.) (pp. xi–xlviii). Soveria Mannelli: Rubbettino.

Courtine, J.-F. (1990). *Suarez et le Système de la Métaphysique*. Paris: PUF.

Cristin, R. (1994). Presentazione. In E. Husserl, *Meditazioni Cartesiane* (F. Costa, Trans.) (pp. vii–xvi). Milano: Bompiani.

Esposito, R. (2012). *Living thought: The origins and actuality of Italian philosophy* (Z. Hanafi, Trans.). Princeton: Princeton University Press.

Ey, H. (1963). *La Conscience*. Paris: PUF.

Gentili, D. (2012). *Italian theory. Dall'Operaismo alla biopolitica*. Bologna: Il Mulino.

Habermas, J. (1995). *Postmethaphysical thinking: Philosophical essays* (W. M. Hohengarten, Trans.). Cambridge UK: Polity Press.

Hadot, P. (1995). *Qu'est-ce que la Philosophie Antique?* Paris: Gallimard.

Husserl, E. (1959). Erste Philosophie (1923/24). Zweiter Teil: Theorie der Phänomenologische Reduktion. In R. Boehm (Ed.), *Husserliana* (Vol. 8). The Hague, Netherlands: Martinus Nijhoff.

Husserl, E. (1970). *The crisis of European sciences and the transcendental phenomenology* (D. Carr, Trans.). Evanston, IL: University of Northwestern Press.

Husserl, E. (1973a). Analysen zur passiven Synthesis. Aus Vorlesungs- und Forschungmanuskipten (1918–1926). In M. Fleischer (Ed.), *Husserliana* (Vol. 11, pp. 336–346). The Hague, Netherlands: Martinus Nijhoff.

Husserl, E. (1973b). Statische und genetische Phänomenologie. In I. Kern (Ed.), *Zur Phänomenologie der Intersubjektivität, Dritter Teil (1929–1935)* (Husserliana) (Vol. 15, pp. 613–633). The Hague, Netherlands: Martinus Nijhoff.

Husserl, E. (1989). *Ideas pertaining to a pure phenomenology and to a phenomenological philosophy. Book II. Studies in the phenomenology of constitution* (R. Rojcewicz & A. Schuwer, Trans.). Dordrecht: Kluwer Academic.

Husserl, E. (2012). *Ideas: General introduction to pure phenomenology* (D. Moran, Trans.). Abingdon UK: Routledge.
Jonas, H. (2001). *The phenomenon of life. Toward a philosophical biology*. Evanston, IL: University of Northwestern Press.
Kant, I. (2007). *Critique of pure reason* (N. Kemp Smith, Trans.). New York: Palgrave Macmillan. Retrieved February 5, 2017, from http://www.phil.pku.edu.cn/resguide/Kant/CPR/contents.html.
Landgrebe, L. (1961–1962). Husserls Abschied vom Cartesianismus. *Philosophische Rundschau, 9*, 133–177.
Landucci, S. (2006). *La Doppia Verità. Conflitti tra Ragione e Fede nel Medioevo e nella Prima Modernità*. Milano: Feltrinelli.
Lovejoy, A. O. (1964). *The great chain of being: A study of the history of an idea*. Cambridge, MA: Harvard University Press.
Maturana, H. R., & Varela, F. J. (1980). *Autopoiesis and cognition: The realization of the living*. Dordrecht: Reidel Publishing Company.
Nietzsche, F. (1968). *Will to power* (W. Kaufmann & R. John Hollingdale, Trans.). New York: Vintage Books Edition.
Scheler, M. (1954). *Der Formalismus in der Ethik und die materiale Wertethik. Eds.* Maria Scheler and Manfred Frings. *Gesammelte Werke 2*. Bern – Munich: Francke.
Ströker, E. (1972). Edmund Husserl's phenomenology as foundation of natural science. *Analecta Husserliana, 2*, 245–257.
Torjussen, L. P., et al. (2008). An interview with Anna-Teresa Tymieniecka. *Phenomenological Inquiry: A Review for Philosophical Ideas and Trends, 32*, 25–34.
Touraine, A. (2006). *Le monde des Femmes*. Paris: Fayard.
Tymieniecka, A.-T. (1971). Die Phänomenologische Selbstbesinnung *Analecta Husserliana, 1*, 1–10.
Tymieniecka, A.-T. (1972). Phenomenology reflects upon itself. *Analecta Husserliana, 2*, 3–17.
Tymieniecka, A.-T. (1986). *Tractatus Brevis*: First principles of the metaphysics of life charting the human condition. *Analecta Husserliana, 21*, vii–ix, 3–73.
Tymieniecka, A.-T. (1988). Creative experience and the critique of reason. In *Logos and life, Book 1*: (Analecta Husserliana) (Vol. 24). Dordrecht: Kluwer Academic.
Tymieniecka, A.-T. (1994). La condizione umana all'interno dell'unità-di-tutto-ciò-che-è-vivente. In M. S. Sorondo (Ed.) *Perì Psyché. De Homine, Antropologia, Nuovi Approcci* (A. Ales Bello, Trans.) 441–456. Roma: Herder Università Lateranense.
Tymieniecka, A.-T. (1996). The golden measure: Self-individualization of life bringing to fruition the ideal for a new epoch. *Analecta Husserliana, 49*, 3–25.
Tymieniecka, A.-T. (1997b). Phenomenology of life (integral and "scientific") as the starting point of philosophy. *Analecta Husserliana, 50*, ix–xiii.
Tymieniecka, A.-T. (1998). The ontopoiesis of life as new philosophical paradigm. *Phenomenological Inquiry, 22*, 12–59.
Tymieniecka, A.-T. (2000). Impetus and equipoise in the life-strategies of reason. In. *Logos and life, Book 4*, (Analecta Husserliana) (Vol. 70). Dordrecht: Kluwer Academic.
Tymieniecka, A.-T. (2002). A note on Edmund Husserl's late breakthrough. In A.-T. Tymieniecka (Ed.), *Phenomenology world-wide. Foundations, expanding dynamics, life-engagements. A guide for research and study*, (pp. 685–687). Dordrecht: Kluwer Academic.
Tymieniecka, A.-T. (2005). The logos of phenomenology and phenomenology of the logos. *Analecta Husserliana, 88*, xiii–xl.
Tymieniecka, A.-T. (2007a). Human development between imaginative freedom and vital constraints. *Phenomenological Inquiry: A Review for Philosophical Ideas and Trends, 31*, 7–16.
Tymieniecka, A.-T. (2007b). The great metamorphosis of the logos of life. In A.-T. Tymieniecka (Ed.), *Timing and temporality in Islamic philosophy and phenomenology of life* (Islamic philosophy and occidental phenomenology in dialogue) (Vol. 3, pp. 11–71). Dordrecht: Springer.

Tymieniecka, A.-T. (2009). The case of God in the new enlightenment. In *The fullness of the logos in the key of life, Book 1* (Analecta Husserliana) (Vol. 100). Dordrecht: Springer.

Tymieniecka, A.-T. (2011). Transcendentalism overturned. Life's geo-cosmic positioning of beingness. *Analecta Husserliana, 108*, 3–10.

Verducci, D. (2010). The development of the living seed of intentionality from E. Husserl and E. Fink to A.-T. Tymieniecka's ontopoiesis of life. *Analecta Husserliana, 105*, 19–37.

Weber, M. (1978). *Economy and society. An outline of interpretive sociology* (G. Roth & C. Wittich, Trans.). Berkeley, CA: University of California Press.

Weber, M. (2004). *Science as vocation* (R. Livingstone, Trans.) (The vocation lectures, pp. 1–31). Indianapolis: Hackett Publishing Company.

Wundt, M. (1992). *Die Deutsche Schulphilosophie im Zeitalter der Aufklärung*. Hildesheim: Georg Olms Verlag.

Wundt, W. (1910). *Psychologismus und Logicismus* (Kleine Schriften) (Vol. 1, pp. 511–634). Leipzig: Engelmann.

Otherness and the Case of the Animal-Human: Interrogating the Post-human from Tymieniecka's Ontopoiesis Multilayered Organization of Life

Sergio Labate

As with every bottomless gaze, as with the eyes of the other, the gaze called animal offers to my sight the abyssal limit of the human: the inhuman or the ahuman, the ends of man, that is to say the bordercrossing from which vantage man dares to announce himself to himself, thereby calling himself by the name that he believes he gives himself
(Derrida, 2009: 49).

1 Introduction

Anna-Teresa Tymieniecka introduces an immanentist critics to Levinas's positions in a famous article published in 1977 in the closing volume of *Analecta Husserliana*, which was opened with a text by Emmanuel Levinas. I have chosen this thread as the main theme of my article (Tymieniecka, 1977: 151).

It is useful to remember that this intervention—originally held in Levinas's presence in 1975—is almost contemporary with the publication of Levinas's most challenging book, *Autrement qu'être ou au-delà de l'essence* (Levinas, 1974). In his opening speech at the 1975 conference, Levinas concisely develops the theses published in this book.

The phenomenological core, which Levinas highlights in his speech, is the recognition—well outlined in the "wrong" interpretation of Husserl's phenomenology in his early works—that the real meaning of phenomenology takes place within a dimension which cannot be understood by the immanence of consciousness, but it breaks with the essence of knowledge itself (with the essence itself *as* knowledge).

In Mikel Dufrenne's initial quotation "... *l'immediat, ce qui se dit dans le poème/ en deçà d'une censure ou d'un contrôle./(Restera à savoir si tout contrôle est/exclu,*

S. Labate (✉)
University of Macerata, Macerata, Italy
e-mail: sergio.labate@unimc.it

C. A. Hornbuckle et al. (eds.), *Phenomenology of the Object and Human Positioning*, Analecta Husserliana 122, https://doi.org/10.1007/978-3-030-66437-4_2

et ce qui distingue alors le sens/du non-sens)" (Levinas, 1977: 3), Levinas announces the fundamental theme of his interpretation of phenomenology: *the immediate*. It could be almost said that this immediacy—announced by Husserl's phenomenology as it was sensational—was finally a sort of *unfulfilled promise* by the German philosopher.

The immediate, i.e. what firstly transgresses the "thèse doxique du logos" (and, therefore, as it comes before it, it is not really a transgression, but a law prior to any beginning) ends up in Husserl's thesis by being brought back to order (and, in this way, it becomes a philosophical transgression at most, not a new beginning). Levinas writes:

> La place faite par la phénoménologie husserlienne à l'intentionalité non-représentative, à une rationalité pré-predicative ou se passant de prédication, promettait une signifiance, c'est-à-dire une sagesse qui ne procède pas du savoir. Mais comme pour attester le caractère insolite d'une solution de continuité et pour retrouver aussitôt le contexte perdu de l'Occident, la promesse n'a pas été tenue jusqu'au bout. L' "intention" dite affective ou pratique—le plaisir ou le désir—se montre prise de position en tant qu'intention précisément (*Ideen* 241), et la thèse de toute position recélerait une "thèse doxique", une croyance exprimable en proposition prédicative. De sorte que Husserl enseignait, en fin de compte, que "tout acte ou tout corrélat d'acte, eveloppe en soi un facteur logique implicite ou explicite." (*Ideen* 244) (Levinas, 1977: 3)

Husserl's "unfulfilled promise" would be linked to an idea of wisdom which cannot be identified with knowledge and is moved by an intentional emotional or practical order: it has to do with "pleasure or desire," as Levinas writes. Even Tymieniecka's attempt tries to recover Husserl's unfulfilled promise, starting from the increasingly urgent reference to the life-world (*Lebenswelt*). This reference would finally allow us to escape "to the sovereign rule of intentionality" and to its "misleading focus: human consciousness" (Tymieniecka, 1988: 6).

Is it, therefore, sufficient to acknowledge that Levinas's proposal, as well as the one by Tymieniecka, may be simply recalled as new original elaborations for this phenomenology of the immediate, attempts to compensate phenomenology of Husserl's "unfulfilled promise"? It's not so easy. In fact, the philosophical notion of immediacy is controversial. In fact, it can acquire a negative, neutral, positive meaning—depending on how it is used by each author. And there is no doubt that the role of "immediacy" is very different in Levinas or Tymieniecka's thoughts.

What we could say is that the theme of immediacy becomes a fundamental way of distancing oneself and elaborating the same phenomenological source again (which is explicitly Husserl of the *Krisis* for Tymieniecka, while it is more marginal for Levinas, although this reference is explained on these specific pages).

Indeed, Tymieniecka's immanent critics against Levinas—the starting point of our paper—enters at this level of the matter. In fact, as we know, Levinas gives a significant epistemic twist to immediacy. In fact, his critics is not limited to emphasise the constant "revenge of the theoretical logos" in Husserl's proposal, but it also avoids the risk of an exclusively deconstructive phenomenology of the immediate (the last pages of his intervention are specifically devoted to this defence) through a tormented breakage of emotional immediacy (Levinas, 2009: 264) by the transcendence of the *Autrui*. This is the point. Which is indeed the vital background where to place the emotional breakage which from pleasure ends to responsibility? This background is certainly the universe of life. But it is the world of *human* life. Even the transcendence of the Others only takes place if there is this fundamental reduction: ethics is a philosophy which is first of all *human*. According to Levinas, even the evocations of animality are undergone to this regime of humanity reduction: the dog, which is the protagonist of many famous pages (Levinas, 1976: 213–216), is evoked as a symbolic transfert which is very meaningful as it preserves the meaning of humanity lost by human beings. Then, it would deal with evoking the name of transcendence without passing through this shortcut of reduction to anthropocentrism. Therefore, I think that Tymieniecka's immanent critics may be precisely read in this way:

> but with this understanding of man as advancing himself within his life-processo by unfolding from within a network of meaningful synthesis, do we not penetrate to the radical borderline between—not as Levinas would say—"the same and the other", that is, man's "sameness", his fundamental identity shaping itself over against the circumambient circumstances, and these circumstances appearing in this processo as the "otherness"? (Tymieniecka, 1977: 151)

We can fully understand the meaning of Tymieniecka's critics, on the one hand, and the reason why I refer to such criticism as "immanent," on the other hand.

First of all, the overall limit of Levinas's phenomenology is what is here effectively defined as "the man's samenesses." Thus, the transcendence of the *Autrui* allows us to escape from the capture of monotony by the Same (the same's sameness), but it remains stifled within the coils of an anthropological reduction, which is unable to account for the otherness within the "life-process."

Thus, it could be said that the transcendence of the Autrui is not given excepting as the preliminary condition of a radical immanence where even ethics as a philosophy must be firstly placed. Circumstances are showed as fundamental *Levels of otherness* in this radical immanence. If I am not mistaken, this is precisely the fundamental value of the proposal of the phenomenology of life (Verducci, 2007: 20): replacing the other man's transcendence within a more complex phenomenological scene and a more irreducible summary of what remains at the end of a purely anthropological reduction.

To be even clearer: the philosophical force of Tymieniecka's proposal is not simply to save immediacy from anthropocentrism—thus correcting Levinas's mistake—through the recourse to Lebenswelt. But it is rather to do it through a positive and "immanent" phenomenological description. The question is no longer if there is a relationship between the immediate and the conscience, but rather to instruct "a radical investigation of the universe of human existence in the unity of all" (Tymieniecka, 1988: 6). The primacy of constitution is thus traced back to the one of an "experiential evidence that is capable of yielding a rationale rooted with certainty in necessary ties within the web of all beingness" (Tymieniecka, 2004: XX).

This initial comparison between Levinas and Tymieniecka helps to contextualise the following notes. They represent a first approximation—as the enunciation of two theses and a final summary—for a fruitful relationship between the phenomenology of life and the *Posthuman Studies*.

My intent is not to disown or to minimise or even to deny the irreconcilable differences between these two philosophical approaches. They belong not only to different fields of study, but they even respond to different questions, different methods, different objectives. In order to immediately end up the comparative question, we could say that the critics to anthropocentrism is an opportunity to delegitimize the function of the human being for *Posthuman Studies*, while the whole complex organization of life, its spheres and its levels has as a condition the recognition of the human being as a code of access to life itself, more precisely as "vortex" or "knot" in Tymieniecka's phenomenology (Tymieniecka, 1986: 10)—as we will be soon able to examine. Thus, the critics to anthropocentrism becomes an opportunity to enhance the function of the human being in the ontopoietic organization of life at the same time.

Always keeping all this in mind, I think it is useful to recognize that if there is a point of essential convergence, it is precisely defined by the attempt to assert a certain work of immanence against "the man's sameness." So, this point of convergence becomes a cultural threshold, which probably defines not only a limit of Levinas's philosophy, but also a wide limit in the inheritance from the philosophy of the twentieth century. In short, it offers an opportunity to think about a change in cultural paradigm, where philosophical disagreements and agreements should be rethought. And I precisely devote the theses of the following pages to this paradigmatic change.

2 First Thesis: The Need to Deconstruct the Primacy of Anthropocentrism (Pars Destruens)

In 1986, when the position sustained in the *Tractatus brevis* already stated that the phenomenology of life must be thought on an anthropological basis, albeit in a cosmic sense (Verducci, 2004), there was, however, a reorganization of the classical position of man at the center of the cosmos by the phenomenologist: "[man] appears caught up in the turmoil of a generating progress, and thus as one of innumerable moments of the immensurable stream of life" (Tymieniecka, 1986: 10). However, the essential phenomenological point is to temper the needs of this reorganization (or decentralization) with the need to recognize that the human being is not simply "a moment among the others within the stream of life. But rather he/she has a special status, for which he/she is precisely "the vortex of the universal sense"" (Tymieniecka, 1986: 10).

Therefore, this thesis is difficult: in fact, it must hold two apparent contradictory aspects together in a rigorous way. The fact that the human being is no longer the demiurge of the cosmos (the critics to anthropocentrism) together with the fact that he/she represents the vortex through which one could access the universal sense.

How is it possible? I think that the answer is twofold. On the one hand, it is a question of recognizing the essence of man in the fact that he/she is an ontopoietic living being, and no longer an autopoietic one. Therefore, it is through the constant "unity-of-everything-is-alive" (Tymieniecka, 1998: 34) that man is no longer a master of himself or his own conscience, through the support of any transcendental subjectivity. On the contrary, man is already inhabited by the complexity of the cosmos or, more precisely, by its dynamism. It is precisely through the immanence of this cosmic dynamism that the human being takes shape. The very dynamics of the human being is thus stratified in functions and movements that follow the ontopoietic rhythm of natural life. Man is nothing but an implementation of cosmic dynamism. On the other hand, man is a "living agent", that is he is located in such a way as to act as a spark of life, thanks to the power of pure consciousness and creative imagination.

In short, it could be summarized by recognizing that the human being remains "transcendentally positionated"—that is, he/she assumes an irreplaceable place in the economy of life, but only on the condition of recognizing that this transcendentality is not a priori and it does not rise from self of conscience but rather from the ontopoietic immanence to which man belongs and of which he is made: "the world of life that man projects around himself is indeed transcendental but not in its fundamental origins in constitutive consciousness/mind—with its specific centrality—but rather with respect to its positioning within the dynamic web of the geo-cosmic architectonics of life. It is life-transcendental" (Tymieniecka, 2011: 17).

For these reasons, in the last phase of her thought, Anna-Teresa Tymieniecka expressly excluded that there might be a reduction of the phenomenology of life to an anthropology of life. This is because a human being cannot be considered in itself as such, unless it is placed in the middle of other things almost by chance (Verducci, 2012: 92). Similarly, a certain post-human line of thought thematizes "the dissolution of man as an epistemological object" (Janicaud, 2002: 23). In this way, he claims the need to think about the future epistemological horizon as it is emended by all anthropocentrism ("anthropo-de-centrism": Marchesini, 2002: 14).

This convergence should be recognised on the basis of various objectives: in fact, the Posthuman thesis challenges anthropocentrism as having the decline of humanism as its final goal (Marchesini, 2009: 186); instead, in the context of a phenomenology of life, questioning anthropocentrism is necessary in order to reposition the human being within the cosmos (not at the centre of the cosmos). After all, the phenomenology of life aims at going beyond Husserl's lesson in comparison to that. If Husserl's unsurpassed gain consists in having thematized the *Lebenswelt*, his limit is not to have thought about it in *natural* terms (not to have thought about "further stretches of pre-predicative "Nature"" [Tymieniecka, 2002: 685]), also abandoning the reflection concerning the primacy of "one's conscious body", in order to focus on the analysis of intentional consciousness (Tymieniecka, 1986: 27–76).

Therefore, what is the phenomenological value for the deconstruction of the primacy of anthropology?

I think it's the recognition of a complex sphere of life where all the meanings and the intentions of the human being's self-understanding are defined in its immanence. In this sense, the deconstruction of anthropocentrism is a classically "genetic" matter. Every "anthropocentric reduction" is unable to grasp the *original donation* of the human being in its worldly appearance. It is not by thinking to oneself that man can really understand himself.

The original donation of the human being does not occur in "such" a pure form, but it takes place during the irreducible genesis of a time which was already self-established within the primitive sphere of the world and, therefore, was already given in the complex stratification of otherness, beginning from the animal otherness which creates the human identity.

I realize that the theme of animality cannot immediately be drawn from Tymieniecka's reflections. But, at the same time, there is no doubt that the phenomenological description of the human being cannot leave an original point out of consideration, where the human being is, first of all, as "an ecological fruit. […] He is formed by the earth and sucks the juices of the earth" (Tymieniecka, 2001: 3). As we will shortly see, this starting point must be placed within a phenomenological constructivism which stratifies the relationship between nature and culture, tightening the links between the intentionality of consciousness and the natural dimension of the human being. But, we can be satisfied with stating up to now that any attempt to trace the essence of man collides with this irreducible exposure of humanity to his animality.

Therefore, the first result of our thesis is the recognition that every anthropological reduction may not be important, but it deals with *Levels of Otherness* which occur starting from the essential condition of animal-man. However, I would like to outline once again what a methodological distance is between a purely post-human approach and a phenomenological one. The deconstruction of anthropocentrism does not aim at abdicating the argument concerning essences. It is not simply a matter of recognising that the human being does not *exist* as such, but it only *exists* within a set of circumstances where it is located. Instead, it is about recognising that any discussion regarding the essence of man is made of the necessary recursion of the onto-phenomenological structure of the world of life. In this sense, the fact that man always comes *after* the animal (which he is too) signals an excess of philosophical discourse regarding the essence of man. The dis-identification of man with his simple humanity allows us to rethink about the relationship between the human essence and the immanence of the world in structural terms. Therefore, we leave any oppositional temptation which aims at establishing the primacy of the human subject on the epistemological separation from the object of knowledge from Descartes onwards: world, things, other human beings, himself as an object of knowledge. If anthropocentrism breaks the world down in order to define it, a non-anthropocentric phenomenology does not give up questioning the essence of phenomena (that phenomenon which we call man: admitting that anthropology is a simple local phenomenology is not a scandal), but it focuses on analysing the correlated mis-identifications, in order to carry out a complex operation for reuniting man with cosmos.

3 Second Thesis: Animal Otherness Between Phenomenology of Life and Posthuman Studies

> *Man becomes animal at the same time in which the animal becomes… but what does it become? Does it become man or something else?* (Deleuze & Guattari, 2010: 292).

Tymieniecka suggests human beings should be referred to starting from the "*human condition within the unity of everything there is alive*" (Tymieniecka, 2006: XIII–XXIII). Similarly, in his latest writings, Derrida points out the need to connect humanity and animality: "man is after the animal. He follows him. This "after," that determines a sequence, a consequence, or a persecution, is not in time, nor is it temporal; it is the very genesis of time" (Derrida, 2009: 55).

My intent is to acknowledge that there is a reservation of hidden meanings commonly used in the category of *otherness* even in the *Posthuman Studies*, a reservation which allows us to approach the cultural discourse of the *Posthuman Studies* with the philosophical discourse of the phenomenology of life.

We could say that the first mediating element between Posthuman Studies and phenomenology of life is getting out of the oppositional logic of nature-culture. This is what Daniela Verducci notes about the relationship between Cassirer and Tymieniecka:

> Ernst Cassirer had stated that precisely as an *animal symbolicum* […] man was continually endangering and at least partially losing the safety and the warmth of his natural existence. There would no longer be any possibility for him to escape or to return to immediate life from the magic circle of mediaticity and the artificial world, once it is produced by a symbolic function. The painful sense of parting from "natural life" and the loss of "immanence to reality" would indeed represent an anthropological constant, inevitably implying the symbolization of leaving the "heaven of immediacy" which belongs to animal existence. […] On the contrary, a renewed phenomenological examination allowed us to ascertain that its apparent cool transcendental conscience is found to be alive and existing or, at least, with connected and compatible life and existence; even the creative and symbolizing function can therefore no longer be considered a prerogative of human spirit in spite of its organic dimension and in dis-communication with it. Tymieniecka's original philosophy begins from this phenomenologically gained discovery: this thinker appeals to it to reform the classical phenomenology and to make it progress beyond itself, through the new phenomenology of life. (Verducci, 2012: 89–90)

Following this perspective, symbolization is not thus an interdiction to the phenomenological journey towards the paradise of immediacy, but it represents an additional tool available to conscience. Every cultural artifact is not opposed to immediate life, but it takes part in it in depth, even for post-human Studies. And it does it as much as it also represents a form of immanent otherness to the order of life where the human being, the concrete living individual who takes part in the ontological order of the world of life, takes shape.

What I would like to do is simply this: using some gains of the *Posthuman Studies* in order to try to dig into the notion of otherness and, of course, to legitimate an irreducible anthropological value. As we shall see, these gains are in one way unsettling and in another way completely insufficient. Nevertheless, it can be useful to try to question the *Posthuman Studies* as if they were a (paradoxical) description of Levels of otherness which necessarily also act in relation to the other human being.

Taking a long road is also necessary here: in order to reach the intersubjective nature of the human being, one must firstly consider a nature contaminated by otherness which are not—immediately—reducible in terms of subjectivity or humanity yet. An otherness which digs down to the bottom of our nature as such—and could be a rule in order to differently think about interpersonal relations as relations of otherness, even before we think about them as relationships with a *particular kind* of otherness.

In order to clarify it, I will hereby use a passage from a text by Derrida, where the matter of the "border crossings" between man and animal becomes a phenomenological analysis of the irreducible twining between humanity and nature, as if the objective of essentialising man's nature would lead to an excess, to a phenomenological projection of man's nature over his own humanity (Derrida calls this project: "limitrophy"). As if man's nature was not simply *itself*. A phenomenology of borders which always show themselves together with their transgressions. An excess of otherness within the heart of humanity, a self-violation which redefines the boundaries of all the discourses concerning man. Derrida writes:

> In what sense of the neighbour [prochain] (which is not necessarily that of a biblical or Greco-Latin tradition) should I say that I am close or near to the animal and that I am (following) it, and in what type or order of pressure? Being-with it in the sense of being-close-to-it? Being-alongside-it? Being-after-it in the sense of the hunt, training, or taming, or being-after-it in the sense of a succession or inheritance? In all cases, if I am (following) after it, the animal therefore comes before me, earlier than me. […] The animal is there before me, there close to me, there in front of me—I who am (following) after it. And also, therefore, since it is before me, it is behind me. It surrounds me. And from the vantage of this being-there-before-me it can allow itself to be looked at, no doubt, but also—something that philosophy perhaps forgets, perhaps being this calculated forgetting itself-it can look at me. It has its point of view regarding me. The point of view of the absolute other, and nothing will have ever done more to make me think through this absolute alterity of the neighbour than these moments when I see myself seen naked under the gaze of a cat. (Derrida, 2009: 47)

First Observation The animal (that animal there: a cat. And even more: that cat there, which has a name and, therefore, it is close to us) becomes the phenomenological pretext to define the boundaries of what is close (to us). I.e., in this case, the animal allows us to avoid any *objectification* of the world (and, at the same time, any human *subjectivity*). Therefore, even the animal is not just in front of me, but it is, above all, the evidence *of my surroundings*. In fact, as it is in front of me, it is also before me. Animal *precedes* man. It is the genetic issue which I referred to above and that, in my opinion, clearly shows the need for a preliminary phenomenological examination. Here, the genesis of man is his essence, but not as his primacy or succession. There is no separation or dualism. His essence is the original donation of man within the unit in which he lives. The essence is the man's original donation within the unity where he lives. What surrounds me would be the essence of man.

But, at this point, such essence represents going beyond the order of immanence at the same time.

Second Observation Derrida evokes a prevalent philosophy method (or rather: a philosophy identity), which would have been built as a "calculated oblivion" of all this. Philosophy as a view of the world and a "calculated oblivion" of the view of the world. In this sense, philosophy is often a human science, which is too human. Man also often seems to be conscious of himself as the one who looks at the world and considers the oblivion for the world as "what surrounds us" or "a unity of the whole which we belong to". In other words, there is a continuous obsession concerning the purity of human nature in the recognition of the "calculated oblivion" of such a philosophy that *Posthuman Studies* very effectively deconstruct it:

> Man subsumes different characteristics of animal othernesses. Therefore, animal othernesses become figures through which it is possible to describe man, where the ontic plurality of othernesses is translated into an ontic partiality. This phenomenon is particularly evident in the metaphorical, semiotic, symbolic use of animal othernesses with their implicit, iconic and stereotypical translation, in order to make them functional to explain some human characteristics. This process leads to a simplification of animal otherness as, whilst man may get co-extensive to all the othernesses, each of them can only approach to a man's quality. (Marchesini, 2009: 64, 66)

Therefore, this calculated oblivion would shrewdly generate subsumption processes of otherness under human order. There are two relevant aspects of these subsumption processes.

First of all, the fact that there is a false movement of enlargement towards the sphere of the world of life. In fact, there is a co-extension, through which man thinks of himself as "together with the othernesses." But such a co-extension is one-sided. This aspect also seems to me decisive, in order to rethink about human nature from a phenomenological point of view: a one-sided co-extension basically corresponds to the eternal temptation of idealism. The world of life does not become the sphere where man recovers the ante-predicative belonging of the transcendental ego, but rather the set of the phenomena which are subsumed within the power of the ego. On the contrary, avoiding this oblivion is the only way to phenomenologically recognize the otherness as it is located in the natural heart of the human being and, thus, to inaugurate the long way for the layers of otherness which start from recognizing themselves as "ecological fruit."

The second aspect also seems to confirm this strange relationship between *Posthuman Studies* and phenomenology of life. Here, Marchesini refers to a recurring semiotic and symbolic translation of the animal otherness into human codes. Taming seems to confirm this increasingly widespread contamination: treating our animals like ourselves, connecting the communication system with a set of codes, meanings and signifiers which belong to human communication. *Empathising* animal otherness, not *identifying ourselves* in it. Isn't this subsumption of animal communication to human semiotics a real idealistic presumption? Let us return to the phenomenological question of the genesis. How may animal otherness be preserved in its eventuality? How may it be possible to avoid this idealism? I think it is possible by specifically recognising that the *genetic* moment of otherness (in this case, a specific animal otherness, but I think we can acknowledge an eidetic structure of otherness as such here) must phenomenologically be attributed to its self-giving. Otherness is that phenomenological event which takes place as the pure form of an *original donation*. On the contrary, the otherness, which is reduced to a mere human symbol, is a dual phenomenon, a phenomenon lacking in its genesis, which is imagined without its original donation. Anthropocentrism would be literally nothing more than a *bad phenomenology*, by applying this aspect.

Through these observations, we can try to answer to the various questions I have asked at the beginning of this part of my essay: what can be at the basis of a fruitful dialogue between *Posthuman Studies* and phenomenology of life is, above all, [I] the acknowledgement that the layers of otherness—starting from the animal we are—can be included within the energy of their cosmic immanence only by escaping from the syndrome of philosophy as a "calculated oblivion", *i.e. from anthropocentrism*; [II] the taking charge of otherness as a genetic issue.

Of course, these two answers only allow us a provisional alliance and do not certainly manage to hide some substantial differences. In particular: [I] That Posthuman Studies replace anthropocentrism with another type of reductionism. In fact, the connatural otherness to the human being is exclusively the animal one for them. In short, there is no process or stratification of otherness, as it happens in Tymieniecka. It could be said that Posthuman Studies emerge from the calculated oblivion of anthropocentrism through the reduction of human experiences to internal modifications and, exclusively, to "a process-like nature" (Tymieniecka, 1988: 5). [II] Therefore, Posthuman Studies take charge of otherness as a genetic issue, while the phenomenology of life deals with *the layers of otherness*: in fact, it is a matter of recognizing that otherness is not only the starting point, but is also the way with which the human being experiences the turning point, which leads him to reach the degree of humanity, through its stratifications. This passage of level—which crosses the layers of otherness from physis to culture, also crossing bios and society—happens using the original experience where the human being experiences his animality, a "man's elementary condition." Interpreted in this sense, the questioning of the human being by his own animality is not an umpteenth reductionism, but the restoration of an original stimulus, which is able to restore the human being to the autopoietic flow of life: "man's elementary condition [...] appears to be one of blind nature's elements, and yet at the same time, this element shows itself to have virtualities for individualization at the vital level and, what is more, for a specifically human individualization" (Tymieniecka, 1988: 28).

It is only if we keep these so deeply different conclusions in mind that we can recognize a starting point of contact between Posthuman Studies and phenomenology of life.

In fact, there may be a philosophical alliance among them, which is able to avoid the calculated oblivion of anthropocentrism: "The hunger for purity leads to read the identifying processes as separating events ... In order to become recognisable, identity must break down any bridge of continuity with otherness through multiple denial events: a) denying a common origin; b) denying any credits from othernesses, i.e. identity may have taken on predicates from othernesses; c) denying belongings, that is, there should be some common dimensions between the two terms" (Marchesini, 2009: 52).

4 Conclusions

We have tried to show the fruitful correlation—at least, at the beginning—between *Posthuman Studies* and phenomenology of life. The theoretically crucial inheritance of the *Posthuman Studies* lies in questioning anthropocentrism as the "calculated oblivion" of man towards the world. Such questioning is primarily carried out through the recovery of the intrinsic role of otherness in the processes of identification/differentiation of man in the world. Of course, as we will shortly see, it is not irrelevant remaining on the phenomenological evidence of the "unity of what is a set" or theorising that such processes may be eventually determined as forms of "hybridisation" (Braidotti, 2003: 72), until going beyond the notion of human being (Luciano & Chen Mel, 2015: 183–207).

What changes is—from a philosophical point of view—the meaning which we assign to the category of otherness as such. This is a phenomenological deficit, which the integration with phenomenology of life can precisely help to fill. In fact, the analytical capacity of the *Posthuman* seems to limit to show the genetic function of otherness in relation to the human condition imagined beyond anthropocentrism. The levels of otherness *essentially* concern the twine of the human being with animal and technological otherness. But can a genetic examination of their role and their function be possible without a preliminary phenomenological study of their past within the human condition? A phenomenology of life is the transcendental condition of all the epistemological reflections which attempt to think about human conditions beyond anthropocentrism.

It is thanks to this phenomenological study that different philosophical outcomes of these projects may be evidenced. A difference which is especially revealed in the meaning to be assigned to the category of *otherness*. It is not by chance that we have insisted on the genetic twine between human condition and animal otherness on these pages, without mentioning technological otherness. It is difficult to assume that the way technological otherness genetically works on human nature should be similar to the way animal otherness does it. The philosophical risk is to have a critics of human identity which ends up by disregarding the irreducibility of differences and otherness as such (Labate, 2015: 77). We are at the phenomenological heart of the question again. Recovering the human condition within the sphere of belonging to life and beyond any anthropocentrism means recognising that the way we use to relate with the levels of otherness in the immanence of the world is full of intentional meanings where there is evidence of an irreducible and necessary differentiation. Derrida explains it very well:

> Let's allow that word to have both a general and strict sense: what abuts onto limits but also what feeds, is fed, is cared for, raised, and trained, what is cultivated on the edges of a limit. [...] Whatever I will say is designed, certainly not to efface the limit, but to multiply its figures, to complicate, thicken, delinearize, fold, and divide the line precisely by making it increase and multiply. [...] So it will in no way mean questioning, even in the slightest, the limit about which we have had a stomachful, the limit between Man with a capital M and Animal with a capital A. It will not be a matter of attacking frontally or antithetically the thesis of philosophical or common sense on the basis of which has been built the relation to the self, the presentation of the self of human life, the autobiography of the human species, the whole history of the self that man recounts to himself, that is to say the thesis of a limit as rupture or abyss between those who say "we men", "I, a man", and what this man among men who say "we", what he calls the animal or animals. [...] For there is not interest to be found in a discussion of a supposed discontinuity, rupture, or even abyss between those who call themselves men and what so-called men, those who name themselves men, call the animal. Everybody agrees on this, discussion is closed in advance, one would have to be more asinine than any beast [plus bite que le bites] to think otherwise. Even animals know that. (Derrida, 2009: 67)

It is not difficult to trace Tymieniecka's general sense of criticism towards Levinas, with which we have begun this article, on this rich page written by Derrida. We need to break the spell of "the man's samenesses." And this break may mainly occur through an affective form, by using the immediate condition of the human being within the *Lebenswelt*. An animal among others, this would be man inside the world. Indeed, this break must stand on the edge of the ever-present differentiation between "the one who calls himself man and the one who calls animal the one who calls himself man". *An animal among others, man is not just an animal*. Thus, all the world is also the set of the abysses on which othernesses settle, collide, are defined, destroy each other, go to war and—also—grow tame. *Levels of Otherness*. It will not be an excess of mixture which will break the spell of anthropocentrism ("the man's samenesses"), but rather the acknowledgement of the differences between othernesses within the immanence of the world.

It is precisely at this point that the phenomenological way fills the cultural excavation of *Posthuman Studies* with genetic meanings. Tymieniecka's words appear decisive once again. Once the operation of extending the human world within all the cosmic life is completed, once the immanence of the layers of otherness in his own human being is acknowledged, it is still necessary to understand how the experience of this extension and this immanence shows in accordance with the intentional order of meanings, "how does the world become a lifeworld, how does it become a world?" (Tymieniecka, 2004: XX).

References

Braidotti, R. (2003). *In Metamorfosi. Verso una teoria materialistica del divenire*. Milano: Feltrinelli (Trans. of *Metamorphoses: Towards a materialist theory of becoming*. Cambridge: Polity Press, 2002).

Deleuze, G., & Guattari, F. (2010). *Mille piani. Capitalismo e Schizofrenia 2*. Roma: Castelvecchi (Trans. of *Mille plateaux. Capitalisme et schizophrénie*. Paris: Les Éditions le Minuit, 1980).

Derrida, J. (2009). *The Animal that therefore I am*. New York: Fordham University Press.
Janicaud, D. (2002). *L'homme va-t-il dépasser l'humain?* Paris: Bayard.
Labate, S. (2015). Umanismo senza antropocentrismo. Note sul Post-human. *Filosofia e Teologia, 1*, 72–81.
Levinas, E. (1974). *Autrement qu'être ou au-delà de l'essence*. La Haye: M. Nijhoff.
Levinas, E. (1976). *Nom d'un chien ou le droit naturel* (Difficile liberté). Paris: Livre de Poche.
Levinas, E. (1977). Pensee et predication. *Analecta Husserliana, VI*, 3–6.
Levinas, E. (2009). *Carnets de de captivité et autres inédits, Oeuvres 1*. Paris: Grasset/IMEC.
Luciano, D., & Chen Mel, Y. (Eds.). (2015). *Queer inhumanisms*. Durham: Duke University.
Marchesini, R. (2002). *Post-human. Verso nuovi modelli di esistenza*. Torino: Bollati Boringhieri.
Marchesini, R. (2009). *Il tramonto dell'uomo. La prospettiva post-umanista*. Bari: Dedalo Edizioni.
Tymieniecka, A. (1977). The creative self and the other in man's self-interpretation. *Analecta Husserliana, VI*, 151–186.
Tymieniecka, A. (1986). Tractatus brevis. First principles of the metaphysics of life charting the human condition. *Analecta Husserliana, XXI*, 1–76.
Tymieniecka, A. (1988). Creative experience and the critic of reason. In *Logos and life, Book 1*. Dordrecht: Kluwer Academic.
Tymieniecka, A. (1998). The ontopoiesis of life as a philosophical paradigm. *Phenomenology Inquiry, 22*, 12–59.
Tymieniecka, A. (2001). The passions of the earth. *Analecta Husserliana, LXXI*, 1–12.
Tymieniecka, A. (2002). A note on Edmund Husserl's late breakthrough to the plane of nature-life, completing his itinerary. In A.-T. Tymieniecka (Ed.), *Phenomenology World Wide. Foundations, expanding dynamics, life-engagements. A guide for research and study*. Dordrecht: Kluwer Academic.
Tymieniecka, A. (2004). Ontopoietic chipering and the existential vision of reality. *Analecta Husserliana, LXXIX*, xiii–xxx.
Tymieniecka, A. (2006). The human condition within the unity-of-everything-there-is-alive and its logoic network. *Analecta Husserliana, LXXXIX*, xiii–xxiii.
Tymieniecka, A. (2011). The new enlightenment. Cosmo-transcendental positioning of the living being in the universe. *Analecta Husserliana, CVII*, 3–17.
Verducci, D. (2004). The human creative condition between autopoiesis and ontopoiesis in the thought of Anna-Teresa Tymieniecka. *Analecta Husserliana, LXXIX*, 3–20.
Verducci, D. (2007). *Disseminazioni. A partire dalla fenomenologia della vita*. Macerata: EUM.
Verducci, D. (2012). *La fenomenologia della vita di Anna-Teresa Tymieniecka*. Roma: Aracne.

The Poiesis of Thought

Antonio Domínguez Rey

1 Introduction

This text aims to fix the attention on the poetic phenomenon of thought. By reflecting on the internal process of the mental act or idea, we observe a special type of innovative, creative action. And that novelty is the source of the poetry or germ of the language. It is also the principle of science as original knowledge. Scientific functionalism, however, often forgets this creative principle of thought, especially when converting the result of computing into a paradigm of intelligence and the germ that activates it. The act of thought is first and foremost a vital act. It makes explicit the life and its creative source. To such phenomenon of explicit emergence we call it language. And the concept that best expresses this process we call it *onto-poiesis*. We will focus on its meaning and importance for both philosophical discourse and criticism.

The poetic word refers normally to the creative literature, work, or product of thought in its innermost process of expressive significance. So George Steiner called one of his most beautiful books, *The Poetry of Thought: From Hellenism to Celan* (2011). Other researchers speak of the language of thought, such as Jerry A. Fodor (1975), who is aware that the word Logos contains a principle of formal action that determines the mind's reflective faculty. It is so, while Logos means word and thought, idea, meaning and operational expression. Modern criticism depends on such a concept for the product of its activity and assigns to it, as definition, the computation that makes it become evident. Biochemistry and neuroscience define the term according to the attributes of the computation performed by means of a detailed calculation of its operations and results. What is produced becomes image and

A. Domínguez Rey (✉)
National University of Distance Education, Madrid, Spain
e-mail: adominguez@flog.uned.es

C. A. Hornbuckle et al. (eds.), *Phenomenology of the Object and Human Positioning*,
Analecta Husserliana 122, https://doi.org/10.1007/978-3-030-66437-4_3

model not only of the production, but also of the origin, germ, basis or principle which produces it. The descriptive features summarize the definition.

The algorithm of mental and expressive functions encrypts thought as a symbol and makes its mental frame a model and pattern of the thinking being. In addition, people who reflect this model, for example, Susan Schneider (2011), forget, to our way of thinking, that the algorithm is clotting a structuring of thought, not the life of the Logos. It happens that the same mirage in criticism confuses the general frame of literature with the creative germ of mind. The development of this germ is language itself. The human faculty of thought reaches its peak when it produces or invents the language. Language is, as is said by Steiner (2011, 31, 40), the "auroral force" of thought, or the "the voice of being." That is the reason why the literary creator aims "to reinvent language," to make it new, conscious, as well as being an oriented critic, that this conception of language is already "the defining nucleus of being" (214).

The mind-stream of the algorithm serves as a symbol of inference to determine by its approach what happens in the nucleus of being when its activity is already language. There lies the root of thought as *poiesis* rather than *poetry*. Is it possible to access the moment or germ of being thought instantly? The poet reinvents language because he sees its permanent novelty. The *poiesis of thought* is this constant and actual presence of the germ in the thinking being. Both Anna-Teresa Tymieniecka, and myself, are located in that original moment of permanence and we raise it to split all philosophical reflection (Tymieniecka, 1957, 243). If this is possible, the fund of philosophy, linguistics, literary theory and germinal science acquire meaning in another way. Language contains and manifests the poetic germ of knowledge, its *grammatical nerve*, as we say (Domínguez Rey, 2014, 91–93). And this is what we try to expose in this reflection.

2 The Framing of Ideas

Starting with Plato and Aristotle, the great thinkers of language and poets have always seen a poetic germ in thought. An excess of rationalism, positivism and, above all, functionalism, as well as current computing, reduced the fertility of this germ to the logical, regulative effect of forms or computerized information. Being was interpreted from formal mechanisms of thinking in an Aristotelian-Euclidian symbiosis, as José Ortega y Gasset explained in *The Idea of Principle in Leibniz* (1971).

Aristotle had already predicted the location of terms and concepts in the foundation of relationships that underlies them. However, the ontological dialectic of terms and concepts prevailed and was oriented only to a designated (entity) or mental reality [the *ousía*], or to the logical implications of its representation [*S est p*] in the classical *phantasma* of objectified conscience.[1] Proposition and argument ordering rules (the syllogism) were the result of this dialectic. Hence the domain of such a logic lasted for centuries.

The real movement of life, the everyday bustle of the existence, the acts and human endeavor were outside, i.e. *the life task*, *household chores*, the *dafare*, the *facenda*, *die Hausarbeit*, *die persönliche Sachen*, the life-world or Husserlian *Lebenswelt*, etc. Something was excluded so clear and intelligible in itself that Ockham once set it up against the intellectualism of Aristotelian and Scottish *species*. Understanding experiences facts which are not sensitive and do not even respond to proper sense: intellection and will itself, joy, sadness (Ockham, 1974, 39–43). They are contingent truths, the life of the concept which the apophantic logos, oriented toward logical judgment, undernourishes on rationalizing it.

On discovering that the relationship of proportions was only useful if it determined the relationship of the things, and in these, the contact with the other was mediated by the intervention of man, and with Leibniz's infinitesimal calculus, the discursive Cartesian method, based on doubt and calculation (number) intersecting space (movement), won the Cartesian frame of reality. It is not searching for the essence of the thing anymore but a network of relationships and correspondences. As Ortega notes (2009, 1093), things are equal now to "media relations." According to Slavoj Žižek (2014, 98), the Cartesian *cogito* results in pure empty formalism or tautology. To know is explored in order to get to know what the reality of the world is since Descartes equates truth with being and being with the existence. Thinking and being [*cogito*: *sum*] are reversible: "cette proposition: *Je suis, j'existe*, est nécessairement vraie, toutes les fois que je la prononce, ou que je la conçois en mon esprit," ("this proposition: *I am*, *I exist*, is necessarily true whenever I utter it or imagine it mentally") (Descartes, 1953, 275).

Here is a named—*phantic*—and translucent fact, therefore, the saying is necessarily in itself (*ego sum*) and the existing (*ego sum dicens*: "*I am by saying*"; at the same time I say; as I say). And this is as pure evidence, without paying attention, Ortega points out, without warning the background which is implicit here, the Cartesian generic notion *Je pense, donc je suis* ("*I think, therefore I am,*") (Descartes, 1953, 147) as happens in the syllogism. The involved, successive connections, their logical chains, become evidences of second rank or deductions (Ortega, 2009, 1162). Intuition then reaches involved, connecting loops. And this is so because it is the spontaneous bud of mind or inspiration from nature, Descartes warns (1953, 276), even if the language involved there assumes some inconvenience or logical opacity.

The imperatives of Kantian critical reason opposed the Copernican turn of knowledge to the reality of the thing. The scholastic and Aristotelian *phantasma* of perception amazes felt reality and it is only the illusion of intuition on a space-time

---

[1]The term *phantasma* is the sensitive, intentional image of the object from which the agent of understanding [*nous poietikós*], with its activity [*energeia*], obtains a form or mental image, the concept that allows us to know objective reality. According to Aristotle, "the soul never thinks without a mental image" (Aristotle, 1957, 177, 431a 17–18) which is similar to the object perceived (181, 431b 9–10).

coordinate, where the glows of a possible reality are registered, but unknown. From it, we only reach what intelligence, assisted by reason, planned within its framework. The truth of things is the regulated order or framing of ideas that are aroused by sensitive traces. The scheme, the representative image [*Bild*], the framing (enframing), the synthetic and phenomenal composition [*Gestell*], are foreshadowed there. Reality is now an idea, and the idea, reality. Subject and object are immersed in a dynamism of consciousness that equates being with thinking, according to the Hegelian viewpoint. We return to the Presocratic origin of the *nous*, but plunged into a stream of reflections or multiple mirrors.

Kant, in form and substance, proceeds, as Aristotle does, to establish this three-way access to reality in a single theoretical-critical and practical opposition: the *logos semantikós* (meaning), apophantics (*S est p*, proposition and judgment), and the *logos poetikós-patétikos* (agent-passive intellect).[2] Theory opposed praxis, a duality that infiltrates thinking to the present day. Kant wanted to solve this duality with the aesthetic variant of a value judgment only partly independent of critical reason, but related to the exercise of practical reason. Aesthetic judgment regulates imaginative representation by the "inner feeling of a state of spirit according to a purpose" (Kant, 1978, 201). Imagination depends now, not on the concept that frames it, but also on the impulse of the feeling regulated by an entelechy or form which accompanies it as a principle. Supposedly, therefore, it is a factor *with* operational compliance. And this process is fulfilled even today in the neurobiological theory of pre-linguistic impulses originating in the limbic area and structured in the prefrontal one, whose forms, synthetics and syntactic structures, would not really have been reversible, but informed by those impulses. So Carl G. Jung (2007, 87) also imagines the alternation between the feeling or area of what is known as experienced, and intuition, which leads to the unknown.

The deployment of the first source so glimpsed will be a reason and a groundwork for that origin previously considered. The work—the formalized sense—supports the impulse or germ that promotes it. There is nothing more beyond aesthetic reason for this deployment or conformity of form with the source that drives it. Kant still sees therein a symbiosis and immersion of the work in the regulating principle of nature [*Naturweck*]. On making the end independent of the original feeling, or on justifying this by achieving that, the intellection of the work is now subjected to the circuit of its construct. The object is subject, and vice versa. Any other predicative inquiry will be a circle of this orbit. There is no origin, there are no other meanings or transcendence than those revealed there: orbs of orbs, networks of networks, indefinite succession of coordinates in accordance with cohesion—regulative syntax—and coherence or adequacy of sensations, images, ideas, concepts, rules,

[2] Let us remember, that for Aristotle, intelligence [*nous*] is a mental activity [*energeia*] adaptable, on the one hand, to the form of things and capable, on the other, of producing them with the crowd of images [*phatasmata*] in the mind. Both passive understanding [*nous patétikos*] as well as active [*nous poetikós*], are denominations related to this mental activity. One should therefore not forget that the passive reception of impressions in human sensibility also presupposes a force [*dinamis*] (Aristotle, 1957, 171, 430a 14–20).

laws, parameters, etc. *Falsifiability* of elements and structures as the ultimate truth of being and thinking correlated in invention, production and creation so understood.

3 A Creative Motion

Anna-Teresa Tymieniecka summarizes this reduction of being to the structure of analysis as an exclusive resource of thought in the book *Phenomenology and Science in Contemporary European Thought* (1962). To the simple diagram or picture of cognitive representation (scheme, frame [*Bild*]), its framework (enframing) or mechanized framing [*Gestell*], and from the spontaneity of experience ("genuine experience") which provides contact with the world and things, such as Descartes and Jung do, she opposes a "sincerity without reserve," the immediate presence or intuition of the Self that every man perceives in himself—scientist, phenomenologist, or creator—on inaugurating the understanding of reality (Tymieniecka, 1962, 182–183). She starts with an originating motion or *Imaginatio Creatrix* who invents or discovers "a Nature-Transcending level of her self-interpretation in existence" (Tymieniecka, 1982, 69), a spontaneous intuition "of nature-transcending tension" (181). This driving force is made explicit, then, as creative emotion, the universal *ontopoiesis* culminating in human individualization or the person equipped with language.

She then interrogates the intrinsic foundations of the constitution or make-up of thought from Aristotle's *Metaphysics* and *Poetics*. It faces mental action "from their intrinsic content and form" process. She frames the analysis "in the conditions of 'making' itself" and in the phenomenon of motion, effort, and the follow-up of the creative Act (Tymieniecka, 1982, 69). It is the critical position of her *Poetica Nova* opposed to the phenomenological constitution and literary criticism. The focus is now on analysis and description of the interactive connectivity "of the continuous-discreet purposive segments" of human organization (72). Her background includes "the SUBLIMINAL SPONTANEITIES" of that internal motion and furthers in man a creative action of himself (*creative Self*) as a creative agent (73).

In this way, the *Poetica Nova* continues the teleological orientation that explains, from Aristotle, the sense of knowledge as an emergent self-awareness in connection with the truth of things. It highlights the etymological factor *erg* of the *intellectus agens* energeia, its creative feature and peculiar resulting shape, which is expressive character. And all that as the process advances creating, from its origin, the interplay of form and content. A dynamic agent operates on a "spontaneous genesis" in addition to being considered by the author as a "lyrical element" of a "lyrical sense." The background of this lyricism shows a specific existential *enjoyment*, since it connects to the creation of the universe. Tymieniecka supports this native *pathos* in the phenomenological aesthetics of art, in the style of Moritz Geiger. The affability or pleasure produced by the work of art has an objective groundwork of aesthetic value which lends an "existential interest" or phenomenological empathy [*Einfühlung*] in its process. The analysis carried out this way reveals, then, in the "ontological-cognitive" structure of thought and of work, an "ontopoetic design" with an "ontopoetic nucleus" (Tymieniecka, 2000, 247). Her mind continues investigating the first goal which she previously stated in *Essence et Existence* (1957): to analyze the conditions, background and origin of the configuration of thought structure in its effort to know something.

So she also works out the transparent film on the "entelechial element" which, since Aristotle, directs the internal process of the mind's internal action. This process has concerned philosophers, linguists and poets since the critical period of Humanity. The Polish-American author therefore deviates from the technical formalism of Roman Ingarden, but she closely follows their inspiration based on existential critical positions. Throughout, the *Poetica Nova* revives the backdrop, then, of the *Vita Nuova* of Dante and the *Scienza Nuova* of Giambattista Vico. Eventually it concludes with the vitalism of reason or *Logos and Life* (1988–2000).

Creative emotion flows throughout like the *élan* of Bergson, the cosmic teleology of Teillard de Chardin, or even the creative excitement of Raïssa and Jacques Maritain, the multiple layers of reality ("a many-layered reality"). It is possible to obtain a common structure that justifies the technical description of the phenomena, their "deductive mathematical formulation" and "the immediate experience of the self, the other man and the human world" (Tymieniecka, 1962, 183). Such a structure would be the "constitutive 'authenticity'" that intersubjectivity arrives at "with reference to both the simultaneous constitutive genesis of the ego within the world, and the polydimensionality of human nature." In this structure, she ends up conceiving a buried and simultaneous intentionality in the "laws of the world and those of the mind" (179).

This constituent genesis is the "shaping force" or shaping element in Husserl's terms, of the *ontopoiesis* for us. Its dynamism is not just biological, historical sketches or materials that melt it, Tymieniecka says (57–58). Life continually discovers new ways of experience that lead one to conceive an ecological inscription of man in the universe, "a congenial locus for man" (179). Such a common realm of being, as George Santayana would say, transcends, according to Tymieniecka, in an "existential dimension," also unyielding to the conditions that underlie and interpret it. Otherwise it would be an "ecological niche," a tomb or refuge, a *topos* or commonplace theme. Irreducible knowledge and life are then shown as a spiritual tension revealed only in man. Now it intends to *experience* or the fact of going through what separates by persisting: a given *entity* that consists *of*, through or an objective field of realities, i.e., by living it.[3] This is our experience of things. The preposition *of* (Latin *de*) indicates the genitive of the genesis, but also what is brought to mind from the reality, the *ablative*, and what there is processual, and what is captured inside, coming beneath it or perceived as there: the *intu-sub-ceptum* (what is got inside).[4]

[3] The Latin word *experientia* is compounded and effectively indicates the fact of going [*ire*] through [*per*] what separates distinguishing [*ex*] what persists [*per-sist-ens*] and consists [*cumsistit*] in itself, the entities [*entia*] or things.

[4] The term *intususception* (from the Latin *intussuscipio*: to undertake, to take upon onself within oneself) is used by Amor Ruibal to explain the "successive form" of the idea as the "psychological ideal" of different actions, surroundings, and intellectual relations: "the assimilation or intususception of the contents of some concepts in the content of other broader concepts, and at the same time, the more synthetic production of those that came earlier" (Amor Ruibal, 1995a, 331 [186]).

4 The Spring of Eros

In *Logical Investigations* (1900–1901), *Ideas I* (1913), *Formal and Transcendental Logic* (1929) and *Cartesian Meditations* (1931), Husserl faced the complexity of the prepositive mental action involved in the act of conceiving and meaning reality as an object. This action contains an intentional and meaningful experience [*Erlebnis*]: somewhat present in consciousness which, however, when captured, refers, mentions, leads to something else, means, concludes by this process as a sign. And this is so because awareness has sealed a double understanding (ideal-real) mode of experience as a stating of its correlative, a direct intuition or a well-founded apprehension, this of a second degree with respect to the thing known by intuition, which recalls the connection of conceptual chains in Descartes. The important thing, now, is the identity of the lived or felt and what is perceived in first, second, and third… instances: phenomenal re-coverings. If the perceived traits are identified, they own objectivity. So the objective results from suspending the difference of things, its concrete existential character. In practice, it is upon the act of suspension that it implies existence, as thought thinking. Awareness of this phenomenon led Husserl to reflect on *The Crisis of European Sciences* (1970).

In this process, then, something appears that contains what is mentioned on the occasion of an ultimately existing or possible existence. And this implies an analytical process. Tymieniecka proposed, therefore, and as did Max Scheler, to exceed the representative constitution of the living world by going back to the source where the reduction or phenomenological *epoché* flows. This wellspring is the creative conscience or Eros and the possible worlds growing in it. We moved, then, from the common, universal, spontaneous, and opaque, and from Eros to the respective unique, singular limits of spontaneity, what is significant, and "from Eros to Logos, whose union presided over the passage from the present life-world to possible worlds" (Tymieniecka, 1988, 39). Thus she reverses the process of *epoché*, but starting with the phenomenal constitution, as recovering the interstices of a delayed reality. And the author then wonders what "the conditions of the ultimate powers of consciousness" are (35).

We must take into account a radical, operating phenomenon that works as something unthinkable in the technique of re-covering experiences, their connections, and the reduction of reduction or return (the Heideggerian *Rückker*) to an increasingly immemorial origin. The intellective direction between the ideal (*eidos*) and the real (*hyle*), either really simple matter, body (*Körper*), or matter already alive (*Leib*), as in Husserl, is the relationship whose interstices connote affection—to be affected by, *pathos*—, the estimation and valuation of the subject. The Latin word *aestimare* contains something as real as bronze, copper [*aes*] and the action of a set price [*aes-tummo*], whose participle [*aes-tumatus*] can indicate it, according to some lexicologists—root * *tem*, not present in Latin—and which means "cut." So the estimate of something means the price of a mineral weighed and, perhaps, according to the cut produced and, hence, its significant expansion of value depends on the mental content of what is objectified, i.e. of the joint or limit of the concept. In any case, it is a transaction involving material and intersubjective contact. Tymieniecka (2000, 247) also refers to this material value or "existential' weight" when she proposes, as the poetic research objective, the recovery of human existence in the world.

5 The Shadow and Subtraction of Being

The creative act arouses the emerging and founding depth of consciousness and, in it, the performance of received impressions and the interactivity of consistent mental motions according to reality presented as an object of knowledge. This correlation determines the associations of traits, characteristics, and their interconnections in nuclei, nodules, and the laws of the constant appearance of things. And such a record translates into levels not watertight, but removable, as evidenced continuously by science. Tymieniecka considers, therefore, one "radical leap" beyond what is constituted and starting from the experience of the creative act, mainly of poetry. It is also our starting point for poetic and theoretical work based on the research we dedicated to Antonio Machado's lyrical reflections (Domínguez Rey, 1979). Forms and objects in the constituted world, to this poet and theoretician of poetry, are opaque regarding the quality felt in the consciousness to determine them. Therefore, if what is represented is the way of being, its objective essence, lyrical consciousness will begin with non-being or a nothingness paradoxically creative of the "shadow of the being" (Machado, 1989a, 690). Being and thinking, homonyms until now, dissociate, "do not match, or only by chance," says Antonio Machado. Investigating the heterogeneous from the root of life corresponds to poetry "once again to perform the *derealized*" (691). The nothingness here referred to is a rather methodological foundation since, behind it, looms the great creator, the poetic principle. Thus Antonio Machado reflects on the lyrical phenomenon, following Leibniz, Descartes, Bergson, Scheler and Paul Valéry. The Spanish poet also differs from Leibniz, and therein keeps, as Tymieniecka does, only some structural similarity with him, three degrees of consciousness (Machado, 1989a, 688–689). They are the spontaneity of the vital impulse, the development of its internal motion, which has turned outward, and thirdly, self-knowledge through objective forms as the reverse of being. The second degree is the erotic strangeness of self when leaving the self towards the objective and heterogeneous other, the compulsive otherness of desire and love, Hellenic—Eros—, Christian—Agape—, and a source of knowledge. The combination of these two forms constitutes the lyrical philosophy of man.

Existence is thus shown as the immediate data of consciousness and its genesis is the impulse of the universal force, as in the Monad in Leibniz. An erotic energy or now the universal love of the cosmos is "a new logical force," says the poet through his heteronym, Abel Martín (Machado, 1989a, 680).

Consistent with this, Antonio Machado (1989b, 2030) gets to the bottom of the first term of the famous Descartes proposition, *Je pense, donc je suis*, and inclines toward existence: *I exist, therefore I am.* He glimpses in Husserl, however, the wisdom of considering the relationship as a shape of reality among others, and provides for an aurora of the analytical phenomenon of consciousness in Heidegger. He quotes from him the famous phrase, "Das Dasein is das Sein des Menschen," ("being is the being of man,") adding, "and to penetrate into being, there is no other gap than the existence of man, the being in the world and time... This is the deeply lyrical note that will take poets to the philosophy of Heidegger, like butterflies in the light" (Machado, 1989b, 2366).

The derealization of the objective forms that, according to Machado, lead to a curiously creative nothingness, is also the nihilizing function of knowledge in Heidegger. The objects of knowledge would be the reverse of the being and that of the qualitative substance that subtends them. In the spherby of being there would be something even more real and primary than the objective forms. The creative wealth of being is restricted by the objective.

6 The Poetic Anonymity

Alain Badiou (2016, 20–21) highlights this phenomenon of poetic desobjectivation following two processes of mental configuration that are actually methods of poetic conception. The first is called subtraction and is taken from the restricted action concept ("l'action restreinte") of Mallarmé. The second is the dissemination of meaning by excess of metaphorical translation of the object, whose paradigm is Rimbaud. Subtraction captures features or attributes of reality by the effect of "negative machineries." The object vanishes as the poem enunciates the being, the idea. On the other hand, the dissemination dilutes the objective by "excessive equivalence with other objects."

They are two different visions of the same phenomenon. Machado highlights the lyric quality that subtends the objective form and announces it as the reverse mode of being. Conversely, Badiou retracts the perception of the object to a non-objective presence, deep nihilizing or merely related to another also figurative. The modes of presence are then the defective restriction, the lack [*manqué*], and paradoxically the common world of the experience present in the language: poetic communism (Badiou, 2016, 158). Now, the common thing announced here is, after all, an "anonymous point" (140), that is, the speaker's depersonalization. The gift that the poet makes to language by inducing and educing from it new forms of its own (undoubtedly, an added value) is restricted to the common anonymous.

This poetic anonymity does not explain the singular form that comes to the common language when the poet increases his expressive reality. It can be induced, motivated, as it happens in the Spanish romances when the medieval jugglers sing the communitarian feeling induced by the common experience of the historical events. The ontopoetic ferment of language induces in the community of experience the singular and intense resonance of the individual when he feels immersed in a rhythm of existence. Even the soul entity of the peoples is born there. That intensity is the creative effervescence of the cosmos in man. That's why the poem is thought in act, as Badiou says, but also, at that moment, and against what this philosopher deduces (2016, 18, 22), knowledge. The poem retracts itself into the germinative core of being and it is its vibration opening and inaugurating the presence of the object. It is the objective constitution of thought as language.

7 Ontopoetics

Since then, we have also followed the roots of this phenomenon called *ontopoetics* or the rhythmic dimension of reality in knowledge. The emotional opening of the word founds the company of the world "on the edge of existence," a kind of Heideggerian *Mit-da* (literally: "with-there"), or "time that *in*forms itself" in a vital way, a "suture of time in space through the living body of the language," as I said on the back cover of books of poems published in 1980 (*La voz y su vacío*), 1983 (*Lúrido ocelo*) and 1987 (*Gluma*). We thus outlined a theory of language from the poetic phenomenon and creative thinkers who felt, and continue to feel, like certain poets, the existential anxiety of knowledge.

The operative turn of phrase is revealing and counters, in a way, the division established by Aristotle between art and science, *poiesis* (making) and praxis (doing), philosophy and poetry, *logos semantikós* (simple apprehension, the intelligible), *logos apophántikos* (reason*: S est p* meaning *of*),[5] and the distinction in intelligence between *pathétikos* (receptive intellect) and *nous poietikós*: the intellective agent, a generator of what is eternally the same.

Aristotle foresaw this difference and homology by considering in art the moment of production and, in philosophy, the concept of being [*tó ón*] from the doubly significant function of what is determined—the essence—and the proposition or fact meaning something *of*.[6] What is generated (remember the mental budding of Descartes) to think something contains genesis, and, as such, is objectively made. It institutes and constitutes, and has two important prepositional motions: *in* and *with* (Latin conjunction *cum*). That objectification is *nóema* (thought) and *ónoma* (noun) without leaving the moment being [*tó ón*] of the entity.[7] It is the present at every moment and time, so what is alive in the living.[8] This sustained ontological

---

[5] The *logos apophántikos* or simple proposition "is an emission of a semantized word [*foné semantiké*, word with meaning] referring to the presence or no presence of an attribute in a subject, according to the times divided up" (Aristotle, 1940, 52, 17a 23–24). The concept of a "semantized word" for us is an example of the *elation* of human sound in a linguistic verbal signifier.

[6] This preposition—*of*—describes as well the mental movement that constitutes a name and the essence determined by such motion when relating another concept. It indicates the reference *being* of a possible subject.

[7] We prefer to translate this Greek present participle, *tó ón*, declined as a noun, for what it really is: being.

[8] The Latin noun *viventia* (suffix *-entia*) expresses the *onto*logical difference to *be* at the instant of *being* or the concrete entity.

relation of the *entity* to the *being* is predicate or precedence (preexistent) both internal and external of the *nóema.*

As a term of conceptual process, the name refers to, appoints, it determines something perceived automatically as referring to another determination or mode of the existing movement. It contains genesis, and so achieves a denomination [*Sinngebung*] whose term is the essence [*ousía*] or principle of all attributes: to be essential.

In such a process, a projective reversion of the sensible happens, known by intuition as synthetic motion, unity of concept and the immediate term of the genesis that we behold and which breeds the writing, reading and reflection it consists of. No description of language achieves this intuitive synthesis. The time involved is absent at the very instant of its presence and opening. It is language, *the word in time*, states Antonio Machado (1989a, 703), the poem.

This ontological motion assists the process of art and thought, with a difference. The intuitive vision of the sensitive or singular referring to the class that it evokes but does not totalize, is being a predicate or pre-reflective motion. It attends or assists its own vision. Such an instant is *poetics*. It lives and contains the logical motion of the synthesis evoked on minding in order to make universal expression concrete. *Ontopoetics* refers to the instance of the sensitive or conceptual noun moved to a sentence (predicate) by a pre-reflective concern for sense. The vocal sound sensation is an energy factor of the sense. The feeling of the object then becomes a relational subject of the qualities that comprise it. Its associative correlation involves comparison. So Aristotle (1957, 118, 420b 31–33) says that voice is sound with meaning and that its impact involves "some imagination." This is the said budding of sensation, the sound spelling of sense or the initial utterance (*legein*) of the Logos. It is an aesthetic logos on the basis of ordinary logic, claims Barbara Cassin (1997, 152). The horizon open there never becomes saturated, and, although it involves a leap, as Cassin says (154), it is never taken. Therefore, great poets and thinkers always warn us of a difference between the world occurring creatively and their determinations. Forms and principles that establish what is determined leave behind it a never totally unthinking world. The poem is this world. And this is based on the dynamism of knowledge, art and culture.

8 Art and Science

Let's look at the text by Aristotle where he claims an inductive totalization of experience as the basis of art and philosophy. He focuses the root of the historical difference between empiricism and rationalism and hermeneutical glosses which run through centuries down to the phenomenalism of Kant and Hegel, Stuart Mill's inductive empiricism, Husserl's eidetic reduction, Heidegger's ontological historicism, until deconstruction and its semantic genealogy:

> Thus sense-perception gives rise to memory, as we hold; and repeated memories of the same thing give/rise to experience; because the memories, though numerically many, constitute a single experience. And experience, that is the universal when established as a wholeness in the soul—the One that corresponds to the Many, the unity that is identically present in them all—provides the starting-point of art and science; art in the world of process and science in the world of facts. (Aristotle, 1966, 257–259, 100a, 4–9)

The crux of the matter is in the multiple number of memory and the juxtaposition, almost synonymy, of experience itself and the "universal whole." The one and the other correspond, then, with mutual regard, respectively, for the "the starting point of art and science," the production or making and the ontological domain.

Experience involves a processual faith that always must be given as it is shown: in a discrete and replicated number. This way of being given is *evermore* on the one hand and *nevermore* on the other, as in the poem, *The Raven*, by Edgar A. Poe.

Aristotle's universal concept is founded on intuitive confidence—"but it is impossible to gain a view of universals except through induction"—(Aristotle, 1966, 107, 81b 3–5): that what is now given will continue being given "in such a way", as fifteenth century poet Jorge Manrique (1990, 154) says in the verses dedicated to the death of his father, don Rodrigo (Domínguez Rey, 2013, 103). A natural faith ("animal faith", George Santayana calls it) which exceeds the exhaustion of the memory. Aristotle compares the tension of memory with soldiers fleeing a battle, scattered, running here and there until they regroup to order the troops again. The same happens with the phenomenological constitution of the soul oriented to the exercise of sensitive reunification of their perceptions and experience (Aristotle, 1966, 258, 100a 10–100b).

We extend the field of experience to the horizon opening in memory, induced by the desire to be universal and eternal or simply because we keep living and we know the limit of life. This induction is an intuitive point of the genesis in which it occurs. And the genetic experience evoked here occurs in the interval of generation and corruption of being thus felt. What remains of him in memory repeating itself alternately is really the universal notion, the unitary character of a single experience. And the various units rest calmly in consciousness. It is the union of thought that knows settlement (Aristotle, 1957, 40, 407a 33–34, 1980, 236, 247b 12). As soon as a perceptual sense ceases, immediately the universal begins in the soul—*próton men en té pschijé kazólou*—(Aristotle, 1966, 258, 100a 16–17). The one or the unit of a multiple presence calms, tempers. And so does the phenomenological constitution of the soul. Art responds to such a dynamism and from this principle given intuitively, as is apparent from other parts of the Aristotle text referred to. It is a principle neither innate nor derived from other faculties, but begotten in sensory perception (*apó aiszéseos*). Pure Eros.

The repeated succession of units more or less discrete and grouped together on the rising horizon of thought forms and responds to rhythm. Soul would be rhythm, orchestration of prints whose traces describe something transcendent. Mind contact with reality originates a sensitive numerically *quantic* power, rising and waning, one and multiple at the same time. The rest found in gender, concept and categories, is worried by the continuous range of the sensitive genesis and the state of *being* [*perí tó ón*] of what is happening.

The sensitive contact of perception is also a spread of energy in the sound wave which passes through the air and penetrates the ear to the brain with an alternation also *quantic*, articulated, of presence and absence covering at present the event or the phenomenon *being*. Therefore, we talk about *gramma*, the first specific, previously non-existent vibration of sound mutated into a voice as a world echo by expressive tension.

This process leads us, as Ortega observes, to Stuart Mill's concept of "uniformity in the course of Nature", which the Spanish philosopher considers rather a "fad" than an axiom or "at most a praiseworthy hope" (Ortega, 2004, 624–638). The *epoché* (bracketing) sends us to the "virtual character" of intelligible printing or the world of possibilities (value of suffix *-ble*) emerging, as also happens in Tymieniecka's *ontopoiesis of life*, which searches by delving "into the subterranean strategies of the logos in its manifestation" (Tymieniecka, 2000, 25). And if we assume such possibilities, we are in a domain of interpretations.

9 A Prelogical Notion

As soon as the impression that is received of things folds in on itself, so also are we already interpreting when we suspend the Space-Time of the phenomenon or experience to consider it, as Amor Ruibal says:

> Strictly speaking it is not possible in the human order to isolate experience from a way of knowing; to the extent that there is no line that may be considered the end of experience and the beginning of knowledge; rather, any conscious experimental act is already a psychologically interpreted act. It is a necessary result of the transcendent relationship prior to all cognitive functioning between the subject and the object. (Amor Ruibal, 1995b, 486 [477])

We are moving into a previous, not yet reflexive relationship, when we are immersed in an existential pregiving or sensitive contact with the things of the world. This giving is a prelogical notion, a restless intellectual movement that traces the points, profiles, and relief of the entities and oscillates among them looking for handholds. This is how Amor Ruibal conceives the basic notion of being of concept, image, judgment, word and reason (Logos): prelogical, logical, and psychological moments. Three respective functions of acquisition, processing, and deduction (reason) correspond to these moments. Being is the simple presence of the "first performance of the activity of the mind," that which makes it fertile as "a true spiritual generation." The act of defining it or describing it, already assumes that it is known: "*Being*, *is*, and nothing else" (Amor Ruibal, 1999, 200 [605]). The restless notion of any concept, idea, word, *being* never runs out, because it always unveils a new relationship: it "is something that makes up the real context in the things so that they are perceived" (Amor Ruibal, 1934, 41 [53]). His notion recalls, in a way, the continuous predicate of Charles S. Peirce or the relationship that differs from what is related, telling it, as if the concept distances itself from itself whenever it is conceived or, rather, suggesting that the relationship is inherent to every concept (Domínguez Rey, 2003, 282; Peirce, 1978, 41). Knowledge is then shown as "the *notional* evolution of the being," whose "intensive evolution" is "the individualized entity," the particular entity (Amor Ruibal, 1934, 40 [52]). This is how Amor Ruibal founds a "transcendent relativity," connotations of which we find in ontological aspects of Tymieniecka.

We assume, therefore, the continuation of nature in the subject as a notion of *being* regarding all it knows. And this is something more than impressive or residual *gramma*, the sensitive or suspending footprint of an active-passive motion, or than the intuited leap between being and not-being, nature and being, chaos-cosmos. This notion appears, says Amor Ruibal, "as a principle of the very form of the idea with its differences, and of the categories of ideas that translate the categories of that which is real" (Amor Ruibal, 1934, 40–41 [52]).

The plus of this notion is, for us, ontopoetic *gramma* insofar as it encrypts the moment of existential inscription in the world, which is also a starting point for Tymieniecka. It contains in itself the dynamic, expansive and implosive *quantic* point of beginning, the prescientific seed (Domínguez Rey, 2014, 177, 182). It is the first cognizant tension of life, the be-(not)being point, the prelogical existence that feels and knows of itself without knowing itself, plunged in its own emergence and whose evolution already depends on an environment. This is why Amor Ruibal considers being "as a modality within the many offered by nature." It allows us, however, to know others; it is understood, since "it is not only the principle of primordial objectivity, but also a permanent element of it in all subsequent forms of knowing" (Amor Ruibal, 1934, 40 [52]).

We pass the limit of giving moved by a desire for updatedness and permanent rest, one and the same. We make the end of expectation a current issue. We integrate the present point in the successive line of expectant protention and we reduce Space-Time intersection. And by proceeding thus, we discover a field of resonance that transcends the phenomenological basis.

We see that it is no longer just a matter of the processual and counterpart line *of*, the counterpart *of … to*, between consciousness and reality, characteristic of the phenomenal system, and adequate as a point of reflection (Domínguez Rey, 2014, 14, 19; Tymieniecka, 1988, 431). The discovered resonance makes any line of contact with the world vibrate and transfers us to a singular energy whose mode and performance are given at that time for being what we are there [*Da-Sein*]. And at this point of Space-Time in resonant Mode, the meaning of logical and analytic strategies is remade. And so we reach the dimension of the *nous poietikós*, and from there, we consider the emerging process of potentials that are being built according to the medium in consonance with which they occur and manifest themselves.

Tymieniecka focuses on the imaginative creation of consciousness and to do this, she still turns, like Descartes, to the first point and positional moment of experience "in the *mode of self-givenness*" (Tymieniecka, 1988, 374), but from her imaginative motion, the *creatrix imagination*. This emerges "with a specific orchestration of functions which brings together the mechanisms and forces of the constituent apparatus with those of the complex realm of the passions" (375). And this is what Kant and Husserl postponed, she points out, as irrational within the soul, and they do not recover it in their thinking. We prefer to say, however, that they bridle it, because the feeling of freedom operates as an intuitive engine of critical and eidetic reason.

The process, thus reverted to its source, discovers there that the self resides "in… 'fountain of enjoyments'" as it appears *clad* in a qualitative glow, in the actuality of nature and of the virtual content of acts. The self is accompanied by a kind of enlightenment integrated in the psyche and in the halo of its shadows. It conceives, therefore, a "subliminal genesis of light," which proceeds gradually in an orchestrated shaping of powers and endows man's intelligibility with a wider horizon than that of critical reason (Tymieniecka, 1990, 89–90).

The organic, psychic, and intellectual articulation of life is then shown as a source of knowledge, of which reason is only a regulating part. And what is articulated in gradual correspondences is the dynamism of the movement and biological activity of the human being in the world. Here, knowledge only represents, as in newthomism, and in the words of Fernand van Steenberghen (1965, 144), "*un élément de cette <<activité>> intégrale* ou de ce <<mouvement>> qui doit me réaliser pleinement; elle [the *connaissance*] est entourée d'un <<contexte>> affectif auquel elle ne s'identifie pas" ("*an element of this integral <<activity>>* or <<movement>> that must make me fully; it [the knowledge] is surrounded by an emotional context with which it does not identify"). And as the deep satisfaction [*jouissance*] of the self depends on this affective context, knowledge itself is subordinated to its pathetic function, either joy or suffering. This suggests that acts of knowledge and appetite (the will) that affect the body in its development, are also *biological functions* (146). And what works, somehow illuminates even the darkness of the appetite and the passions, whose basis and end shine with the light of feeling, at the same time sensitive and intelligent. In other words, and with the Tymieniecka's title, *Logos and Life*.

10 The Emergence of Language

This brightness is, to our understanding, decisive in order to explain the emergence of language and to retrieve, with its colourful sensibility, the poietic seed of the abstraction that has been so criticized since Kant and especially since Kierkegaard and Nietzsche.

One of the points in the process of what is separated [*diaíresis*] persisting [*synaíresis*], and as an establishing nodule of Space-Time, is language. And in it, the word as rhythm, as the separation of something that persists assumes a discrete character and also a significant interval. Being-(not)being relations, nature-subject, subject-object, idea-difference, ideal-real, are rhythmic moments of knowledge. Language is a nodal point because all faculties of human life converge in it. In and from it, we discover the genetic architecture that shapes it. And this is so with two considerations. One, the most immediate, as already formed language [*ergon*]; the other, forming itself, *in fieri*, as *energeia* or *perí génesis*, as in Aristotle's intuition. The language that we learn and acquire from our childhood is an already formed form, the being of a specific vital energy. The balanced ratio of sonority, length-duration, intensity (pitch), amplitude, tone (fundamental frequency of sound waves) and quality of tone (timbre), allows us to hear and understand the mental energy contained therein. We can also quantify this energy, but its determination will be only the basis, the intelligible condition, not the intelligence of the phenomenon. The conceptual determination of the sound position (generation—presence—and corruption—absence) is word, stem, lexis, semantic phonation. Its absence evokes, opens the field of virtual imagination, and distends time. It is already impossible to measure this other proportion. Let us move on to the field of interpretation, to being here while considering there, as if we stood in two different and simultaneous dimensions of the present, or being, a *power* that we call existence.

Amor Ruibal makes the distinction between formed language and language undergoing formation in his linguistic work. The original sound of language has "a real basis" or reason based on some aspect of the object that it denominates and names, which indicates that the human being is not indifferent to this process. A human being activates his or her vocalic status based on what he or she knows and names, thus "when language is formed, intellectual action falls on the words" (Amor Ruibal, 2005b, 348), in such a way that these already tie ideas "scientifically and systematically" (Amor Ruibal, 2005a, 61). In the constitution of a language, there is no original arbitrariness, neither need nor free will, but rather an occasion for the natural belonging of object-subject to have and be in each case a suitable, fit sound for that which forms itself mapped in the brain by some implicit relationship (Amor Ruibal, 2005b, 699). Language has a denominative justification. Knowing its already lost source is another thing, but the word, however, maintains a relationship "of origin" (Amor Ruibal, 2001, 23–26), although its semantic value is, for this author, conventional (Amor Ruibal, 2005b, 701). That is why, in his reflection of language, is also affection with the idea (672), and, furthermore, the form of conceiving the thing itself in the concept (338). Some connection exists, therefore, between the production and the original sound reception and the knowing form at the time the language is formed. And this is a result of "the spontaneous movement caused by the diversity of subjective and objective agents who appear in its constitution" (700). The knowing form then assumes a natural aptitude or specific faculty of the human being (656).

Already-formed language [*ergon*] is existential and is maintained by use or custom [*consuetudo*] in an ideal-real relationship between indeterminate-determinate motion, whose alternative also forms, in our view, a kind of rhythm. Nevertheless, the interstitial process of this relationship encourages more language due to the absence which the signifier involves when producing the word. The correlation significant-meaning-(idea)-reality sets rhythmic intervals which deepen the aforementioned interstice and increase the pursuit of more nouns which are already concepts transformed into symbols of the reality that they announce. The intervals are instants of the implied relationship of the sign. That which is implicit is both a shadow, ferment, and, as José Ferrater Mora says, a poetic feature of the language. However, the permanent involvement of poetry, we add—"le langage de la poésie est *toujours* implicite" (Ferrater, 1959, 158)—does not exclude the explicit designation of science also having a poetic seed. Progress in atomic science makes the implicit of the atom explicit, always naming its power differently. It seems to us that physical language goes behind the invention, but the act of invention contains a poetic seed. The neurological and vital stream of consciousness (Husserl) is no stranger to the quantum energy of the universe and to all sound that resonates in it. Husserl (1973, 595) conceived a universal teleology that is a universal intentionality.

The ontopoetic orchestration of powers proposed by Tymieniecka stems from this fluency or "creative inwardness" conceived as "created work," a concept that philosophers since Plato have been after, by different names: "Together with the advance of individual experience and with the advance of collective knowledge, the life-world stays in a perpetual process of constitution" (Tymieniecka, 1988, 156). Poetic work, the poem regarded as an internal creation of the world in the singular itemized existence of the particular human being or *Da-Sein*, is energy whose effect contains it in differentiated degrees of evolution and development: "The essential poem at the centre of things," as Wallace Stevens (2015, 465) states in the first verse of the poem, titled "A primitive like an orb I."

11 The Value of Abstraction

The rhythmic alternation of intervals and the implied relationship of language, which is also an implied relationship of concept, allows us to guess a value and *color* of living reality in the process of abstraction. As we think, reality induces sensitive notes correlated with notions. This correlation determines a value in itself, because the idea "is, rather than a fixed type, a psychological ideal of successive formation," says Amor Ruibal (1995a, 331 [186]). It organizes the sensitive and perceptive (concepts) components (attributes and number) with regard to the relational understanding of each (reason), where it is made. And this implies a process of translation, therefore, an interpretative, hermeneutic process, too, which assumes incorporation of "new objective relationships," (131 [186]) at the same time that it interprets them, assimilating and interpolating [*intussusception*] some in the others (329 [183]).

The trope is based on this, as neither the language nor the concept includes the immediate reality that they evoke. Language is, therefore, a *pseudonym* (Amor Ruibal, 2005b, 702) and in virtue, we may say, of the abstract appearance that it acquires, it's also just apparently outside of the value that constitutes it. The trope condenses the internal relationship that a term or concept establishes with another, both induced by some objective relationship, as in a kind of mutual coalescence of mind and reality.

Amor Ruibal conceives a gradual process of continuous translation of actual and subjective elements as some go towards others (*trans-ducere*) at their respective levels and based correlatively on teleological organization, until the idea that includes them and signifies them as its own expression is formed. Tymieniecka's orchestration of functions and powers proceeds also in gradual and teleological order, with another internal explanation, obviously. Her concept of light, color, experience, synthesis, also resembles the *colorful* one that, according to Amor Ruibal, occurs in the formation and content of the idea as "more real and alive" when it faces "the real position of being to which it corresponds in the set of all beings", making it possible "to evaluate the true destiny of each in the universal correspondence of all, giving the standard of a synthetic conception of all that exists" (Amor Ruibal, 1995a, 328 [181]).

Reality does not go away when it is abstracted; rather, we continue to be oriented and revitalized by it when we conceive and name something, when we form and classify the ideas with which we know and signify it. The same happens in the work of art (329 [183]). And this is also the language for us, a creative spontaneous act of human nature vital ideation.

Determination of conceptual and significant features is set according to the value that appears in a practical and operational function. The idea does not distance itself from the value of the thing, "but rather, conversely, [the thing] dresses in its true colors the idea" which is oriented, as we said before. This even means "subordinating the abstract idea to the concrete form that the thing wears, as an expression of given values" (330 [184]). The meeting of mind with external reality induces a halo or vital value of knowledge. This experience is expression, that which emerges from knowing. The concept is attended by the vital vibration of its conception. This is why ideas vibrate when they are a nest of the novelty that they reveal. This is where poetry is, whose act is the present or the instance *being* of the being which acts thus. We can even say that the internal osmosis of the flow between *nóesis* and *nóema* is poietic, as between *poiesis* and *poiema*. The elation of the body sound to voice or phone should integrate, at some singular moment of existence, the action of the mind in a vital act, unit or common factor of understanding. The sound unit is a value that combines action, event, or product and *nóesis*: *pathos*, *praxis*, *ethos*, and *logos*. The human voice is emitted, goes out into the air (phonation), from whence it came (inspiration), and resonates at the same time in the body in addition to receiving it (hearing) as external.

Here, a phenomenon which Aristotle had already observed appears: "The activity of the sensible object and of the sensation is one and the same, though their essence is not the same" (Aristotle, 1957, 147, 425b 26–28). The current energy of that which is made sensible is the same as the sensation acting. And this *same* turns out to be something new, an internal sonance that did not exist before, something that sounds *with* and, at the same time, announcing the sounding object. This *with* factor that Aristotle did not directly cite is very important. He is assuming it. The proof of this is that the Greek philosopher inferred a *ratio* or internal *logos* of this new object and even named it "symphony." This Greek term contains the conjunction *with* (*sin*) plus the material sound [*phoné*]. That is, the human sound becomes a voice when it is emitted *with* certain tonality. This is why Aristotle compared, immediately and in this sense, the consonance of the "symphony" with the human voice: "If harmony [*sinfonia*] is a species of voice, and voice and hearing are in one sense one and the same, and if harmony is a ratio, then it follows that hearing must be in some sense a ratio" (149, 426a 28–30).

The action of hearing is, therefore, a "ratio," a *logos*. That is why we think the separation between Logos and *poiesis* (Aristotle, 1982, 334, 1140a) is thus resolved in the unity of poem, where motion and that which is moved (the "acting and being acted upon" of Aristotle) become one and the same act. The speaker *makes* something with the oral sound, converting the intentionality or the living impulse included therein into an idea and this new formation is an articulated act of thinking. We are what we do, as we can deduce, following Aristotle in *The Nicomachean Ethics* (74, 1103b 20–25), from the relationship that exists between our activities and the human being's affective aptitudes. And this doing is already, one can say, a making, poetical praxis. We are this act called language.

This vital phenomenon is decisive for thought. The phonic feature [*foné*] inscribes in the mind an *erg* factor which is both active and passive, factual, in such a way that the voice is a *foné engrammatos*, *gramma*. The intellectual motion of the mind returns with a resonance that expands it or it resonates itself with the world vibration of embodied sound. And this happens in the critical period of training and speech development, in line with the epigenesis of the brain, i.e., in the infancy of intelligence. Once this singular period has passed, speech does not develop biologically.[9]

Here is a unique, peremptory experience. It founds the "grammatical nerve of knowledge" when it registers in the mind with its own power, a singular feature that reveals in it a motor power or principle whose product is replicated: the name. And it is, at this moment, substantial, since it conducts the relationship of the mind with the world through the body, although its concrete forms vary, just like one body or person is distinguishable from others (Domínguez Rey, 2009, 91, 2014, 91). Language is, then, the expressive principle that specifics the activity of thought and expands it beyond the senses. It is polyphonic. Here begins another phase and era of humanity.

It is in this process of appraisal, where the priority of qualities or predicates, responding to "previous evaluative acts" that are oriented not only toward *of… to*, but in all directions, happens resounding, like an intense volume of sound, and "to a particular subsisting type" which is, at the same time, "ideal", as Amor Ruibal notes (1995a, 331 [186]): the real subject that links up qualities that are perceived and understood.

Abstraction affects itself because of the value that is established as the real entity of things. And this also justifies the retro-projection of language and thought. Insofar as it is *retro,* it affects that which is given and its reflection or footprint. Insofar as it

9 The implications of sound *elation* are more complex than the reference in the utterance to the time or time not contained respectively in the verb and the noun according to Aristotle (1977, 60, 1457a 10–16). In sound *elation* happens at least an impetus induced by some articulatory reflex. And this means brain development, transit, mental translation. The nomination (the fact of naming something) implies reflects intuition, inner *tempo* named Word.

is *pro*, it also has a pronominal value, like the subject when it materializes itself in I/me or the first-person of the prelogical object-subject relationship. Because it is something *jectum* [*ergon*], it remits to the *jective* motion of conscience. Language is the result and the reflux of this *intussusceptio* and contains, as a value, the ongoing relationship of colorful experience that tends toward the idea and the ideal of cognitive interpretation.

The assumption of sensitive values in the ideal training process also turns the subject and, therefore, the issue along with it, we must understand, into a reflection "of previous evaluative acts." This inversion is important, because the initial enactment of the phenomenon discovers a world or "pre-beat" of conscience which delays the logical figure of self with regard to the interstice and interval in which it is formed. It is an "idea of one or many qualities" before it is an "ideal type that subsists as a subject, which can only be reached by resorting to subsequent elaborations," as Amor Ruibal states (1995a, 331 [186]).

This is especially true, in our view, in the work of art, which reflects and results the perceptual process in such a way that, as Ortega says (2010, 188, 189), poetics is the manner and form of knowledge. From the ontological subject-object correlation, immersed in the same vital substance of nature, the two poles of knowledge are inferred in a field of resonance, "preordered to produce a representative synthesis," that is, "a gnoseological model, or a way of knowing this reality," as Amor Ruibal holds (329 [182], 333 [188]). Given this base of logical, existential order, "The first determination of the spirit, concomitant with perception, comes true under an intuition of *values* into which it immediately translates the content of perception and the perceptual phenomenon" (329 [182]). Here is for us the ontological and poetic form of thought, the *ontopoiesis*.

This is the way Amor Ruibal sensitizes the logical problem of abstraction which continually reverts to the qualities that form the ideas thus "revitalized." Abstraction contains its own reverse side, a reduction of the phenomenological reduction. Once the process is established, the double aspect is the unique act of intuition, because it already really was a unique act of intuition right from the start. The colorful field or "lifehood" is open, in Husserl as well, to specific determinations according to the situation, the environment, physical and mental need, the occasion (Bühler, 1982, 56), etc. The note of existence always refers to something alive or atomic and to a principle of framing, conceptualization, category, that which makes it possible to think about it relationally, even if only as an absence.

So then, language is, in our opinion, one of these values or the inherent translation value of the implied order of "objective relations" in knowledge. The *phonetic type*, shaped with the articulation of sound, already ties up features, elements, and phonic characters with a relational horizon of integration in units which are, in turn, relational, thus setting up a basic sense or principle of meaning, a conceptual value. It is, according to Amor Ruibal (2005a, 75 n1, 2005b, 699–700 n1), the first hint of abstraction: "The meeting of many phonetic expressions of the same nature form a whole which creates an idea shared by all of them." What is shared is the idea that arises or attends in this particular way and in no other way, regarding the phonemic features which are related in this way. The fact of going from one feature to another, to assemble them by categorizing them, assumes an implicit, spontaneous, and unconscious comparison. We can only enter into it from the formed language, analyzing it.

12 Theory Implies Poetics

Here, then, is the factive *poiesis* instant, the instant of a praxis and of the formation of categories. The idea and the moneme (double or triple linguistic articulation) appear embedded within a product and recognizable only based on some degree of organization. It must be said, therefore, that the theory contains *poiesis*. The theoretical formulation process is already *poietic*. The *apophantic logos* implies *nous poietikós*. And internal productive doing continues to be active, bringing new modes of relationship and correlations to the initial process. It maintains the "relationship of origin" mentioned above. This self-reference is specific to language in both poetry and rhythm of thought. It expands, dilates, and relates, but its mode of action is productive even when the already organized forms are repeated. Saying fulfills, as an act, this moment of living action. Speech is always current. This is why languages remake and renew themselves, and assimilate the situation of every significant moment of life. Language, in formation and once formed, activates its *poiesis* in thinking. And then the thought is poetry. Its common source is the entitative realization of being. Thus we understand the almost apodictic conclusion of Heidegger (1977, 328) by identifying thought with the original mode of poetry: "Das Denken des Seins ist die ursprüngliche Weise des Dichtens." This mode keeps the truth of being insofar as it is its original saying: "Das dichtende Wesen des Denkens verwahrt das Walten der Wahrheit des Seins," ("The poetic essence of thinking preserves the ruling of the truth of being") (329). Saying. And this also happens in a way that is different from the construction of a simple product according to a previous intentional plan. The poetic singularity of language is deeper than the *poiesis game* of a praxis or distanced construction, untied "from its individual practical growth," as Karl Bühler (1982, 54) phrases it. The implied distance of the product language [*Sprachwerk*] from the speaker or from itself—metalanguage—is relational. It receives assistance itself in the product as an interpretative comparison. It contains the productive seed. The self-assistance of the signifier or *autopoiesis*, as other thinkers call it, is a peculiar characteristic of language. It remakes itself in each speech act. It contains the effect of the action and goes beyond the document coding, because the position exceeds itself, differentiating itself beyond the simple concepts of *gramma* and essence [*ousía*] conceived, respectively, as an objective line of thinking and presence of itself to itself in the entity [*ens*], according to Jacques Derrida (1972, 77–78).

Self-assistance of language does not reproduce a source which is already lost forever, but rather acts in accordance with a foundation, originating, so that the unknown origin still finds an opportunity to emerge and this encounter continues to be its own current situation. The *phonetic type* is its first characterization. In the phonetic type, the equality of fluid between the physiological and psychological processes, or at least its common base energy—the *energeia*—takes place, processes which, with the phenomenological difference, orient themselves, according to Max Scheler (1974, 68), to a whole [*Ganzheit*] as parts of a vital over-mechanic event.

The evolution of speech and derivation of some languages into others, their differences of parameters with respect to more or less uniform structural types, lead us to think of a seed or genetic *gramma* that assimilates, like an amoeba, certain components of the environment. Or perhaps—it's a hypothesis—a germinative *gramma* whose trait arises singularly in the genetic process and interprets Space-Time as a concrete form of periods or rhythms of existence. This is what languages and the type-logical language that they imply are like. The concept of *ontopoiesis* and the experience or *erg* factor of poetic *making* induce us to formulate this hypothesis. The *gramma* is ergative, the *erg* factor of awareness, in whose dynamism Heidegger intuits the viewing power of the poet, who found being ("der das Seyn >>stift>>"), as he says in *Beiträge zur Philosophie* (1994, 11). There would then be a dimension of permanent presentness in the polymorphic difference that interactivity with the environment induces. The phonic articulation of language is a singular instance of biological interconnection of Space-Time. The fact, however, is that the materiality of the act of writing, linear and positive, discrete and kinetic, is delayed in shaping this present and does not reflect it *in nuce*, but it's his filter. The poem attempts this act in the imaginative source of consciousness.

The product language solves the *aporia* of Aristotle between theory and praxis, as well as the displacement of abstractive reason or phenomenological suspension of the concrete [*epoché*]. As we never get a full saturation or intuitive fulfillment of essence or significant intention, language reveals in its inner core the original tension of the absence, even virtual, unknown, still nothing, and offers, at least, a sensitive inherence of thought in its own *erg*, the strength and sound energy, which moves the breath inspiring the articulation of the air we breathe and generates a specific type of sensation and salivation. Its intuitive fulfillment is the animated body resonance [*Leibblichkeit*] and semantic value of the world [*Lebenswelt*]. And this justifies the mutual qualification of sound and concept—the singular linguistic unity "pensée-son" ("thought-sound") of Saussure (1983, 156)—being the concept quality of the phonic substance and the verbal sound a quality of the concept: "in the language, a concept is a quality of the phonic substance, as a determined sound is a quality of the concept" (144–145) (Domínguez Rey, 2017, 122). Language saturates the thought as poem, as a work that contains its dynamism and expands it in unpredictable mental energy orbs.

In the original tension is where the field of time takes place, as seen by Husserl, Heidegger, Jaspers, Levinas, Ricoeur, Derrida, and a list of glosses on Plato, Aristotle, Plotinus, Porfirius, Augustine… And over time, or even earlier, the verb character of every word shows itself as act and atom of thought. The aforementioned distance, the self-assistance of the signifier, the articulated movement of phonation, intense and qualitative, the space involved in it and the resulting paradigm, all manifest the being of language. This is why Giorgio Agamben (2008, 36) can say that the paradigm has an "ontological character" and that it "does not refer to the cognitive relationship between a subject and an object, but to the being." Relational stress *being* is *verb, word*, whose foundation is the present of life: *tó ón* or the living present. There is language if there is life.

13 Logos and Life

Logos and Life was the title Tymieniecka (1988, 424) gave to the set of her most significant works. In this work, she reproaches Kant and Husserl, essentially, and as we said, for the reduction of the intellectual complex "to the intellectual type of rationality." That which is outside of reason is for them, she says, irrational and does not shine in the constitutional process. The rationality of principles, categories and forms is, on the contrary, "the vehicle of life." Its discovery belongs to the metaphysics of life (429). And the synchronization and synergy of operating functions corresponds to language. Tymieniecka conceives the word as a synthetic eruption of a subliminal light that stains sound and expresses the synchronous completion of intelligence as a "sign symbol" of vital functions. The word implies effort (acquisition, learning), so it is not an automatic eruption, but rather a process, the last sedimentation of the individualized existence that brings this functional complex to light (Tymieniecka, 1990, 84–85, 92).

Language acquires synchronous function when it assigns meaning to the spontaneous signals of the vital process. These then acquire the function of signs. The description that Tymieniecka makes of this phenomenon seems, however, to confer expressive priority to a previous and independent instance of sound, as if there were some kind of demonstration prior to sound the shaping of sound that is, basically, language: "language is embedded in sound and oriented toward 'the other'" (Tymieniecka, 2000, 326). A sound that arises in a subjectivity that is directed or attracted by the presence of another, i.e., in a prior axis of communication? Establishing a special kind of original *saying*? We do not know. We imagine some spontaneity *between* communicating subjects.

This sound formation is already language in itself, a cognitive operation carried out with sound as a phenomenon *being* of the Logos, we would say. It is not something prior that is later embedded. It happens at that very moment even if there is a previous natural sound. And this is *ontopoiesis*. Communicative objectivity becomes language because it is a new object, not an objectivity created to designate *of* rather than to communicate *for*, as Tymieniecka argues (326): "It explicates within its complex subjective form the hidden and otherwise reposing in meaningfulness of the life saga itself (not the meaning but simply of)." However, the prepositions "of" and "for" are language in themselves, parts, therefore, of the signifying process. This inherency is already attributive as well as antepredicative. It not only "[expresses] the subject's dominion over the life script" (325), but it is this very domination insofar as it synchronizes by objectifying. It is therefore not enough to say that "language is clad as it emerges," which is evident. It is the clothing that it processes. It clothes proceeding, as *nous poietikós*. This nuance is decisive for us, because it reveals the singular grammatical function that is implied and inherent to language.

Indeed, language assumes the sensitive ghost or the shadow of the concept in phonetic constitution. It is the *gramma* of resonant immersion in existence. And in resonating it superimposes polar relations that allow a here and there of the here-there [*Da*] as pronominal remission (PRO) or the name tendency of the phenomenon thus lived. Each element contains a PRO relation in the verbal denomination (discursive *verb*) of reality. This is how the semantic inherence of the "fono" that Aristotle quotes is formed. And this is a *novum* previously non-existent. A novelty that increases the power of thought, taking place as a singular moment of existence with a critical period of development. In the first level of synthesis we find the position of sound as voice or sound elation in a tone with a comment on timbre (modal background and shape) or an individualized mark. And this happens by typifying a correlation—Tymieniecka points out (1988, 425), as Jacques Maritain does, the "co-genesis" of world and life—with that which consciousness intuits stirred by the perception of reality. Its presentification allows us to understand that which is thus constituted by recalling it and comparing it with the continuous process of perception, and then categorizing it. This type is a spontaneous mark, a body response depending on the life medium (World): *Lautbild* (Humboldt), *Wortlaut* (Ingarden), a mouth flower (Hölderlin), light (Heidegger and Tymieniecka, each one with a different background and horizon), value (Amor Ruibal) onto-linguistically grounded (Domínguez Rey, 2003, 187).

14 Conclusion

Language reveals the permanent novelty of being. It manifests a presence that updates the germ of the existential Space-Time of man whenever we activate the creative power which drives it. Actually, the language remains active in latent mode—a continuous becoming—once formalized. This is so because verbal learning presupposes a rhythmic formation of thought. Such an act is the word-instant or Logos, the conceptual form of knowledge. It is an *ontopoietic* instant or an integrating unfolding of the creative powers of knowledge. And this is science. Scientific forms are also creations and rhythmic products of thought. They come from the integrating nucleus of existence. That is why they continually remake the novelty that propels and activates them. The language reveals the becoming of the entity, the germ of existence. The basic concepts and categories of science, such as being, space, time, entity, essence, gender, species, relationship, algorithm (matema), infinite, etc., recover the formally active, vital principle (Logos) that grounds them. This dimension is reflected in the great philosophical, scientific, linguistic, critical creators and, especially, in the poets. We have considered several examples in this study. Then we can say that the analytical reduction of being to formal paradigms of computation from whose criticism we started with Anna-Teresa Tymieniecka at the beginning of is philosophical reflection, it also discovers its principle of poetic and objective thought formation. To forget this revelation would be to deviate from the origin and destiny of man in nature. A deadly detour.

Phenomenologically and poetically, we can conclude, language is value that, produced and manifested *a posteriori*, reveals a constituent *a priori*. It is the color or brightness of the intentional, significant and grammatical impulse of the mind. This is how the Logos arises and, in it, the only clarity of life that we know. Its radiance effects and concretes the Space-Time-Movement in Nature's Being Mode. Language makes the cognitive process transparent by opening a new existential value, *the poiesis of thought*.

References

Agamben, G. (2008). *Signatura rerum: Sur la méthode*. Paris: J. Vrin.

Amor Ruibal, Á. (1934). *Los problemas fundamentales de la filosofía y del dogma. Tomo noveno* (10 Vols.). Santiago (de Compostela): Tipografía del Seminario Conciliar. [The numbers in brackets are the paragraphs].

Amor Ruibal, Á. (1995a). In S. C. Blanco (Ed.), *Los problemas fundamentales de la filosofía y del dogma. I. [Problemas preliminares]* (5 Vols.). Santiago de Compostela: Xunta de Galicia. (First edition published 1914). [The numbers in brackets are the paragraphs].

Amor Ruibal, Á. (1995b). In S. C. Blanco (Ed.), *Los problemas fundamentales de la filosofía y del dogma. II. [Problemas histórico-críticos]* (5 Vols.). Santiago de Compostela: Xunta de Galicia. (First edition published 1917). [The numbers in brackets are the paragraphs].

Amor Ruibal, Á. (1999). In C. M. Robles (Ed.), *Los problemas fundamentales de la filosofía y del dogma. V. [El conocer humano]* (5 Vols.). Santiago de Compostela: Xunta de Galicia. (First published 1934). [The numbers in brackets are the paragraphs].

Amor Ruibal, Á. (2001). ¿Lengua o dialecto? In *Álbum literario 1907 da Asociación Iniciadora e Protectora da Real Academia Galega da Habana* (pp. 23–26). A Coruña: Real Academia Galega, Xunta de Galicia.

Amor Ruibal, Á. (2005a). "Introducción". In P. Regnaud (Ed.), *Principios generales de lingüística indo-europea* (pp. 1–137). Santiago de Compostela: Consello da Cultura Galega. (Facsimile of the first published edition 1900).

Amor Ruibal, Á. (2005b). *Los problemas fundamentales de la filología comparada: Su historia, su naturaleza y sus diversas relaciones científicas* (Segunda Parte). Santiago de Compostela: Consello da Cultura Galega. (Facsimile of the first edition 1904–1905).

Aristote [Aristotle]. (1977). *Poétique*. (J. Hardy, Trans.). Paris: Les Belles Lettres.

Aristotelis [Aristotle]. (1940). *Categoriae et Liber de Interpretatione. Recognovit Brevique Adnotatione Critica Instruxit L. Minio-Paluello*. New York: Oxford Classical Texts.

Aristotle. (1957). *On the soul*. (W. S. Hett, Trans.). Cambridge, MA: Harvard University Press.

Aristotle. (1966). *Posterior analytics*. In H. Tredennick (Ed. & Trans.) *Topics* (E. S. Ferster, Ed. & Trans.). London: William Heinemann.

Aristotle. (1980). *The physics* (P. H. Wicksteed & F. M. Cornford, Trans.) (Vol. 2). Cambridge, MA: Harvard University Press.

Aristotle. (1982). *The Nicomachean ethics* (H. Rackham, Trans.). Cambridge, MA: Harvard University Press.

Badiou, A. (2016). *Que pense le poème?* Caen: Nous.

Bühler, K. (1982). *Sprachtheorie*. Stuttgart: Gustav Fischer.

Cassin, B. (1997). *Aristote et le logos: Contes de la phénoménologie ordinaire*. Paris: PUF.

De Ockham, G. (1974). *Comentarii in I Librum Sententiarum*. New York: Instituto Franciscano de San Buenaventura.

De Saussure, F. (1983). In T. de Mauro (Ed.), *Cours de linguistique générale*. París: Payot. (First published 1916).

Derrida, J. (1972). *Marges de la philosophie*. Paris: Minuit.

Descartes, R. (1953). *Oeuvres et lettres* (A. Bridoux, Compiler). Paris: Gallimard.

Domínguez Rey, A. (1979). *Antonio Machado*. Madrid: Edaf.

Domínguez Rey, A. (1980). *La voz y su vacío*. Talavera de la Reina: VI Colección Melibea.

Domínguez Rey, A. (1983). *Lúrido ocelo*. Madrid: Molinos de Agua.

Domínguez Rey, A. (1987). *Gluma*. Madrid: Playor.

Domínguez Rey, A. (2003). *El drama del lenguaje*. Madrid: Verbum.

Domínguez Rey, A. (2009). *Lingüística y fenomenología: (Fundamento poético del lenguaje)*. Madrid: Verbum.

Domínguez Rey, A. (2013). El horizonte rítmico del lenguaje. (Trasfondo fenomenológico en las *Coplas* de Jorge Manrique). In A.-T. Tymieniecka (Ed.), *Phenomenology and the human positioning in the cosmos. The life-world, nature, earth: Book One* (Analecta Husserliana CXIII) (pp. 97–133). Dordrecht: Springer.

Domínguez Rey, A. (2014). *El Gramma poético: Germen poético del lenguaje*. Barcelona: Anthropos.

Domínguez Rey, A. (2017). Aporía(s) de Saussure, valor *gramma* y función retroproyectiva del lenguaje. In V. Martínez Paricio (Ed.), *Cien años después del Cours de Linguistique Générale* (pp. 109–136). València: Universitat de València.

Ferrater Mora, J. (1959). Le Langage de la Poésie. In A.-T. Tymieniecka (Ed.), *For Roman Ingarden: Nine essays in phenomenology* (pp. 147–159). The Hague: Martinus Nijhoff.

Fodor, J. A. (1975). In J. J. Katz (Ed.), *The language of thought* (Vol. 5 vols). Cambridge, MA: MIT Press.

Heidegger, M. (1994). *Beiträge zur Philosophie: (Vom Ereignis)* (Band 65, 2. Auflage). Frankfurt/Main: Vittorio Klostermann.

Heidegger, M. (1977). *Holzwege* (Vol. Band 5). Frankfurt am Main: Vittorio Klostermann.

Husserl, E. (1970). *The Crisis of European Sciences and Transcendental Phenomenoloy: An Introduction to Phenomenological Philosophy* (David Carr, Trans). Evanston, IL: Northwestern University Press.

Husserl, E. (1973). *Zur Phänomenologie der Intersubjektivität*. In I. Kern (Ed.), *Texte aus dem Nachlass. Driter Teil: 1929–1935*. Den Haag: Martinus Nijhoff.

Jung, C. G. (2007). *Sobre el fenómeno del espíritu en el arte y en la ciencia* (C. García Ohlrich, Trans.). Madrid: Editorial Trotta.

Kant, I. (1978). In W. Weischedel (Ed.), *Kritik der Urteilskraft* (3rd ed.). Frankfurt/Main: Suhrkamp.

Machado, A. (1989a). In O. Macrí (Ed.), *Poesía y prosa. Tomo II. Poesías completas* (4 Vols.). Madrid: Espasa Calpe-Fundación Antonio Machado.

Machado, A. (1989b). In O. Macrí (Ed.), *Poesía y prosa. Tomo IV. Prosas completas (1936–1939)* (4 Vols.). Madrid: Espasa Calpe-Fundación Antonio Machado.

Manrique, J. (1990). In M. Á. Pérez Priego (Ed.), *Poesías Completas*. Madrid: Espasa Calpe.

Ortega y Gasset, J. (1971). *The Idea of Principle in Leibniz and the Evolution of Deductive Theory* (M. Adams., Trans.). New York: W. W. Norton & Company.

Ortega y Gasset, J. (2004). Sobre el concepto de sensación. In *Obras completas. Tomo I_1902/1915* (Vol. 10 Vols, pp. 624–638). Madrid: Santillana Ediciones Generales, Fundación José Ortega y Gasset.

Ortega y Gasset, J. (2009). La idea de principio en Leibniz y la evolución de la teoría deductiva. In *Obras completas. Tomo IX. (1933–1948). Obra póstuma* (Vol. 10 Vols, pp. 929–1174). Madrid: Santillana Ediciones Generales, Fundación Ortega y Gasset.

Ortega y Gasset, J. (2010). El hombre y la gente [Curso de 1949–1950]. In *Obras completas. Tomo X_1949/1955. Obra póstuma. Índices generales* (Vol. 10 Vols, pp. 139–326). Madrid: Santillana Ediciones Generales, Fundación Ortega y Gasset.

Peirce, C. S. (1978). Deuxième lettre a Lady Welby. In *Écrits sur le Signe* (G. Deledalle, Trans.) (pp. 35–57). Paris: Seuil.
Scheler, M. (1974). *Die Stellung des Menschen im Kosmos*. München: Nymphenburger Verlagshandlung.
Schneider, S. (2011). *The language of thought: A new philosophical direction*. Cambridge, MA: MIT Press.
Steiner, G. (2011). *The poetry of thought. From Hellenism to Celan*. New Directions Book: New York.
Stevens, W. (2015). A primitive like an orb I. In *The collected poems of Wallace Stevens* (corrected edition by J. N. Serio & C. Beyers) (Second Vintage Books Edition). New York: Vintage Books.
Tymieniecka, A.-T. (1957). *Essence et existence: Essai sur la philosophie de Nicolai Hartmann et Roman Ingarden*. Paris: Aubier.
Tymieniecka, A.-T. (1962). *Phenomenology and science in contemporary European thought*. Straus and Cudahy: Farrar.
Tymieniecka, A.-T. (1982). Poetica Nova. The creative crucibles of human existence and of art, a treatise in the metaphysics of the human condition and of art. Part I. The poetics of literature. In A.-T. Tymieniecka (Ed.), *The philosophical reflection of man in literature* (pp. 1–93). Dordrecht: D. Reidel Publishing Company.
Tymieniecka, A.-T. (1988). *Logos and life: Creative experience and the critique of reason. Book 1* (Analecta Husserliana XXIV). Dordrecht: Kluwer Academic.
Tymieniecka, A.-T. (1990). *Logos and life: The passions of the soul and the elements in the onto-poiesis of culture. The life-significance of literature. Book 3*. Dordrecht: Kluwer Academic.
Tymieniecka, A.-T. (2000). *Impetus and equipoise in the life-strategies of reason. Logos and life. Book 4* (Analecta Husserliana LXX). Dordrecht: Springer Science+Business Media.
Van Steenberghen, F. (1965). *Épistémologie*. Louvain: Publications Universitaires.
Žižek, S. (2014). *Event: Philosophy in transit*. London: Penguin Books.

Part II
Transcendental Idealism: Investigation Continues

Ecce Zarathustra: Nietzsche's Answer to the Human

Marcella Tarozzi Goldsmith

Zarathustra announces the coming of a new type of human being, the *Übermensch*, whose qualities are partly left undisclosed. The human being—leaving aside the herd—is unpredictable; the *Übermensch* even more so. Nietzsche has a specific vision that coincides with Zarathustra's role as an Oracle. Zarathustra, while trying to avoid his solitude, says: "Nobody tells me anything new: so I tell myself—myself" (1966: 196). Is he human? Or non-human in his solitude? By his own admission, Zarathustra is not the over-human being he advocates. The mere humanity of the herd constitutes a hindrance that interferes with the requirements of higher values, whose aim is the realization, not of a utopia in the classical sense (as would be the case if believing in progress were a possibility), but of a *Übermensch* capable of ecstatic nihilism and who can embrace the cosmos in such a manner that it coincides with the vision of the Eternal Recurrence.

How are we to understand what Nietzsche had in mind while embracing this Dionysian mode or vision, esoteric and mysterious, whose power is such that it can change the life of a person? The idea of the Eternal Recurrence indicates the transition from the merely human to the over-human, to the exclusion of non-human animals (Ponton, 2007: 319–320). I maintain that the essential meaning of the *Übermensch* rests on the complete acceptance of a cosmic reality, which Nietzsche called "The Eternal Recurrence of the same." It is a real, or imagined, vision of an eternity that returns—the beginning, which, to paraphrase Hegel, is an end; and an end, which is a beginning. A chiasmus that points to a paradoxical infinity which Hegel described by evoking the moment of birth as the moment of death. The beyond human is one example of human *hybris*, since it ventures into a new territory that is not delimited by the already experienced, the already understood (Del Caro, 2004: 135). Nietzsche's exceptional experience of the Eternal Recurrence at

M. Tarozzi Goldsmith (✉)
Independent Scholar, New York, New York, USA
e-mail: mftarozzi@gmail.com

C. A. Hornbuckle et al. (eds.), *Phenomenology of the Object and Human Positioning*, Analecta Husserliana 122, https://doi.org/10.1007/978-3-030-66437-4_4

Sils-Maria is a personal, esoteric experience indicating that the human can be overcome by embracing the totality of the cosmos in its magnificence—only then the-world-to-be is experienced in all its relevance.

In order to understand how Nietzsche thinks of the human and the non-human, it is crucial to take into account not only his most mysterious and arcane belief, the idea of the Eternal Recurrence, but also his descriptions of human types—the free spirit, the bound spirit, the last Man, the herd, and the *Übermensch*. These all demand a different philosophical approach and, consequently, a different terminology. As for Zarathustra, in my view, he is not strictly speaking a type; rather, he is an individual with a proper name. Yet he is the mythical figure who condenses myth, symbolism, poetry, and farce. He is the Oracle who awaits the coming of the *Übermensch*. Zarathustra is a hybrid, not exactly a type. A type does not have a proper name. He is half man and half Overman, always in search of a future. To be merely a human type means that individuality is secondary. As Nicholas More tells us, Nietzschean Zarathustra wears three masks: he is simultaneously a "singing poet, a warrior, and an independent thinker" (More, 2014: 165). As such, Zarathustra is also a free spirit, and, in that sense, he is a human type. But, as a symbolic and allegorical figure, Zarathustra stands alone with his non-human animals—the serpent and the eagle. Nietzsche specifies: "I found life more dangerous among men than among animals" (1966: 25).

Two levels of discourse are present here. One is the descriptive; the other is the prescriptive level, which qualifies certain human types as being close to or far from the ideal (Haar, 1985: 7). The ideal is the *Übermensch*, but Nietzsche thinks that few will ever attain that state, so he approaches the issue indirectly by dismantling old truths, which are nothing but accepted falsities. In an unforgettable moment of "revelation," Zarathustra qua *Übermensch* reaches the fundamental idea (which is not a concept) of the Eternal Recurrence, an idea that is hardly communicable to the herd. This idea makes Zarathustra—as close as it's possible to be—an *Übermensch*.

A different mode to introduce the truth or falsity of a given moral and ethical belief consists of rejecting the old truths that have become obstacles to the emergence both of the free spirit and of Zarathustra, the new Oracle who announces the *Übermensch*; that is, a new human, or even non-human, capable of responding to the deficiencies of the past and present, and who would bring forth a superior worldview. Nietzsche's prescriptive level of discourse presupposes that to be human means to be able to change and enhance life in order to attain the highest possible worth. Yet the prescriptive level of discourse is immanent to the first level, the descriptive one. The will to power, which acts as a factor of change, is the welcome "obstacle" to the homogenization of different human types into a single category. Such new "pluralism" becomes a truth indicating whether or not the will to power shows signs of decadence.

In Nietzsche's *Human, All Too Human*, the prescriptive level of discourse is one with the descriptive level, as together they discuss those free spirits who somehow have broken the spell of unquestioned old truths. To be human is human and necessary, to be *too* human is an indication of having rejected what one could have become. As is the case with the *Übermensch*, who is spoken of in a variety of utopian or mythical languages, the free spirit also remains an elusive but lively human type. Rightly so, since Nietzsche refuses to give specific rules of conduct instead of accepting what could resemble a Kantian "ought." What is decisive, instead, is Nietzsche's conviction, as he writes in *Human, All Too Human*: "Men can *consciously* decide to develop themselves forward to a new culture, whereas formerly they developed unconsciously and by chance" (Nietzsche, 1984: 30). One-sided as the voluntarism invoked in this passage may be, Nietzsche also tells us: the will to power allows us to change the old, stagnant conditions that made the human *too* human, that is, too inept vis-à-vis becoming.

In an uncompromising philosophical criticism, Nietzsche discusses the human beings that emerge from the confining space of moralism and conformity. He calls them "the herd"; that is, undesirable and detrimental modes of being that pose the question of the human and the non-human one more time. It is clear that Nietzsche did not believe in the idea that human beings are perfectible, certainly not on a large scale, and that progress eludes us because history is in the hands of fate or chance, not of providence. On this point, the words of Emil Cioran are more than pertinent as applied to Nietzsche: "The true prophet is he who suffers from the obsession of the future without believing in 'progress'" (Cioran, 1997: 223).

As opposed to the humanity of the free spirit, the humanity of the bound spirit is more questionable. The last Man is also considered by Nietzsche as a kind of happy slave who is content in his utter mediocrity. He is not in search of a higher destiny, of a life worthy of the over-human and of the Dionysian passion. According to Gooding-Williams, the last Man is the modern man, indifferent to the impending nihilism and, therefore, indifferent to the typical dangers of modernity (2001: 84–90). The last Man is, in fact, human to the point of disappearing.

It is indicative that Nietzsche, the philosopher of the human and of life, had to venture into the "more than human" to make his interest in the human plausible, and to justify it philosophically. The prescriptive recommendation to aim at a higher human level tells us to leave behind what is detrimental to human health. To reach this end, to experiment, even with oneself, becomes essential. In order to break the vicious circle of the merely human, Nietzsche gives us the merciless descriptions of what is not so human, including its physiology, passions, fears, and weaknesses. The human body will tell us what it means to be human and not a non-human animal.

The obstacles to arriving at the extraordinary vision of the Eternal Recurrence, and to understanding what it takes to live an over-human life, are by now well known. The conditions in which the human finds itself are dangerous to the point of indicating a diagnosis of a serious form of devitalization that limits experiencing new possible outcomes. It is a philosophical, and not merely a psychological, task whereby diagnosis and therapy are identified in the overcoming of the merely human and of the non-human. However, Nietzsche is cautious concerning a prognosis, so much so that the *Übermensch* is, for Nietzsche, not an event foreseeable in the immediate future.

For Nietzsche, to propose a return to nature would be a regression, even considering that human life certainly belongs to nature. Nature can be reassuring, witness the aphorism that proclaims: "We like to be out in nature so much because it has no opinion about us" (Nietzsche, 1984: 237). In fact, nature, in itself, does not have a specific mission. Moreover, there is no such thing as a "natural human being." To enhance human life, as Nietzsche insists it must be done, is to adopt forms of comportment that are not specifically natural (instincts notwithstanding). To enhance life means to put into "nature" something more or to push artificiality to its extreme limit, backed by the will to power. To determine what this "more" is, raises the parallel difficult question of what or who the *Übermensch* is.

Yet Zarathustra is surrounded by non-human animals. He loves them, particularly the serpent and the eagle—those symbols whose meanings speak to us. The serpent as earth animal; the predatory eagle, its exact opposite. This is an indication that Zarathustra subsumes the whole of the planet Earth in its dual aspects. A first approach to the cosmic idea of the Eternal Recurrence.

In the passage "On the Three Metamorphoses" in *Thus Spoke Zarathustra,* we find an illustration of how non-human animals add to the symbolism of Nietzsche's work. Zarathustra mentions two more symbolic animals that speak of the two human conditions. First, as mentioned above, the camel; an animal whose burden comprises all the moral norms of the herd and all the unnecessary values that burden human beings. Left to themselves, human beings stagnate, are prey to envy, pity, and impotence vis-à-vis their burden—victims of what Nietzsche calls "The spirit of gravity." I quote from *Thus Spoke Zarathustra*: "'Yes, life is a grave burden.' But only man is a grave burden to himself! That is because he carries on his shoulders too much that is alien to him. Like a camel, he kneels down and lets himself be well loaded" (1966: 193). The second metamorphosis involves the wild lion and, finally, a human animal—the child, when we have a new beginning, an innocent being (Loeb, 2010: 106–107) who is perhaps the precursor of the *Übermensch* at the almost embryonic state.

After having explored the possibilities of what it means to be human, with all the limitations he describes, Nietzsche is ready to put forth his extraordinary "revelation"—namely, his idea of the Eternal Recurrence, the significance of which is willfully kept obscure. But if we explore further what the Eternal Recurrence means, we must consider that Nietzsche has prepared the route for a clearer explanation when he introduces in his oeuvre the relevance of the different human types and non-human types, including the role of the different animals, in particular the serpent and the eagle. Additionally, I maintain that the idea of the Eternal Recurrence is not to be taken literally. It is, rather, a vision; an experience so rare and intense that it cannot stand any comparison to what has been held in the historical past. It is not to be excluded that Nietzsche's personal experience qua unique evokes an intense rapture inclusive of both the idea of the Eternal Recurrence and of the *Übermensch*. But this is just the beginning of a spiral of what is to happen in an eternal future.

The theme of the Eternal Recurrence of the same does not require a transition from rational concepts. In Nietzsche we do not find transitions of a dialectical kind that unite the human types such as the free spirit and the last Man. What is peculiar to the "last phase" of the Nietzschean "evolution" is that the Eternal Recurrence occurs suddenly, and that fact has the character of a revelation, similar to a spiritual revelation. What it does reveal, though, is hard to ascertain. As a revelation, it has a content; as a vision, it is a worldview complete in itself. Its own completion does not belie the fragmentary nature of the earlier Nietzschean views, but it does not contradict them. The idea of the Eternal Recurrence does not introduce a different worldview, the world will eternally remain the same in its cosmic, terrifying aspects.

Temporal considerations are not extraneous to this view, and yet they remain marginal, since time is the first "victim" of a view that presents itself almost like a *fresco* in which human types and non-human types remain what they are ... as they have always been. A simplistic conception of an evolution connecting humanity's past, present, and future would make the philosophical idea of the Eternal Recurrence, and with it the *Übermensch*, a general way to solve the issue of placing these two crucial views in a historical framework with temporal implication extraneous to the typological relevance Nietzsche's methodology.

The pressing question instead now is to determine whether the *Übermensch* is actually a type or a being who goes beyond such generality. If he were a type—as, for instance, the free spirit—a future *Übermensch* would lose all the characteristics evoked by this idea. Everything would lose the peculiarity of the idea of the Eternal Recurrence. Everything would collapse into a senseless multiplicity, and the mysterious, unique, and eternally futural "over" would be reduced to a banality. What would be lost is the meaning of the Eternal Recurrence, which would be reduced to a subjectivistic ecstatic and sublime emotion not too different from those of the last Man declaring the death of God, rather than being the cosmic and terrifying meaning elevating Zarathustra to a timeless life moment.

References

Cioran, E. (1997). *Cahiers: 1957–1972*. Paris: Gallimard.

Del Caro, A. (2004). *Grounding the Nietzsche rhetoric of earth*. Berlin: Walter de Guyter.

Gooding-William, R. (2001). *Zarathustra's Dionysian modernism*. California: Stanford University Press.

Haar, M. (1985). Nietzsche and metaphysical language. In D. B. Allison (Ed.), *The new Nietzsche: Contemporary styles of interpretation* (pp. 5–36). Cambridge, MA: The MIT Press.

Loeb, P. S. (2010). *The death of Nietzsche's Zarathustra*. New York: Cambridge University Press.

More, N. D. (2014). *Nietzsche's last laugh: Ecce Homo as Satire*. Cambridge, MA: Cambridge University Press.

Nietzsche, F. (1966). *Thus spoke Zarathustra: A book for all and none* (W. Kaufmann, Trans.). New York: The Viking Press.

Nietzsche, F. (1984). *Human, all too human: A book for free spirits* (M. Faber, Trans.). Lincoln, NE: Nebraska University Press.

Ponton, O. (2007). *Nietzsche: Philosophie de la légèreté*. Berlin: Walter de Gruyter.

Phenomenology and Formal Ontology: A Theoretical Model of Max Scheler's Early Phenomenology of Sense Perception

Martina Properzi

The topic of this paper is a contemporary reading of the ontological dimension of Max Scheler's early phenomenology of sense perception, which will be performed within the rigorous framework of the so-called Formalized or Formal Ontology (FO. See § 2). As a first step in this direction, it is important to give some general information about Scheler's philosophical work. To this end, in Fig. 1 I have schematically reconstructed the three different periods of Scheler's philosophical production. For each period, I have taken into account the main historical sources, fields of interest and works. Here I would like to briefly recall Scheler's reception of contemporary scientific research.

However, to understand the theoretical meaning of Scheler's work it is necessary to include another important aspect: the comparison with contemporary sciences. More precisely, in order to introduce the topic of this paper, it should be highlighted that, in the length of time during which Scheler works out his early phenomenology of sense perception, that is, between 1908-1909 and 1914, and 1914, he delves into the study of consolidated and emerging scientific paradigms, disciplines and theories as well as their empirical results (1993). His research follows two main directions:

1. A negative direction focused on the criticism of the "mechanist ideology" that tries to expand the application domain of the physical theory known as classical mechanics to other disciplinary fields (e.g. physiology) and/or scientific areas (e.g. life sciences)
2. A positive direction focused on the investigation of the theoretical basis of the emerging biological paradigm in the contemporary life sciences

As a main consequence, this research leads to the production of an original epistemology (Scheler, 1986a, b; Properzi, 2018a) and a phenomenological philosophy of life (Scheler, 1993; Bosio, 2000; Verducci, 1997).

M. Properzi (✉)
Pontifical Lateran University, Rome, Italy
e-mail: martinaproperzi@alice.it

C. A. Hornbuckle et al. (eds.), *Phenomenology of the Object and Human Positioning*, Analecta Husserliana 122, https://doi.org/10.1007/978-3-030-66437-4_5

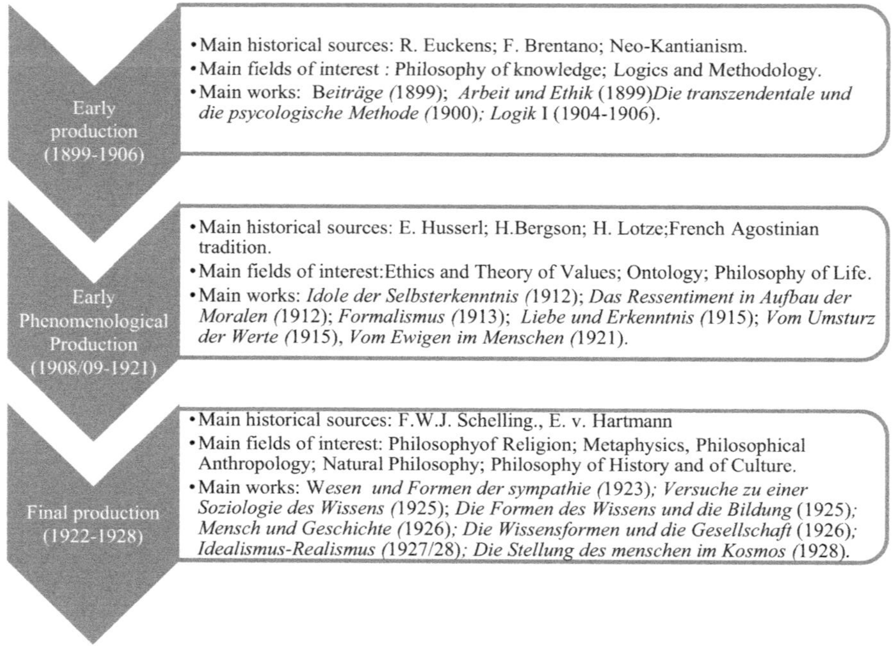

Fig. 1 Schematic representation of the three different periods of Scheler's philosophical production. For each period the main historical sources, fields of interest and works are taken into account (Hartmann, 1963; Avé-Lallemant, 1975)

1 Max Scheler's Early Phenomenology of Sense Perception

The preliminary remarks mentioned above allow me to approach directly the topic of this section: Max Scheler's early phenomenology of sense perception. In order to examine this subject matter, let me sketch beforehand the background of Scheler's proposal.[1]

Strongly influenced by the French philosopher Henri Bergson and the German Baltic biologist Jakob von Uexküll, Scheler compares himself critically with the main proponents of the philosophical schools of neo-Kantianism, empiriocriticism and American pragmatism (Scheler 1986a; 1993) as well as of the contemporary scientific theories of psychophysics and associative psychology (Scheler, 1916; 1993). Scheler criticizes also ancient and modern philosophies of sense perception, such as the influential philosophies of Aristotle and of modern empiricism and rationalism (Scheler, 1955). However, the main target of Scheler's criticism is Immanuel Kant, with which the author establishes a close comparison in his most famous work, *Der Formalismus in der Ethik und die materiale Wertethik* (1916, 49–58; 139–161).

Clearly, in the 1910s Scheler's phenomenology of sense perception is already a well-articulated theory. Indeed, the more mature theory of the 1920s, which was systematically presented in the work Erkenntnis und Arbeit (1960), depends either work-genetically[2] or theoretically on the research done during the previous decade, even if some relevant modifications occurred.[3] Scheler elaborates a non-representationalist theory of sense perception, whose core aspect is a selective interpretation of the synthesis or identification of the perceptual object based on a minimal, pre-reflective kind of motor consciousness that Scheler usually names "dissociation" (Dissoziation) (Properzi, 2018b)". I reckon that it is possible to expose the core of Scheler's theory, i.e. its ontological dimension, by breaking it up into five points. It is important, however, to keep in mind that my attempt does not claim to be a complete reconstruction of Scheler's proposal but only to provide a textual basis for the formalization pursued in section 3.

Point 1. Sense perception as a cognitive component of the natural worldview. As a cognitive component of the natural worldview (natürliche Weltanschauung), sense perception brings phenomenal contents to common sense, interpreted by Scheler as a sound practical judgement originating from the intellect (Menschenverstand) which is shared by ('common to') all human beings. The author looks at the worldview as to a collective attitude of experiencing the world, which is objectified in the natural languages used inside what he called

[1] In this contribution I shall leave this point almost unexplored. A good introduction can be found in Cusinato (1986/1987) and Properzi (2018b).

[2] In spite of the fact that in the *Formalismus* Scheler refers to *Erkenntis und Arbeit* as a soon-to-be-published book (1916, 153), the book's first draft is datable only between 1923 and 1925.

[3] These modifications presuppose a deep rethinking of ontological categories and scientific data and mainly concern the interpretation of the concepts of "corporal thing" (Koerperding) and "perceptual form" (Form) (Scheler, 1979).

the "people's living community" (Lebensgemeinschaft des Volkes. See Scheler, 1986a, 451–452). Thus, for Scheler different worldviews are expressed by different natural languages. They share however the same set of conceptual forms (categories) that phenomenology brings to light by virtue of the method known as the eidetic reduction (eidetische Reduktion). The eidetic reduction allows the phenomenologist to trace phenomenal data back to the laws of meaning that regulate their essence, i.e. the set of their invariant properties, grasped as a self-evident object of an ideal or eidetic intuition (1986a, 447–448).

Point 2. Sense perception as a practical kind of knowledge that emerge from behavioral adaptations. Scheler gives a practical interpretation of sense perception also from the point of view of the individual (1968). American pragmatists are right to underline this aspect: their considerations can help overcome the theoretical bias guiding the tradition of Western philosophy (1993, 311). However, the description offered by American pragmatists has a main limit, that is precisely in the fact that the practical nature of sense perception has been understood on psychological basis (making reference for example to the concept of need). For Scheler, instead, it emerges from the behavior that humans and other living beings perfom to be able to continue to exist in their environment. The author was strongly influenced by Uexkull's interpretation of the behavioral adaptive dynamics, where the organism and its surroundings constitute a coupled interacting system it comes instead from Uexküll's idea of a "double" matching of organism/environment. Figure 2 shows Uexkull's original representation of the so-called functional cycle (funktional Kreis) that models the organism-environment coupled interacting system.

Scheler describes the interaction of the organism and its surroundings in the following way:

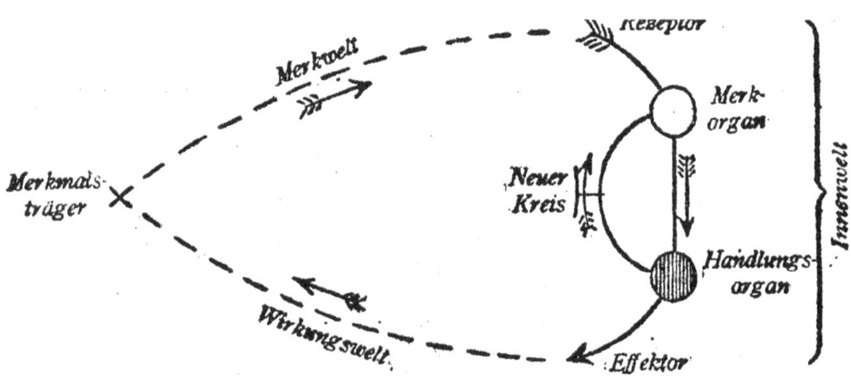

Fig. 2 Uexkull's original representation of the functional cycle with negative feedback control (Uexkuell, 1928, 117)

> Sind die wesenhaften Akte, die zu einer Erkenntnis führen können, aber gegeben, so verstehen wir dagegen sehr wohl, daß die Auswahl zwischen ihren besonderen möglichen Inhalten für Lebewesen–sofern Lebewens Träger diser Akte sind–durch spezifische Lebensfunktionen und durch für sie eingerichtete spezifische Organe hindurch erfolgt, nämlich durch die Sinnesfunktionen und die Sinnesorgane ... Dann ist aber auche dises "Aktionssystem" eines Lebewesens das leitende für die Ausbildung der Sinnesfunktionen, wogegen die Sinnesorgane erst innerhalb des Spielraumes dieser und ihrer Eigengesetzmäßigkeit auf Grund der Besonderheiten der realen Welt und ihrer Wirkungen auf das Lebewesen sich ausbilden. Das "Aktionssystem" bestimmt dann sowohl die anatomische Struktur des Organismus wie sein "Milieu", von denen keines die Wirkung des anderen ist–obzwar sie streng einander angepaßt sind. (Scheler, 1986a, 439)

The personal contribution of the author is the detection of an action system (Aktionssystem) as the mover of the adaptive dynamics on the organism's side. Clearly, this contribution depends on the assimilation of Bergson's interpretation of the body as a 'centre d'action that is "capable d'exercer une action réelle et nouvelle sur les objets qui l'entourent" (Bergson, 1986, 17). However, Scheler's action system can not be described by making reference to a "zone of indetermination" that depends on the organism's anatomical and physiological structure, and, more precisely, on the number of sensory cells placed between the terminal branches of the centripetal fibres and the motor cells of the Rolandic area, as suggested by Bergson (1986, 27–31). Indeed, for Scheler, there are well-defined foundational relationships between the action system and the lived body (Leib), i.e. the body-as-subjectively-experienced, considered in both its configuration, namely as Leibseele and Leibkörper (Scheler, 1916, 416). Furthermore, because of this dynamical correlation, Scheler's "Aktionssystem" always exercises its activity according to a prearranged set of possible dynamical configurations, and not in accordance with motor possibilities determined by a given anatomical and physiological structure.

Point 3: The selective synthesis of the perceptual object. Scheler describes closely the structure that underpins the perceptual synthesis. A strong influence on Scheler's description was exercised by Uexkull's biosemotic theory of animal (external) sensibility. As T. Ruting explains,

> This theory 'had to pave the ground for further research into the problem of access to the *Umwelt* of animals. This basic research could only reveal a first outline of the realizable *Umwelt* of the animal. However, by investigating the animal's ability to perceive and discriminate different physical stimuli, Uexküll tried to get initial indications of their significance for the animal's behavior–firstly, ideas about the signs that could possibly constitute the animal's *Umwelt*. For Uexküll this was the basic methodology to analyze the 'subjective space' ("der subjektive Raum") of the animal. Uexküll and his assistant Friedrich Brock tried to give the reader a vivid demonstration of the results of basic *Umweltforschung* in the new laboratory and published illustrations of the different *Umwelten* of different living beings. Normal photographs represented the human *Umwelt*. By using grids with different pitches of the matrix, the resolution of the compound eye of a fly (*Musca*) or the eye of a mussel (*Pecten*) was emulated. The pitch of the raster was corresponding to the frequency of sensory elements within the eyes of the animals. These dots were called "Sehorte," 'visual locations' in the visual space. In order to eliminate the artifacts of the grid, water-

color paintings of the supposed *Umwelten* were made. However these pieces of art were based on scientific grounds and were later reproduced in the famous *Stroll through the Umwelten of animals and humans*. The pictures helped to make the viewer aware of how differently humans and animals perceive the world they share. Therefore they served as a method for intersubjective and interspecies understanding–as the first fascinating steps into the *Umwelten* of other organisms. (2004, 56–57)

Uexkull's account of animal (external) sensibility is explicitly linked to his proposal of expanding Kant's interpretation of the synthesis or identification of the perceptual object beyond human perception (Uexkuell, 1928, xv–xvi). Clearly, this point could not be shared by Scheler. Instead of making reference to Kant's interpretation, the author stresses a core aspect of Bergson's analysis: the characterization of the perceptual synthesis as a selective unity. To quote the French philosopher:

On pourrait résumer, en effet, nos conclusions sur la perception pure en disant qu'il y a dans la matière quelque chose en plus, mais non pas quelque chose de différent, de ce qui est actuellement donné. Sans doute la perception consciente n'atteint pas le tout de la matière, puisqu'elle consiste, en tant que consciente, dans la séparation ou le "discernement" de ce qui, dans cette matière, intéresse nos divers besoins. Mais entre cette perception de la matière et la matière même il n'y a qu'une différence de degré, et non de nature, la perception pure étant à la matière dans le rapport de la partie au tout. C'est dire que la matière ne saurait exercer des pouvoirs d'un autre genre que ceux que nous y apercevons. Elle n'a pas, elle ne peut receler de vertu mystérieuse. Pour prendre un exemple bien défini, celui d'ailleurs qui nous intéresse le plus, nous dirons que le système nerveux, masse matérielle présentant certaines qualités de couleur, de résistance, de cohésion ..., possède peut-être des propriétés physiques inaperçues, mais des propriétés physiques seulement. Et dès lors il ne peut avoir pour rôle que de recevoir, d'inhiber ou de transmettre du mouvement. (1896, 79)

Point 4: The lived body as the form of organic data. Scheler describes the lived body as the 'form' of organic data (1916, 417-419). This position is developed by the author in close connection with the criticism of Conrad Avenarius, Ernst Mach and Gustav Theodor Fechner's interpretation of the body-as-lived. For Scheler, indeed, Avenarius and Mach's empiriocriticism and Fechner's psychophysics share a common mistake when they describe the relationship between Leib and organic sensibility. Both the theories invert the "order of giveness" (Gegebenheitsordnung) that structure the experience of the body-as-lived. In the empiriocriticim of Avenarius and Mach a not-yet-differentiated cluster of sensory and organic data is deemed as originary, and hence genetically prior to the individual's experience of the own body (1916, 419–424). Fechner's psychophysics assumes instead mental connections or associations between sensory and organic data (1916, 414–417). Against these theories, Scheler suggests that:

"Leiblichkeit" eine beßondere materiale Wesensgegebenheit (für die pure phänomenologische Anschauung) darstellt, die in jeder faktischen Leibwahrnemung als form der faktischen wahrnemung fungiert (wir können in Sinn unserer früheren Charakteristik alles Kategorialen auch sagen: als Kategorie). Dies schließt ein, daß sich diese Gegebenheit also wedere auf eine solche äußerer Wahrnemung, noch auf eine solche innerer Wahrnemung, noch auf eine Zuordnung von Inhalte beider Wahrnemungen züruckführen läßt; gleichweige gar auf einen Tatbestand induktiver Erfahrung d.h. der Wahrnemung eines beßonderen Einzeldinges. Und es schließt auch ein, daß ebenso umgekehrt der Leib niemals als eine primäre Gegebenheit anzusehen ist, auf deren Fundament ein allerst als

> psychophysisch indifferent "Vorgefundenes" sich durch ein verschiedenartiges Verhältnis zum Leib als "Psychisches" und "Physisches" differenziere und abhebe. (1916, 413–414)

Point 5: space and time as the forms of sensory data and the enviromental thing as the synthetic unity of dissociative consciousness. Scheler delves into an analysis of space and time as forms of sensory data. Differently from the Kantian view, where the so-called 'forms of sensible intuition' are framed as components of a cognitive experience, Scheler's focus is on the way they organize the experience of the organism interpreted as an agent that interacts with a surrounding environment:

> Auch hier ist die Entfernung der Punkte immer zuerst ins Auge gefaßt; nicht wie bei der Lebensbewegung, wo die Verkleinerung der Entfernung im Nachlassen der tendenz als Langsamkeit, im Wachsen als Geschwindigkeit und im Erlebnis des wachsenden Spannens als Funktion der wachsenden Spannung erlebt wird. Da ich hier auch bei Annäherung der Punkte, sich verkleinernder Entfernung doch auch eine Entfernung ins Auge fassen muß (negative Entfernung), so scheidet die Langsamkeit aus. Ist in gleichen Zeiten die Stecke kleiner, die er durchquert, so ist die Geschwindigkeit nun kleiner geworden. Hinsichtlich der Zeit bestehet ähnliches: in der Lebensbewegung erlebe ich jede Phase als die Erfüllung der vorausgehenden Bewegungsintention und die ganze Bewegung al seine einheitliches Akt meiner selbst. (Scheler, 1993, 331–332)

This position is strongly influenced by Bergson, a point to which Scheler himself paid attention (1972, 330–331). Following the French philosopher, Scheler describes a minimal, pre-reflective kind of motor consciousness whose function is to dissociate those elements of the network of organic and sensory stimuli that are relevant to increase behavioral adaptations. Maybe the most interesting aspect of Scheler's descriptive analysis of dissociation is the concept of enviromental thing (Milieuding). This concept refers to the synthetic unity of the dissociative consciousness that embraces and identifies both the lived body and the external environment. In the following text Scheler draws attention to Bergson's contribution to the description of sense perception interpreted as the cognitive structure of dissociative consciousness

> Wenn jemand einen Baum wahrnimmt, so fügt sich nach Bergson nicht an die Endstelle des sensorischen Reizprozesses im Gehirn irgendwie ein neues, durch den Gehirnvorgang bestimmtes "psychisches Gebilde" zeitlich an; vielmehr setzt sich der nervös geleitete Reizvorgang in einen motorischen Impuls um, während gleichzeiting aus der Umwelt, die der Perzeption stets als daseiend in allen möglichen Stufen der Vagheit und Helle mitgegeben ist, sich irgendeine Seitenansicht des Baumes selbst "aufleuchtend" heraushebt und den Charakter eines "Lebhaftigkeitswert" über Null gewinnt, ohne den e seine bewußte "Perzeption" nicht gibt. (1972, 334)

To sum up the contents of the five points described above, Scheler's early phenomenology of sense perception fundamentally depends on the characterization of a minimal, pre-reflective kind of motor consciousness named by the author dissociation (Dissoziation). Dissociative consciousness is at play during adaptive behavior and allows for the identification or synthesis of the enviromental thing as a perceptual object structured by material components (i.e. organic and sensory data) and formal components (i.e. the lived body and the forms of space and time). This kind of individual consciousness is part of a linguistically mediated collective structure of practical experience, i.e. the natural worldview, according to which the world is pre-given to (the individual experience of) all human beings.

2 What Is Formal Ontology?

In order to provide a brief introduction to FO, it is necessary to sketch its disciplinary background. FO originates from the close cooperation between the scientific disciplines of cognitive engineering and computer science, which has been established during the last decades to advance the simulation of semantic properties distinctive of natural intelligence and knowledge-acquisition in human beings (Studer, Benjamins, & Fensel, 1998). Following previous work (Gruber, 1993; Studer, Benjamins, & Fensel, 1998), Guarino and colleagues provide for a definition of the concept of ontology as it is used by cognitive engineers and computer scientists, i.e. as Computational Ontology (CO) (see Fig. 3):

> **Definition 3.4 (Ontology)** Let **C** be a conceptualization, and **L** a logical language with vocabulary **V** and ontological commitment **K**. An ontology $\mathbf{O_K}$ for **C** with vocabulary **V** and ontological commitment **K** is a logical theory consisting of a set of formulas of **L**, designed so that the set of its models approximates as well as possible the set of intended models of **L** according to **K**. (1998, 11) (Guarino et al., 2009, 11).

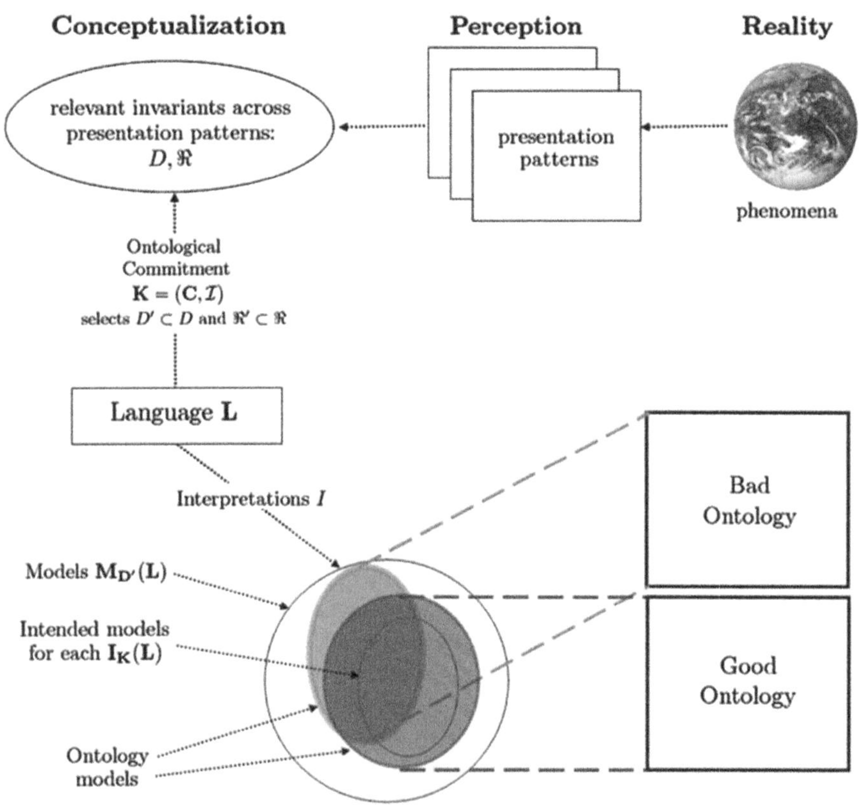

Fig. 3 Diagrammatic representation of the relationships between phenoma occuring in reality, their (time-variant) perception and abstract conceptualization, the language used to talk about such conceptualization, its intended model and ontologies. See Guarino et al., 2009, 9

As observed by the authors themselves, this definition shows some problematic aspects

> Indeed, one may correctly argue that it is not possible to share whole conceptualizations, which are private to the mind of the individual. What can be shared are approximations of conceptualizations based on a limited set of examples and showing the actual circumstances where a certain conceptual relation holds (for instance, actual situations showing cases where the *cooperates-with* relationship occurs). Beyond mere examples it is also possible to share meaning postulates, i.e., explicit formal constraints (e.g., the relationship *cooperates-with* is symmetric). Such definitions, however, presuppose a mutual agreement on the primitive terms used in these definitions. Since however meaning postulates cannot fully characterize the ontological commitment of primitive terms, one may recognize that sharing of conceptualizations is at best partial. For practical usage of ontologies, it turned out very quickly that without at least such minimal shared ontological commitment from ontology stakeholders, the benefits of having an ontology are limited. The reason is that an ontology formally specifies a domain structure *under the limitation* that its stakeholder understand the primitive terms in the appropriate way. In otherwords, the ontology may turn out useless if it is used in a way that runs counter to the shared ontological commitment. (Guarino et al., 2009, 14) (1998, 14)

In other words, the problem here is how to bring into harmony the so-called 'shared-condition' of the conceptualization seen as a structure of meaningful relations with the impossibility of obtaining 'a mutual agreement' on its primitive terms. These letters are used to reproduce the condition of meaningfulness distinctive of human natural intelligence and knowledge-acquisition. To understand why the problem arises, it is useful to bring to light a very general assumption of Guarino's position, i.e. the distinction between CO and philosophical ontology formulated in terms of the opposition between (the ontological commitments of) formal and informal conceptual structures (Guarino et al., 2009, 1). As a main limit, this formulation overlooks the progress made in the so-called Formal Philosophy (FP), a comprehensive approch to philosophical studies whose primary aim is to formalize through the axiomatic method traditional philosophical concepts and theories to make them usable in contemporary multi-, inter- and trans-disciplinary research (Hendricks & Symons, 2005). In this context, several formal definitions of ontology have been produced, which can be compared with the one formulated by Guarino and colleagues from the perspective of CO.

As G. Basti pointed out: The necessity of formal philosophy became particularly urgent in TCS, because of the emergence of the semantic problem in CS, in the two key-domains of semantic programming and of functional programming. Both of them, indeed, have in the algebraic version of category theory and of modal logic—traditionally, two main formal tools of the philosophical logic—the core of their amazing developments in TCS, during the last twenty years. (2015, 3). In this contribution I do not inted to reconstruct the history of Philosophical Formal Philosophy (For an historical contribution see Properzi, 2015). My aim is to tackle the topic of PFO from a theoretical perspective. To this end, I will follow the interpretation of PFO elaborated by N.B. Cocchiarella (2001; 2007). In comparison with other approaches, Cocchiarella's interpretation reveals a strong foundational basis, as I am going to show. There are three main aspects in Cocchiarella's interpretation of PFO: 1) the distinction between logical languages (**L**) and natural languages (**NL**); 2) the detection of the general structure of PFO; 3) the formulation of the idea of a compared formal ontology (2001, 2007). Let me present these three aspects in more details.

In regard to the first aspect, Cocchiarella distinguishes between **L** and **NL** by making reference respectively to their logical and ontological nature. He earns this distinction through a criticism of Frege's notion of propositional function focused on the reduction of the ontological concept of predication (i.e. 'being predicated of') to the set theoretic concept of membership (i.e. 'being a member of'). In order to go beyond this Frege's reductionism, Cocchiarella analyzes both the concepts, i.e. the concept of predication and the concept of membership. Membership is defined by self-identity of its members:

Membership. *Def.*: $a \in \mathbf{V} \equiv \hat{a}\ (a = a)$.

Predication is defined by a "mutual saturation" of referential and predicative concepts (2007, 142–144), as displayed in Fig. 4.

Cocchiarella detects three basic theories of predication, on the basis of which it is possible to classify historically developed theories (2001).

Considering the second aspect, Cocchiarella describes the general structure of PFO as composed of

– a logical system, i.e. a deductive apparatus or a logic that consists of axioms (or axiom schemata), derivation rules used to derive theorems of the system and a semantics (tipically, the model-theoretic semantics based on the concept of model or frame, i.e. structures that satisfies all the axioms of the system);

– ontological forms or categories;

– an ontological grammar, i.e. a set of rules that allow to form ontological formulas from ontological categories;

– ontological axioms and ontological rules.

The third aspect of Cocchiarella's interpretation of PFO is the idea of a compared formal ontology that furnish a sort of platform to compare different ontologies as far as their soundness and completeness are concerned.

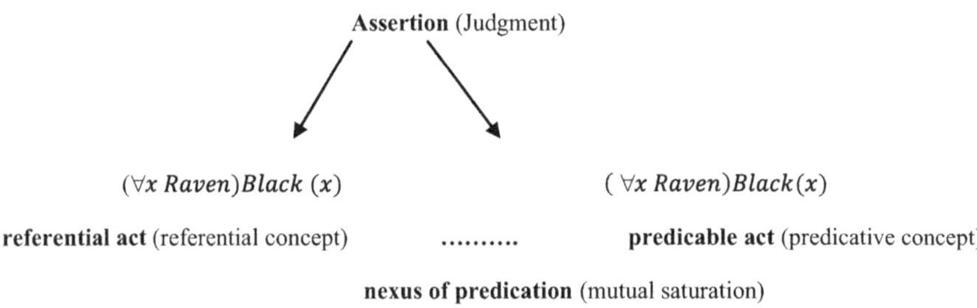

Fig. 4 Representation of the nexus of predication as "mutual saturation" of a referential concept expressed by a referential act, and a predicative concept expressed by a predicable act in the judgment's unity expressed by the assertion (Cocchiarella, 2007, 144–145)

> We do not always have to decide in advance whether or not there must (or even can) be a final completeness to the categories or of the laws of logic before undertaking such a research program. In particular we can try to establish restricted or relative notions of completeness for special area of a formal ontology, and then we can compare and evaluate those results in the context of a comparative formal ontology. The construction of abstract formal system and model theoretic semantics within set theory will be especially useful in carrying out and comparing such research programs. In other words, set theory is an ideal framework within which to carry out comparative analyses of different formal systems proposed either as a formal ontology or a subsystem of such. Set theory is not itself a formal ontology. (2007, 19)

Starting from Cocchiarella's work, in the following section I will attempt to interprete the ontological aspect of Scheler's early phenomenology of sense perception within the rigorous framework of PFO.

3 A Theoretical Model of Max Scheler's First Phenomenology of Perception

The reconstruction of the ontological dimension of Scheler's early phenomenology of sense perception carried out in section 1 has shown the central role played by the ontological dimension. This centrality depends on the way in which Scheler describes the minimal, pre-reflective kind of consciousness that enables perceptual experience, i.e. dissociation. In order to formalize this core dimension of Scheler's theory in the context of (Cocchiarella's interpretation of) PFO, it is first of all necessary to find a suitable deductive system.

To this end, let me focus on the logical nature of dissociation. From a logical point of view, dissociation may be represented as a relation (selection) which is reflexive. I.e.: $\forall p\ (select)p \rightarrow p$. A deductive system which is sufficiently rich to express reflexive relations and the (ontological) necessity thereof is the modal system **KT**. As a modal system, **KT** includes the axiom and derivation rules of the classical propositional calculus **k** plus the so-called Necessitation Rule **N**. This rule may be expressed as follow: 'if from a set of formulas it is possible to derive a formula, then from the same set of formulas it is possible to derive also that the formula is necessary'. **T** is the so-called axiom of reflexivity: 'if a formula is necessary, then it is', where the existence of a formula means that it may be inferred in the deductive system without contradition. To sum up, the modal system **KT** consists of axioms and derivation rules

D(KT) = D(K) + T

where **K** is **D(K) = D(k) + N** and **T**.

Another more suitable way to logically represent dissociation is as a relation (selection) which is Euclidean. I.e.: $\forall p, q, r\ (select)p, q \wedge (select\ )p, r \rightarrow (select)q, r$. A deductive system which is sufficiently rich to express Euclidean relations and the (ontological) necessity thereof is the modal system **K5**. The definition of **K5** is similar to the definition of **KT**, save for that **K5** includes the axiom **5** instead of **T**. **5** may be expressed as follow: 'if a formula is possible, then it is necessary that it is possible".To sum up, the modal system **K5** consists of axioms and derivation rules

D(K5) = **D(K)** + **5**

where **K** is **D(K)** = **D(k)** + **N** and **5**.

From a logical standpoint, therefore, dissociation is a relation (selection) which is reflexive and Euclidean. A deductive system which is sufficiently rich to express reflexive and Euclidean relations and the (ontological) necessity thereof is the modal system **KT5** or **S5**:

KT5 with **D(KT5)** = **D(KT)** + **5**.

The next step in the direction of interpreting the ontological dimension of Scheler's early phenomenology of sense perception within the rigorous framework of PFO is to introduce ontological categories, an ontological grammar, ontological axioms and ontological rules. For the sake of simplicity, in what follows I will focus on ontological axioms, showing how they are satisfied by a version of the model-theoretic semantics of **KT5**, the so-called possible words semantics.

According to the reconstruction conducted in section 2, I suggest to introduce two non-logical or ontological axioms. That is, an onto-epistemic axiom that guarantees the non-reducibility of the objectual form to the in-formed data, in our case organic and sensory data, and an onto-dynamic axiom that guarantees the spatio-temporal variability of the synthesis or identification of the object, in our case the perceptual object.

Now, we are able to concentrate on the (onto-)epistemic-dynamic interpretation of **KT5**. Let then u be the percipient and v, w, z relevant data either organic or sensorial. Now the Euclidean relations uRv, uRw, uRz represent their dissociation from the network of potential enviromental stimuli. Euclidean property of R involves secondary transitive relation (vRw, wRv, vRz, zRv, wRz, zRw), secondary symmetrical relation ($vRw = wRv$, $vRz = zRv$, $wRz = zRw$) and secondary reflexive relation (vRv, wRw, zRz). These relations represent the integration and mutual stabilization of organic and sensory data in function of the percipient's motility, that is, a mutual invariance of organic and sensory data obtained whith the constitution of the perceptual object, and hence the organization of the data by means of the (objectual) form. This letter is represented in the frame as an equivalence relation given by secondary reflexive relation + secondary transitive relation + secondary symmetrical relation (see Fig. 5).

The reflexive relation (uRu) represents the status of the percipient before stimulation. Together with the Euclidean relations, reflexive relation can be used to represent the percipient as a boundary condition of the perceptual experience.

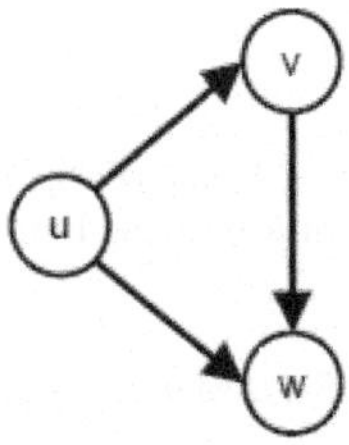

Secondary transitive relation: (om *u*) (om *v*) (om *w*) (*uR*v et *uRw*⇒ *vR*w)

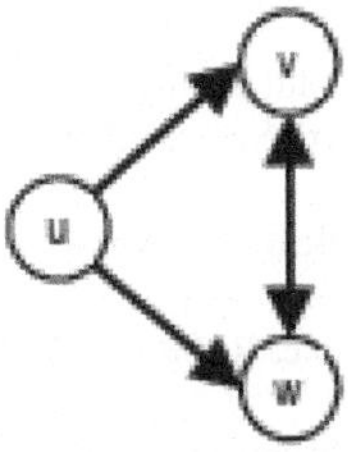

Secondary symmetrical relation: (om *u*) (om *v*) (om *w*) (*uRv* et *uRw*⇒*vRw* et *wRv*)

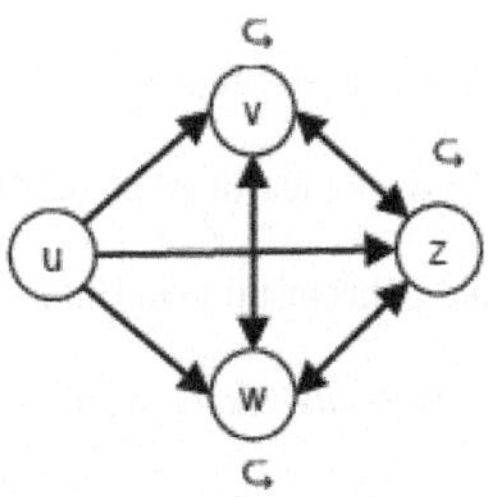

Formation of an equivalence class (secondary reflexive relation + secondary transitive relation + secondary symmetrical relation).

Fig. 5 Eulero-Venn diagrams show the formation of an equivalence class

This a point which is in line with Scheler's descriptive analysis, which concentrates on the ontological dimension of sense perception. If the elements of the frame are indexed, the frame may be used to represent the spatiotemporal variability of the perceptual object and of its boundary condition, i.e. the percipient, interpreted in this case as a sensorimotor agent. R can therefore represent the dynamic structure of the objectual synthesisis in the dissociative consciousness, according to Scheler's description.

References

Avé-Lallemant, E. (1975). Afterword. Bio-bibliographischer Anhang. In P. Good (Ed.), *Max Scheler im Gegenwartgeschehen der Philosophie* (pp. 267–284). Bern-München: Francke Verlag.
Bergson, H. (1896). *Matière et mémoire*. Paris: Alcan.
Bosio, F. (2000). *Filosofia e scienza della natura nel pensiero di Max Scheler*. Padova: IlPolignafo.
Cocchiarella, N. B. (2001). Logica e Ontologia. In *Science, theology and the ontological question*. Rome: Pontifical Lateran University. (Trans. of F. Marcacci. Logic and ontology. *Axiomathes* 12: 117–150).
Cocchiarella, N. B. (2007). *Formal ontology and conceptual realism*. Dordrecht: Springer.
Cusinato, G. (1986/1987). Intuizione e Percezione. Bergson nella prospettiva di Scheler. *Annali di Discipline Filosofiche dell'Università di Bologna, 8*, 117–145.
Gruber, T. R. (1993). A translation approach to portable ontology specifications. *Knowledge Acquisition, 5*(2), 199–220.
Guarino, N., Oberle, D., & Staab, S. (2009). What is an ontology? In S. Staab & R. Studer (Eds.), *Handbook of ontologies* (pp. 1–17). Berlin: Springer.
Hartmann, W. (1963). *Max Scheler. Bibliographie*. Stuttgart-Bad Cannstatt: Frommann.
Hendricks, V., & Symons, J. (2005). Formal Philosophy. Copenhagen: Automatic Press.
Properzi, M. (2014). Il linguaggio religioso è traducibile in uno analitico? In C. Canullo (Ed.), *Differenze e relazioni* (Vol. 3). Roma: Aracne Editrice.
Properzi, M. (2018a). Il giovane Scheler e l'epistemologia. Dialegesthai, 20, https://mondodomani.org/dialegesthai/mpr01.html.
Properzi, M. (2018b). Materia e Forma nella prima estetica fenomenologica di Max Scheler. Rivista internazionale di filosofia e psicologia, *9*(2), 162–177.
Rueting, T. (2004). History and significance of Jakob von Uexküll and of his institute in Hamburg. *Sign Systems Studies, 32*(1/2), 35–71.
Scheler, M. (1916). *Der Formalismus in der Ethik und die materiale Wertethik. Neuer Versuch der Grundlegung eines etichen Personalismus*. Halle: Niemayer.
Scheler, M. (1955). Die idole der Selbsterkenntnis. In M. Scheler (Ed.), *Gesammelte Werke* (Vol. 3, pp. 213–292). Bern-München: Francke Verlag.
Scheler, M. (1960). Erkenntnis und Arbeit. Eine Studie über Wert und Grenzen des pragmatischen Motivs in der Ernenntnis der Welt. In M. Scheler (Ed.), *Gesammelte Werke* (Vol. 8, pp. 9–423). Bern-München: Francke Verlag.
Scheler, M. (1968). Vom Wesen der Philosophie und die moralischen Bedingungen des philosophischen Erkennens. In M. Scheler (Ed.), *Gesammelte Werke* (Vol. 5, pp. 61–101). Bern-München: Francke Verlag.
Scheler, M. (1972). Versuche eine Philosophie des Lebens: Nietzsche–Dilthey–Bergson. In M. Scheler (Ed.), *Gesammelte Werke* (Vol. 3, pp. 311–341). Bern-München: Francke Verlag.
Scheler, M. (1979). Manuskripten zur Metaszienzien. In M. S. Frings (Ed.), *Gesammelte Werke* (Vol. 9, pp. 125–184). Bern-München: Francke Verlag.
Scheler, M. (1986a). Lehre von den drei Tatsachen. In M. Scheler (Ed.), *Gesammelte Werke* (Vol. 10, pp. 431–475). Bern-München: Francke Verlag.
Scheler, M. (1986b). Phänomenologie und Erkenntnistheorie. In M. Scheler (Ed.), *Gesammelte Werke* (Vol. 10, pp. 377–431). Bern-München: Francke Verlag.
Scheler, M. (1993). Biologievorlösung. In M. S. Frings (Ed.), *Gesammelte Werke* (Vol. 5, pp. 257–367). Bonn: Bouvier.

Studer, R., Benjamins, R., & Fensel, D. (1998). Knowledge engineering: Principles and methods. *Data & Knowledge Engineering, 25*(1–2), 161–198.
Uexkuell, J. (1928). *Theoretische Biologie*. Berlin: Springer.
Verducci, D. (1997), La phénoménologie de la vie et la philosophie selon Max Scheler. Analecta Husserliana, *50*, 165–180.

Puzzles in Phenomenology

Lucian Delescu

1 Preliminary Considerations

The challenge phenomenology must overcome is the integration of conscious intentional experiences within a scientific framework. In order to do that, two painfully entangled aspects must be clarified. The first is the *epistemological aspect* which is concerned with setting out a proper theoretical framework. The second is the *ontological aspect* and is concerned with the *explanation* of conscious intentional experiences. Ideally, one should begin with the ontological aspect. However, one becomes aware of conscious experiences when noticing the ontological implications of the "peculiar pattern of combination of the concepts, and truths which form the ideal unity of a particular science" (Husserl, 2001a, 2001b, 115). To be sure, higher-order propositions are not be taken as sole hints to the existence of consciousness. One is also right if arguing that consciousness can be inferred from the pattern of the simplest propositions such as "I am" or even from the pronoun "I." In other words, consciousness is involved in generating propositions with lower and higher epistemic values and anything in between at the same time. This is to say that, in order to explain consciousness, it is required to set out the proper epistemic conditions. By that, I mean conditions for unraveling the ontological implications of lower and higher-order propositions beneath which runs the conscious stream. In the last decades, there was a lot a debate regarding this matter. The result is a growing number of phenomenologists attempting to explain consciousness from an emotivist perspective. By "emotivism," a term first coined by A. J. Ayer, I mean the view according to which all mental activities arise at the intersection between senses and the external reality. The quest of emotivism is to realistically solve the problem of

L. Delescu (✉)
St. Francis College, Brooklyn, New York, USA

C. A. Hornbuckle et al. (eds.), *Phenomenology of the Object and Human Positioning*, Analecta Husserliana 122, https://doi.org/10.1007/978-3-030-66437-4_6

consciousness. But in doing away with cognitive processes, emotivism ends up from an ontological point of view by denying a fundamental feature of mind without accounting for sense-making (traditional emotivism) or by redefining it from a sense-interactivist perspective (phenomenological emotivism) which is the same. From an epistemological point of view, emotivism (re)commits itself to subjectivism which goes against its own realist claims. My claim is that reason is "the key feature of conscious life" (Husserl, 1970, 338). It is by explaining reasoning, no matter how difficult that might be, that we can begin to unravel the mystery of consciousness. In what follows, I deal with some differences between reason and emotion within phenomenology and argue for the need to reinstate reason as a key feature of consciousness.

2 Some (Onto-)Epistemological Puzzles

Let us talk about phenomenological emotivism which is the most recurrent since the early Sartre (2000) and Merleau-Ponty (2005). The question is: are there reasons to consider that there is compatibility between Husserl's phenomenology (I take as a cognitive theory of consciousness) and the early Sartre and Merleau-Ponty reinstatement of the role of emotion? Now, on the one hand, Husserl argues that "sensory-feelings" have no intentional character because they are of strictly "physical order." On the other hand, Husserl states that "pain and pleasure are of psychical order" and display intentionality (Husserl, 2001a, 2001b, 110). Supposing that "sensory-feelings" are peripheric extensions of their more complex pain pleasure arousal one may argue that Husserl have endorsed a form of emotivism. However, later in his work Husserl argues that "sensory-feelings" generate a "cogito" which job is to generate meaning about contents. Husserl describes the "cogito" as "essential for the phenomenology of intentionality" and "cogitatum" as, say, an application of the ability to make sense about individualized objects (Husserl, 1969, 135). Husserl also argues that "cogito" generates propositions such as "I perceive," "I remember," or "I expect" as "modes of consciousness," which belong to the doxic sphere thought although have nothing to do with "the sphere of predicatively thinking" (Husserl, 1969, 135)? Does Husserl want to say that (*a*) *all propositions are perceptual*, but some are tighter to perceptual contents (propositions with limited qualitative insight), some remote (propositions with relative qualitative insight), some remoter (propositions with absolute qualitative insight) and yet all perceptual? Or that (*b*) *some propositions belong to the perceptual and emotional sphere while some belong to the predicative sphere (cognitive)*? Or, finally, that (*c*) *all propositions are ontologically predicative*, hence generated by the ability to think which extends itself to the level of perception and emotions without interfering with their ontology? In here, one needs to figure out where Husserl stands with respect to the ontology of consciousness via the understanding of what different propositional levels have to say about the ontology of mind in general.

Let us briefly consider the first (*a*) hypothesis. In here, perception and feelings (taken peripherally and as part of their upper arousal) are considered necessary and sufficient ways to get acquainted with the outer world. It is also considered that the continuity between propositions of any level of complexity are spontaneously generated and without ontological gaps. To some extent, this description fits the view of phenomenology as a first-person theory many contemporary phenomenologists endorse. Some of its roots are to be found in the Gestalt school, the early Sartre, and Merleau-Ponty who remains the most influential champion of *continuous-like* (intersubjective) solutions. Merleau-Ponty follows the underlying assumption that the mind-body problem can be solved by reducing the ontology of propositions to perception and emotions and their relationship with the environment. For Merleau-Ponty, propositions aren't the expression of a prior "cognitive intuition" or of an "inner thought" (Merleau-Ponty, 2005, 206). Rather, they are the expression of a "system of equivalents" between mind and reality (the *out-there*) shaping the mind via perceptual/emotive processes and which on their turn contribute to a broader "bodily" intersubjective experience. Thus, propositions are generated via a "blind recognition which precedes the intellectual working out and clarification of the meaning" (Merleau-Ponty, 2005, 215–216). It seems that for Merleau-Ponty humans are emotional non-cognitive entities verbalizing their environmental mingling. But does this view account for the transition from one propositional level to another without silently ditching the *predicative sphere* and, inevitably, breaking away from Husserl's account of consciousness? Is it indeed (and independently from what Husserl might thought about this matter) by an intersubjective perceptual and emotive mingling with reality that the mind generates propositions?

Let us consider the second (*b*) hypothesis in which *some propositions are of perceptual order and some of predicative ontological order*. One might find this approach suitable if the purpose is to pacify the division between perceptual and cognitive features of mind in order to ensure that there is no ontological gap between commitment to knowledge about the external world and knowledge about the inner world. However, propositions can be either of perceptual order or of cognitive order. To argue for dual ontological origins and yet to use the same conceptual characterization is to commit a contradiction in terms in a sense that if reason and emotion have different ontological status one of the two propositional kinds must carry a different label in order to recognize its distinctive ontological origin. That is not what Husserl claims.

Finally, the third (*c*) hypothesis according to which *all propositions are of predicative ontological order*. In the *Crisis* Husserl states that "reason is the characteristic of conscious life" (Husserl, 1970, 338). For Husserl: "Only human beings have language and reason, only human beings can carry out psychic acts such that they are subject to the normative regulations of reason, at least this is the general conviction" (Husserl, 2001a, 2001b, 10). But what is ontologically speaking "reason"? Husserl states that "thinking includes every mental process appertaining in this fashion to chief function of the expression (namely to express something) and going on simultaneously with speaking: that is to say, every mental process in which the sense that is to become expressed becomes constituted in the manner peculiar to consciousness—the sense that, if it does become expressed, is called the signification of the expression, particularly of the location as used on the particular occasion. The process is called thinking, whether it is a judging, a wishing, a willing, an asking, or an uncertain presuming" (Husserl, 1969, 23–24). We also know that thinking or reasoning provides "qualitative insights," which are the mark of knowing (Husserl, 2008a, 2008b, 13). To that extent the proposition "I perceive a horse galloping" is a (cognitive-) qualitative insight into a perceptual content while the act of seeing (visualization) a horse galloping (content) has no intentional value. Eyes, as organs, are transporters or collectors of bioelectrical signals while intentionality is that ability of conscious mind to intend toward something. Thus, intentionality becomes *visible* only when the proposition "I perceive a horse galloping" is disclosed because the proposition is, again, a qualitative insight is the mark of the knowing itself generated by the ability to reason/think. Husserl argues that "real insight only consists in the moment of this actual performing of the thought processes involved in inferring and proving," thus "phenomenology is the descriptive discipline of thinking. It is the essence description and essence analysis of experiences of thinking" (Husserl, 2008a, 2008b, 377). Husserl continues: "thinking is a kind of mental process obviously bound in human nature, its patterns obviously psychological and like all psychology having their place in biology" (Husserl, 2008a, 2008b, 407). Finally, a rather unexpected claim that "the most universal validity of logical laws is explained by the fact that they express the most comprehensive conditions of biological adaptation" (Husserl, 2008a, 2008b, 407). Thus, reason is, phenomenologically speaking, an embodied, natural non-emotional process which ensures the producing and the disclosure of propositions.

Let us make some additional remarks with respect to the qualitative insights. Qualitative insights need not to be of strictly propositional order which doesn't change the fact that a predicative act is always behind. One can read a book or listen to Dvořák's "Symphony No. 9," it still involves a predicative act (Husserl, 2001a, 2001b, 11). This is to say that qualitative insights are ontologically different from their disclosures in a sense that reasoning is ontologically wider than its expressions. As Husserl argues,"conceptual thinking om general prior to the norm however does not encompasses all thinking in general" although the first sign or the "outset" of thinking is language (Husserl, 2001a, 2001b, 9, 12). Thus, it is important to distinguish between the variety of propositions and expressions and their ontological roots. That's why qualitative insights can be considered hints toward the possibility of an even more special feature of mind Husserl describes as "reason." In here, the proposition "I perceive a horse galloping" is the indication of the qualitative insight taken as the effect of the ability to predicate/reason/think while seeing the horse galloping is an act of insulation or a perceptual pinching. More precisely, there is the propositional embodiment that must be considered as the indication of a hidden (remote) feature of mind, and there is visualization which is to be described as the biological non-intentional activity of perceptually pinching objects from the external reality. There is nothing inside the very process visualization which could be employed to explain the origin of propositions but visualizing grounds the predicative acts into the external reality (Husserl, 2001a, 2001b, 153). Yet again, propositions, including propositions concerned with objects from the external reality, aren't the enactment of "a mere sequence of presentations, but a judgement, a peculiar unity of consciousness, that bind these together" (Husserl, 2001a, 2001b, 155). In other words, propositions are expressions of the natural ability "to execute judgement, and to be conscious of a state of affairs, in this synthetic positing of something as referred to something, are one and the same (Husserl, 2001a, 2001b, 155–156). If so, the proposition "I see horse galloping" should be read as "I think (believe) I see a horse galloping." It follows that all propositions have an inferential character since they "posit something" about something by "putting-together" or (predictively) synthetizing the meaning of different objects without the "putting-together" to occur at perceptual level (Husserl, 2001a, 2001b, 184). Or it is from the "peculiar unity" or the "putting-together" that the existence of a predicative process can be inferred. To that extent "intersubjectivity" is not an emotivist mingling with external reality as the early Sartre and, especially, Merleau-Ponty claim, but a predicate relation with objects from reality upon which an entity can act as long as can make sense of them. Thus, there is nothing in the making of a proposition of perceptual or emotional order, rather of predicative order where perception and emotions serve only to pinch/tag contents (Husserl, 2001a, 2001b, 109–110). If all propositions are of predicative order, it follows that it is only when one *acknowledges* that something has been perceived or felt, that he/she may claim he/she had a conscious experience (Husserl, 2008a, 2008b, 184). As Husserl argues, "in these we discover as a common circumstance the fact that certain objects or states of affairs of whose reality someone has actual knowledge indicate to him the reality of certain other objects or states of affair, in the sense that his belief in the reality of the one is experienced as motiving a belief or surmise in the reality of the other" (Husserl, 2008a, 2008b, 184). The belief in the external reality cannot be of perceptual order because it cannot generate judgements about the perceptual contents: one cannot say "this is a chair" without passing a judgement about what is perceived. Therefore, beliefs about the external reality must be too of predicative (cognitive) order in a sense that beliefs are possible if and when a basic sense-making occurs. By extension, what qualifies an experience as "conscious" is, as mentioned above, the ability to "peculiarly" acknowledge which shall not be confused with the ability to disclose although the two are inextricably linked. *To acknowledge* means to have the type of awareness enabling one to judge as "true" or "false" (and for that matter of any word employed to characterize something) that which has been tagged by perception. "True," "false," "chair" are qualifiers of emotional/perceptual contents which cannot be derived from the content of sensing and visualization. If so, *believing* is an *act of predication* generated by a non-sensorial feature of mind which qualifies a process as a "conscious experience" (something can be described as an "experience" if and only if conscious), and if willing to disclose what has been acknowledged, it becomes a full *conscious intentional experience* where the hint to intentionality is the disclosure of proposition. However, the quality of "conscious intentional" of an experience is not to be treated strictly as an inference. More precisely, an entity doesn't become conscious and intentional simply because one states so. Between what is disclosed and what makes the disclosure possible there is an *asymmetrical relation*.[1] Thus, the ontology of conscious intentional experiences should be considered apart from their propositional disclosures although that doesn't change the fact that a conscious intentional experience is built upon *acknowledging* (mainly) and *disclosing* (and many additional complex processes in between)—an entity is first aware that something occurs within the mind of that entity which is an independently act from the content of experience and then to decide to disclose or not whatever is aware of. Being conscious and disclosing what one is conscious of can be turned inward or outward. For Husserl, a "descriptive unity which is not to be conceived as a mere form-quality founded upon our acts of judgement, for it is in their unity that the essence of indication lies. More lucidly put, the "motivational" unity of our acts of judgement gas itself the character of a unity of judgement, before it as a whole an objective correlate, a unitary state of affairs, parades itself, is meant in such judgment, appears to be on and for the judgement" (Husserl, 2001a, 2001b, 184). In other words, descriptions are non-emotional acts of judgement of what is out-there. That means, again, that behind all propositional disclosures there is reasoning which confers unity or coherence to any proposition. If so, it follows that the third hypothesis is correct.

3 Some (Epistemic-)Ontological Puzzles: Cognitivism or Emotivism?

Now, it is clear that from a classical phenomenological perspective the origin of all propositions is *reasoning* under which judging, wishing, willing, asking, or presuming are included. Moving back to the early Sartre, Merleau-Ponty et al. (post-Husserlian era), several questions must be posed: (*a*) how did we get from a phenomenology of consciousness in which reasoning was considered a key feature to a phenomenology of consciousness in which emotions are the key features of consciousness? (*b*) how to reintroduce reasoning in a contemporary phenomenology dominated by emotivism and (*c*) how to unravel the ontology of reasoning while keeping in mind that this is a problem beyond strictly theoretical considerations? With respect to the first question, the answer is most likely, because emotivism is an easier way out. There is an old pre-phenomenological assumption according to which describing something is to comply with the empirical standard of objectivity upheld by Merleau-Ponty who argues, roughly, that language emerges as combination between external reality and emotions/perception which would ensure the objectivity of propositions (Merleau-Ponty, 2005, 206). This would explain the preference for an emotivist phenomenology which cannot be considered anymore the extension of Husserl's argument for a cognitive phenomenology (Husserl, 2001a, 2001b, 184). To that extent, Husserl's cognitive phenomenology

[1] My assumption is that the asymmetrical relation between generating and disclosing propositions has much to say about consciousness.

becomes after the early Sartre and Merleau-Ponty an emotivist phenomenology delivered as a phenomenology of emotions in which the problem of objectivity is solved following non-traditional phenomenological assumptions. From where the forth (*d*), and probably the most important question for our current purpose, *does the cognitive character of conscious experiences entail subjectivism?* The short answer is "yes." Regardless how one interprets the ontology of consciousness there is something inherently "subjective" about any claim on consciousness. As Husserl argues, "all knowledge is realized as a subjective act, and the subjective act must harbor within it what pleads and warrants its claim to legitimacy. Only in this is the justification to be sought" (Husserl, 2008a, 2008b, 129). To be clear, a form of subjectivism can be found in any kind of propositions and regardless how one understands the ontology of consciousness, but I believe that cognitive subjectivism is better if granted that the validity of externally focused propositions is not provided by data alone which is not to claim that reality plays no role whatsoever, yet only to recognize that all propositions are expressions of an act(s) of judgement. To that extent, an act(s) of judgement is part of the ontology of "reasoning" which is the only one able to lift up certain processes to "conscious intentional experiencing." Inevitably, another question is: (*e*) is indeed emotivism preferable to cognitivism? The answer, briefly, is that emotivism relies upon the premise that a valid theory of mind, and by the default of objectivity, must admit that mind emerges from a combination between emotional and perceptual experiences entangled with environment. I think this is embellished determinism sold out as pluralism.

Now, Husserl rejected subjectivism in all its forms although he acknowledged that a form of subjectivism is inherent to any proposition in a sense that while the intrinsic character of conscious intentional experiences cannot be disputed, the modality in which they are described affects how we relate to them. On the one hand, this is to say that there is also an intrinsic limit of the epistemological power of any description. On the other hand, this is to say that if one admits the ontological intrinsicness of conscious intentional experiences, he/she must deal with the gap *between ontological claims* and the actual ontology of conscious intentional experiences. Finally, this is also to say that propositions (regardless of their object) are subjective since shaped by the modality in which an entity is able to predicate about its object. Notice here another difference between emotivism and cognitivism: the emotivist believes in continuity between emotional content and meaning while the cognitivist admits only a correlation between content and meaning. There is, however, a moment when emotivism and cognitivism meet: when it comes of recognizing that the intrinsic reality of an object from within the external reality is not propositionally-dependent. So, no matter which perspective we adopt, we need to keep in mind the boundary between their intrinsic dimension and their propositional interpretation. That is to say that speaking of them as "experiences," or for that matter as "occurrences," is to appropriate them and in that sense to "subjectivize" them in a sense that each description is a predictive act or an act of judgement (weather one understands it from an emotivist or cognitivist perspective). If so, a proposition is the constitution of the meaning of what is described, hence *a meaning of*, but also *a meaning for the subject which confers meaning*. To put it differently, their meaning, yet not their intrinsicness, is dually attributed *toward outside and toward inside*. Let us call this *the duality of the meaning-conferring* which is experienced when conferring a meaning to something and when the meaning is (self-reflectively) experienced. We can argue that it is out of the dynamics of this process that conscious-intentional experiences emerge in a sense that meaning-conferring is accompanied by a conscious experience.

Now, we have seen that there is a difference between Husserl's account of mind and the post-phenomenological accounts of mind. To that extent, the modality their attempts to solve problem of objectivity is fundamentally different. In post-phenomenology (if by that we understand the early Sartre and Merleau-Ponty), the solution is reminiscent to empiricism where the assumption is that in order for propositional objectivity to be established one needs to eliminate reason from the ontology of mind or to reduce reason to the ontology of emotions. In classical phenomenology, we have the cognitivist model of mind where reason is the key feature of conscious life irreducible to emotional processes and yet non-dualistically related to the other psychological processes. It might be that the post-phenomenological preference for emotions over reason to be a consequence of the absence of a clearly formulated phenomenology of cognition. But it might be also the case of a fundamentally different assumption about mind—in that case all attempts to compatibilize classical phenomenology with post-phenomenology are doomed.

3.1 Can Subjectivism Be Minimized?

In order to maintain an *a* as possible as non-biased epistemological relation with objects from external reality, one must acknowledge that real occurrences aren't ontologically determined by propositions. Occurrences (objects from the external reality) are always and only *out-there* while the description is *in there* (or *out-there* to them). So there is an onto-epistemological gap between them. The rule stands for emotivism and cognitivism. To that extent occurrences are "transcendental" in two senses: (1) in a sense that they are intrinsic, hence beyond grasp, and (2) in a sense that all one can do is to make a judgement about them where the value conferred is predicative. In other words, one always adopts a special positioning or an "attitude toward" what one wishes to describe. Certainly, adopting an attitude toward, means methodologically speaking little. Yet without it, there is no way to understand that regardless how sophisticated a method can be, it is still biased. In the past of philosophy, Plato and Descartes were among the first to point out the importance of a meta-methodological analysis. Husserl, I believe, took this task to the next level by attempting to design a "transcendental logic" which will solve once for the problem of objectivity.[2] Husserl pointed that "the consciousness of intersubjectivity, then,

[2] To this day, we still don't have a "transcendental logic" or a method (if by "method" one understands a "textbook") to design absolute objective statements—it would be interesting to imagine how a "transcendental logic" might look and how such "logic" could enhance the possibility of objective propositions.

must become a transcendental problem, but again, it is not apparent how it can become that except through an interrogation of myself, [one that appeals to] inner experience, i.e., in order to discover the manners of consciousness through which I attain and gave others and a fellow mankind in general, and in order to understand the fact that I can distinguish, in myself, between myself and others and can confer upon them the sense of being "of my kind"" (Husserl, 1970, 202). The way I see it "the consciousness of intersubjectivity" is another way to speak about the problem of objectivity where both the puzzle of truth and the puzzle of the first-person-cognitive perspective are included. This is also the reason that I earlier mentioned that epistemological and ontological problems cannot be treated separately. But even if one acknowledges the limitations of propositional objectivity, it is not clear to what extent he/she could avoid bias. Propositions, especially those dealing with the nature of mind, are *naturally* inferential and methodologically ascriptive. Regardless how one understands ontology, propositions are the expression of a belief which on its turn is a predicative/cognitive insight. If so, all descriptions are cognitive insights into the independently given reality. To put it differently, propositions are the embodiment of a predicative act. Thus, when one describes something, he/she cannot avoid prescribing a (bias) meaning (he/she has always a perspective) about what he/she is about to describe. This is also a puzzle for those who believe that philosophical phenomenology can lead toward a psychology able to integrate the predicative sphere of mind without collapsing it into the ontology of emotions. This is the reason, I think, Husserl argues that a psychology in which emotions are considered the origin of thought cannot assist in solving the problem of objectivity (Husserl, 1970, 202). For Husserl, the conditions of universal truth must supersede circumstantial truths. To be clearer, from a phenomenological point of view all propositions are of predicative order but the assumption which lies beneath may or may not enhance their epistemic amplitude. To that extent let us say that even if the classical phenomenological defense of the predicative sphere of mind is valid, no proposition can be entirely deemed as fully objective. It follows that the only way to increase the level of objectivity is to "transcendentalize the consciousness of intersubjectivity," namely to recognize that a future "logic" (science) of consciousness should contain only universally-valid propositions. For critics, this is to adopt a dualist or a nominalist view. I will not discuss these issues here but let me point out that, again, for Husserl occurrences from within the field of reality (where the predicative sphere is included) aren't constructs. The status of occurrence is exclusively determined by its own inner structure while their disclosure belongs to the predicative sphere a conscious intentional entity projects in order to label them. The occurrence is a piece of reality, independently given of puzzling, and it becomes a "theme" only at propositional level. The puzzling ("thematization") about occurrence is of predicative order—conscious-intentionally designed with the purpose to deliver a meaning to the entity involved in the act of predication. Therefore, the best one can do is to put together, if possible, a meta-phenomenological analysis of occurrences or to provide a "transcendental logic" which is supposed to be more powerful than the initial phenomenological analyses and able to preclude biases.

4 More (Epistemic-)Ontological Puzzles: Cognitivism Reinforced

Now, I said earlier that the difference between Husserl's phenomenology and post-phenomenology shows a deep division within the general field of phenomenology which might be either (*a*) the consequence of the lack of a fully articulated theory of reasoning (transcendental logic included), or (*b*) of the fact that post-phenomenologists have a different assumption about the ontology of mind. If so, (*a*), an exciting but a very difficult job is in front us. I have elsewhere argued that classical phenomenology could become a science of consciousness in which reason is the key feature of consciousness if:

(a) *intentionality is a special aspect of reasoning impossible to explain without explaining reasoning which is possible if:*
(b) *reason is considered a higher order ontologically intrinsic feature of conscious mind, therefore*
(c) *it must be distinguished between the act of reasoning which can be called indeed "reasoning" and propositional correctness which means that*
(d) *making errors cannot be used to discard the intrinsic ontological reality of reasoning.*
(e) *Most importantly, reasoning cannot be described in sensorial terms because perception cannot be substantiated* (Husserl, 2008a, 2008b, 9).

This is an incomplete basic guidelines list for a science of consciousness in which reasoning is taken as a key feature. Keep in mind that in here my aim is not to address the ontology of reasoning in all its details, yet only to point out some of the basic epistemological problems that have to be addressed in order to get there. One of the modalities to identify the reality of conscious intentional experiences is to analyze the disclosure of statements, in particular of statements with explicit cognitive value. Notice though that the disclosure of meaning must not be evaluated exclusively based upon the accuracy of the statements or according to the relation between statements and events to which they refer. One may state: "The bus arrives at 6.30 pm" and be wrong either because the person did not properly generate the meaning or because something occurred and altered what could have been a correct statement. It is important to understand the relevance of the *asymmetry* between the ontology of sense-making and its implications for a future proper theory of reasoning. However, once the information is disclosed, we can assume there is conscious intentional intent accompaniment. The correctness of a statement can be altered by circumstances, but as I mentioned earlier, correctness alone cannot be employed to determine the reality of conscious accompaniment. If so, deducing the conscious intentional character of a proposition strictly based on the relation between what is felt/perceived and their reference is not possible. To put it differently, a proposition needs not to be evaluated based upon its matching with the external reference in order to be considered a conscious-intentional expression although it must be verified against the reference in order to determine if it accurately grasps it. Of course, the problem is how do we know that the references have been properly identified and much depends of the type of reference. There are more "accessible" references or "layers" of those references and there are less "accessible" references, i.e. "mind," "consciousness," etc.[3] In any case, this is to say that we must distinguish between making sense of the *out-there* and the attempt of determining its conscious intentional roots. If one is asking, "Do you think that is an interesting book?" to answer the question, the respondent must *think* what precisely from his/her point of view makes a book interesting. It is that particular effort that can be qualified as a conscious-intentional experience.

5 Conclusion

We know that Husserl considered reasoning a key feature of consciousness independently given of propositional disclosures. We also know that in order to pave the way toward a proper science of consciousness emotivism (along with variants of traditional naturalism) must be reformed or entirely abandoned. Setting a difference between reason and emotion is not to give up to reason as "key feature of conscious life" yet only to repositions emotion as data collector. The task of phenomenology is to provide insights into the ontology of reason without reducing it to non-cognitive processes and without recommitting to Cartesian dualism. It remains to see if besides asymmetry I very briefly mentioned earlier, there are other relations between the ontology of reasoning and propositional disclosures and to what extent these relations can be described in a manner which doesn't compromise the intrinsicness of reasoning. Most importantly, it remains to see what relations exist between reasoning and consciousness and how precisely the intentional aspect emerges.

References

Cairns, D. (2000). Reason and emotions. *Husserl Studies, 17*, 21–33.

Husserl, E. (1969). *Formal and transcendental logic* (D. Cairns, Trans.). Leiden: Martinus Nijhoff.

Husserl, E. (1970). *The crisis of European sciences and transcendental philosophy* (D. Carr, Trans.). Evanston, IL: Northwestern University Press.

Husserl, E. (2001a). *Logical investigations* (Vol. 1) (J. N. Findlay, Trans.). London: Routledge.

Husserl, E. (2001b). *Analyses concerning passive and active synthesis. Lectures on transcendental logic* (A. J. Steinbock, Trans.). Amsterdam: Kluwer Academic.

Husserl, E. (2008a). *Logical investigations* (Vol. 2) (J. N. Findlay, Trans.). London: Routledge.

[3] Following Husserl, Cairns argues that "believing is something as having a red surface is a *prima facie* rational believing when it is a believing-seeing of the thing as having a presented red surface" (Cairns, 2000, 24).

Husserl, E. (2008b). *Introduction to logic and theory of knowledge. Lectures 1906/07* (C. O. Hill, Trans.). New York: Springer.
Merleau-Ponty, M. (2005). *Phenomenology of perception* (C. Smith, Trans.). London: Routledge Classics.
Sartre, J.-P. (2000). *Sketch for a theory of the emotions* (B. Frechtman, Trans.). New York: Citadel.

Part III
Politics/Social Issues/Question of Universality

Freedom and the Human Positioning in the Lifeworld: The Transcendence-Immanence Contrast in Simone de Beauvoir's Existentialist Feminism

Natasha Kiran and Abdul Rahim Afaki

Woman's situation of being the Other of man is, according to Beauvoir, a result of men's chauvinistic attitude throughout History. Men have always been intimidating women in terms of power and so the later have failed to claim a position of human dignity as liberated and independent beings along with the former. Since her adolescence, the time when in her mind the idea of individualism was firmly rooted making her believe that each individual was responsible for securing his own, woman had been of the view that if she being a woman had accepted a secondary status in lifeworld as compared to that of man, she would have been a mere parasite degrading her own humanity. Identifying herself as such a woman, Beauvoir was clearheaded that she was suffering from the same problems because she 'happened to be a woman' and she could control the situation if she was to attempt 'qua individual' (*not qua woman*) to resolve it. This justifies her deviation from the nominalistic abstractness of the meaning of womanhood to the existentialist notion of human being that define her as a 'concrete' existent always a singular, separate individual. Her interest in existentialism sets for her the task of resolving the pivotal and ultimate problem of her feminist discourse: 'why is woman the Other?' And she knows that this resolve is not possible until and unless she is able to reply to a more fundamental question: 'What is a woman?'

In Beauvoir's analysis, woman's existence is a human existence whose socio-historical progress in the lifeworld has to be interpreted in its entirety (rather than with reference to one particular dimension like biological, psychoanalytical or economical etc.). Thereby, she finds existentialism as the most appropriate framework for this analysis, as it affords one the transcendence from the one-dimensionality of life leading one to the overall human situation which can be explained on the ground of its

N. Kiran
Bahauddin Zakariya University, Multan, Pakistan
e-mail: natashakiran@bzu.edu.pk

A. R. Afaki (✉)
University of Karachi, Karachi, Pakistan
e-mail: arahim@uok.edu.pk

C. A. Hornbuckle et al. (eds.), *Phenomenology of the Object and Human Positioning*, Analecta Husserliana 122, https://doi.org/10.1007/978-3-030-66437-4_7

'ontological substructure' defined by the of human nature. For Beauvoir, existentialism is the only paradigm that can show the most transparent picture of the human life, as it encompasses all categories of defining human life that separately unsatisfactorily attempts to attain the same task in the forms of the biological science, Freudianism and Marxism. As regards the utility of the existentialist framework in order to address the problem of feminism, Beauvoir refers to the transcendence-immanence contrast perspective for the resolution of the perennial issue of women's subordination as the Other of man who has been the superior being in all terms. From the existentialist point of view, what Beauvoir tries to establish regarding woman's defining trait as an existent it is that when woman was to accept her biological fate to be a secondary contributor to the socio-economic life she was to do that in bad faith, as she was in fact an existent like man who could transcend her givenness being the Other of man. In the face of this facticity of being in bad faith, she is very much capable of showing her aptitude as an existent being free to engage in those life projects that could bless her with new frontiers in her future life. It is all an attempt to get rid of her bad faith as "woman" who could only be a biological being-in-itself. Instead, she needs to realize that she is a free individual being-for-itself that can improvise the life projects to make her own what she is as an existent. Under the yoke of her being a traditional consciousness shaped through the effective history, woman in bad faith accepts her role as a weak, inferior and secondary being-in-itself (which is to say that 'one is not born, but rather becomes a woman' in the process of effective history). But she always has the aptitude of getting rid of her bad faith by transcending the facticity to realize that she is a being-for-itself who can freely deliberate to develop her own life projects.

1 Defining the Nomenclature of Simone de Beauvoir's Philosophy

The nomenclature of Simone de Beauvoir's philosophy is defined by her attempt to address lopsidedly the question of woman as the perpetual other in comparison with man in the perspective of Sartre's phenomenological existentialism. One may properly term her philosophy as phenomenological existentialist feminism: feminism as it pivots around the question of woman and phenomenological existentialism as this pivoting takes its place mainly in relation to the composite perspective of phenomenology and existentialism.

1.1 Life-Experience and the Philosophical Meanings: Phenomenology, Existentialism and Marxism

Feminism, existentialism and phenomenology all forms of her thought emerge from her life-experiences, and so it becomes an essential trait of her philosophy that it is rooted in her life-practice rather than intertextual reading. In the third volume of her autobiography, *Force of Circumstance*, Beauvoir confesses that even until the post-war scenario of her life in France she 'had no philosophical ambition.' It is the influence of Sartre's specifically of *Being and Nothingness* that she gave care to developing her philosophical insight through her mutual life-praxis with Sartre. Although she does not deny her own influences upon him, what she received inversely from him was so stronger and deeper that whatever she perceived about the world with all of its 'problems' and 'their subtlety' was through his presence with her. And this impact of her life-experience with Sartre is so forceful that she directed herself to talking about that phase of her life which was mutually practiced by her and Sartre as such a 'realm' that must not be taken as of theirs mutually but only of Sartre's. She confesses:

> [Sartre] found himself committed to action in a much more radical way than myself. We always discussed his attitudes together, and sometimes I influenced him. But it was through him that these problems, in all their urgency and all their subtlety, presented themselves to me. In this realm, I must talk about him in order to talk about us.[1]

She recalls her memories about their youth when they found themselves anarchists and so felt themselves close to the Communist version of 'negativism.' It seems that it was their youthful romantic longing to Communism rather than a serious thought out philosophical instance, as she explicitly clarifies that they mutually 'wanted the defeat of capitalism, but not the accession of a socialist society' which would have possibly 'deprived' them of their 'liberty.'[2] This clarification also reflects that they were more strongly committed to existentialism rather than Marxism, as they were unable to sacrifice their individual freedom for the expected economic betterment of their collective lifeworld. In support of their mutual adherence to the existentialist meaning of individual freedom against their abhorrence to the Marxist meanings of the collective economic betterment, she cites from Sartre's notebook, the entry on 14 September 1939:

> I am now cured of socialism, if I needed to be cured of it.[3]

Beauvoir interprets this socialism-liberty contradiction of their thought referring to their existentialist commitment to experiencing *authenticity* of moral life. Under the dictates of circumstances in postwar France, they were to become face to face with problems of poverty, injustice and deprivation which determined them to be against capitalistic structure of their social order that might be further suppressing for them. Out of this fear of economic insecurity and social injustice, they found Marxist ver-

[1] Simone de Beauvoir, *Force of Circumstance*, trans. Richard Howard (Middlesex: Penguin, 1963), 12.

[2] *Ibid.*

[3] Sartre did not totally abandon his commitments to Marxism though this notebook entry reflects such meanings. Beauvoir on the same page clarifies about their confusion regarding the meanings of socialism and liberty: "Yet in '41, when [Sartre] was forming a Resistance group, the two words he brought together for its baptism were: socialism and liberty. The war had effected a decisive conversion." See *Ibid.*

sion of socialism to be an urgent solution of those problems.[4] But this urgency of finding solution might damage the continuity of the social order to which they traditionally belonged and in the nexus of which they wished to prosper as creative writers for which the value of liberty was a prerequisite. Liberty and socialism were for them like two horns of a dilemma, and they found themselves hanging between the illusions of the former and the deceptions of the latter. At that point, the shield of protection from this forceful attack of these two horns came from the existentialist ethics—'the morality of authenticity.' The circumstances of life force one to submit to the facticities without leaving any room for transcendence; but from the existentialist point of view, one's freedom makes every action a project of salvaging whatever problems one is facing in life situations. They were not ready to be living with the 'absolute' meanings of facicticity—whether socialistic or capitalistic; rather, they were interested in the 'transitory' spheres of life-experience wherein they 'had to renounce *being* and resolve to *do*.'[5] Being existentialists, they rebelled against 'bourgeois humanism' characterized by the reverence of a specific human nature that determines every act of man. Instead of this essentialist approach towards human life and act, they appealed to the existentialist creativity of human action based upon man's being condemned to be free. Out of this freedom, man does not only accept the given situation subjectively, but he modifies the situation objectively 'by constructing a future in accord with his aspirations.' In this regard the phenomenological intuition would be heuristically significant, as to it everything in the lifeworld is immediately shown as it exists in itself and so an aspirant soul can constitute its own lifeworld freely. But still there is a difference between phenomenology and existentialism as regards how the subjective consciousness relates to the objective world. Phenomenology defines consciousness as consciousness of something, and when it does so it takes the full presence of the world as *noema* being a correlate of *noesis*—immanently the active pole of the transcendental subjectivity. That is to say, it is the pure consciousness or the transcendental subjectivity that constitutes the world immanently within itself in the paradigm of the *noema-noesis* correlates which Husserl terms as the structure of intentionality. As compared to phenomenol-

[4] It reflects their old romance with Marxism and their perpetual detestation for capitalism. Beauvoir in the second volume of her autobiography recalls those memories of their youthful days when they were to dream of the ruining of capitalism. She says: "We counted on events turning out according to our wishes without any need for us to mix in them personally. In this respect our attitude was characteristic of that general euphoria affecting the French Left during the autumn of 1929. Peace seemed finally assured: the expansion of the German Nazi party was a mere fringe phenomenon, without any serious significance. It would not be long before colonialism folded up: Gandhi's campaign in India and the Communist agitation in French Indo-China were proof enough of that. Moreover the whole capitalist world was, at that time, being shaken by a crisis of the utmost gravity; and this encouraged the assumption that capitalism as such had had its day. We felt that we were already living in that Golden Age which for us constituted the secret truth of History and the revelation of which remained History's final and exclusive objective." See Simone de Beauvoir, *The Prime of Life*, trans. Peter Green (New York: Paragon House, 1992), 18.

[5] Beauvoir explains their attitude referring to the influences they experienced at that time through reading both Heidegger and Saint-Exupéry who taught them the 'meanings came into the world only by the activity of man, practice superseded contemplation.' Beauvoir, *Force of Circumstance*, 13.

ogy, the case of existentialism regarding the consciousness-world relationship is a little different. For Sartre, consciousness is similarly defined as consciousness of something, but consciousness does not act in the field of fantasy rather within the realm of facticity. This factual consciousness receives impressions as subjective plenitude through perception of the things in the external world. And in doing so, this factual subjectivity that cannot transcend itself to posit the world; rather, it negates itself for the assertion that the world exists concretely as being-in-itself. Out of this assertion about the concrete world as being-in-itself, what consciousness realizes about itself is that it is always of phenomenal world and without this phenomenon consciousness is only a void or an emptiness whereby consciousness implies in its being a non-conscious being 'the nature of which is to be conscious of the nothingness of its being.'[6] This experience of nihilism becomes original of existentialist conception of creative freedom that guarantees authenticity of one's moral life. Sartre and Beauvoir share the notion of freedom not as autonomy of thinking or doing with certain *a priori* meanings rather the creative freedom—freedom as will to act *ex nihilo*. This affords an absolute guarantee to experiencing the existentialist authenticity of one's moral life. How in their youth Sartre and Beauvoir were to experience it she describes:

> We had no external limitations, no overriding authority, no imposed pattern of existence. We created our own links with the world, and freedom was the very essence of our existence. In our everyday lives we gave it scope by means of an activity which assumed considerable importance for us—private fantasies...We embraced this pursuit all the more zealously since we were both active people by nature, and for the moment living a life of idleness. The comedies, parodies, or fables which we made up had a very specific object: they stopped us from taking ourselves too seriously. Seriousness as such we rejected no less vigorously than Nietzsche did, and for much the same reason: our jokes lightened the world about us by projecting it into the realm of imagination, thus enabling us to keep it at arm's length.[7]

But from the ethical point of view, one should not take this existentialist practice of life as merely nihilistic though this paragraph may reflect such meanings. Sartre and Beauvoir were accused of being quietists or nihilists, but they refused to accept such labeling. Beauvoir clarifies that instead of 'being a quietism or nihilism, Existentialism' was to define man in terms of action. Although it condemned man 'to anxiety it did so only insofar as it obliged him to accept responsibilities. The hope it denied him was the idle reliance on anything other than himself; it was an appeal to man's will.'[8] An existentialist does not act in accord with moral principles but in the light of ends. Beauvoir while recalling her memories when she started

[6] Jean-Paul Sartre, *Being and Nothingness,* tr. Hazel E. Barnes (London: Methuen, 1972), 47.

[7] Beauvoir, *The Prime of Life*, 20.

[8] This clarification is not of Beauvoir's rather of Sartre's. Actually, certain Marxists at that time were criticizing Sartre for being influenced by Heidegger and so gone astray being a Marxist. Francis Ponge who ran cultural section of *Les Lettres françaises* told Sartre and Beauvoir about a huge number of articles against Sartre that he was receiving for publication. When he published some of those articles, Sartre was to reply 'with a *Mise au point* (Definition of Terms).' This clarification is a part of that reply to the Marxists. On this see Beauvoir, *Force of Circumstance*, 16.

publishing as a writer and Sartre was contributing to the cinema and the theatre explains this trait of existentialist ethics. She justifies that she and Sartre had always pooled their earnings, and so she was not obliged to bother about her daily expenses. This act of her seems to be against her feminist orientation, as she being a feminist advises women to be independent of their male cohabiters and that independence begins with economic freedom. She explained this attitude appealing to existentialist meanings of morality. She had taken a leave of absence from the University in order that she could focus her reading and writing. She could assure her economic autonomy 'since if the need arose' she could always get back to her teaching position in the University. To her it seemed 'stupid and even criminal' that in order to prove her economic freedom she would sacrifice her precious time that she was spending in her creative work. So in that sense, her act might be in aberrance with the principles of feminism, but it was in accord with the existentialist commitment with the ends of act that motivate her for that action. Being a writer, she found creative writing as a 'demanding task' that motivated her to do plenty of things and she could not afford to spend her time in making money. Thereby, she guaranteed her 'moral autonomy' in existentialist sense; 'in the solitude of risks taken, of decisions to be made,' she made her freedom more real than by accommodating herself to 'any money making routine.' For her, her reading and writing were a genuine satisfaction, and as such they freed her 'from the necessity to affirm' herself in any other way.[9] Being authors and thinkers, Sartre and Beauvoir deeply related their consciousness to life-experience, and that phenomenological trait of their intellectual orientation was so significant for them that at times they found themselves ready to repudiate the label of existentialism for the sake of their affinity with life-experience. When Beauvoir published her second novel, *Blood of Others*, it was an instant success. Critics labeled it an 'Existentialist novel' which was not astonishing, as an affixing of such a label on works of Beauvoir's or of Sartre's was more than obvious. But surprisingly Sartre was to refuse out of irritation to allow Gabriel Marcel to label him with the adjective—existentialist during a discussion which the Cerf publishing house was to organize for Beauvoir's novel. Sartre said abhorrently and Beauvoir shared his irritation: 'My philosophy is a philosophy of existence; I don't even know what Existentialism is.' Beauvoir furthered this abhorrence by adding that she had written that novel long before she had come across with the term—Existentialism. She explained that for that novel her inspirations came from her own 'experience, not from a system'[10] whether philosophical or social.

In the face of their mutual irritation and protest against the epithet—Existentialists—which people were using for them, it became a readymade label available to be put on everything came from their mouths or their pens. After

---

[9] *Ibid.*, 21.

[10] *Ibid.*, 45–46. On another occasion Beauvoir expresses her unqualified faith in life experience as the most important trait of the art of writing. She said: "I want to write: I want to put down phrases on paper, to take elements from my life and turn them into words." She further clarifies her ambition as an author more precisely: "I shall never be able to give myself to art excepting as a means of protecting my life." On this see Beauvoir, *The Prime of Life*, 26.

Beauvoir's novel, during the course of a few months, Sartre published *The Age of Reason* and *The Reprieve* and gave a lecture—Is Existentialism a Humanism? Beauvoir also gave lectures on her novel and on metaphysics as well as her play—*Les Bouches inutiles* opened for public, and simultaneously the first few numbers of *Les Temps Modernes*[11] also appeared. And so, they caused a sudden uproar in cultural and literary circles of France. They both were pushed out into the limelight: Sartre was vehemently flung into 'the arena of celebrity' while Beauvoir was identified as an 'associate' of his. The newspapers and the magazines discussed their works and thoughts, and there appeared gossips about their life and particularly about their cohabitation everywhere. The *paparazzi* started to take their candid shots intrusively, and the strangers rushed up to talk to them. They were so much popular that when once Sartre was invited to give a lecture, so many people gathered at the place that they all could not enter the lecture hall, and there was that much rush that some women fainted. This cultural uproar created by their philosophy and literature is what Beauvoir negatively terms as an 'Existentialist offensive.'[12] Beauvoir analyzes how Sartre suddenly turned out to be an existentialist hero in the post World War II France and why he was welcomed as a new ideologist by not only the literary people but by the public and not in only France but the whole world. According to her, the social scenario of the post World War II France happened to be in favour of Sartre's philosophy, as there was a 'remarkable' symmetry between what the public wanted and what Sartre was offering to them. The French middle class, which was the main addressee of Sartre's works, had lost its faith in 'peace' and 'progress,' and they felt tiresome due to the permanent givenness of 'unchanging essences.' They needed an ideology which could guide them to surging up these problems without denouncing the traditional meanings they adhered to. Sartre's Existentialism was striving to establish a harmony between the facticity of life and what was morally required in order to transcend the unwanted elements of factual life. Striving for the compatibility between the historicity of life and morality, Existentialism authorized the people 'to accept their transitory condition without renouncing a certain absolute, to face horror and absurdity while still retaining their

[11] Beauvoir and Sartre mutually published this periodical as an organ of existentialism. Its first number appeared in October 1945. The title of the journal was inspired by the Chaplin film—Modern Times. The editorial committee was comprised of Raymond Aron, Michel Leiris, Merleau-Ponty, Albert Ollivier, Jean Paulhan, Sartre and Beauvoir. See *Ibid.*, 22. This magazine was to play the major role in making Existentialism a worldwide movement in culture and literature; this new ideology of liberation and individualism was projected by Sartre and Beauvoir right from the first number of this periodical. While writing its preface he showed how that new ideology would dwell 'not only on responsibility in literature, but on the concept of each man as a totality. By implication, not solely in France and its citizens, but people everywhere were to be the concern of the new existentialist periodical. This program [had] been carried out by the magazine to such a degree that literature [had] never attained the importance accorded to political, economic, and sociological matters, both in France and abroad.' On this see Kenneth Cornall, Les Temps Modernes: Peep Sights Across the Atlantic, in *Yale French Studies: Foray through Existentialism* (No. 16: Winter 1955), 24–28.

[12] *Ibid.*, 46.

human dignity, to preserve their individuality.'[13] The people thought that, through the Existentialist heuristics, they could educate themselves how to surge up their problems and that surging up seemed to be closed to something they dreamed of. But it might be their bad faith under the yoke of which they thought so, as according to Beauvoir, Sartre's Existentialism did not offer such heuristics. She saw an 'ambiguity' between what the new ideological recipe was offering and what the people 'were starved for.' She found an element of intellectual seduction in that offer, as the world he was creating in his novels or presenting in his philosophical writings afforded certain space in the nexus of which man being an individual had to maintain a particular level of morality. They could not accept the Existentialist morality, as it was altogether different from the morality they were practicing in their facticity; and so they rejected Sartre's offer and 'they accused him of sordid realism, of 'miserabilism.' The moral choice Sartre offered them was grounded upon the freedom that implied tedious responsibilities that might turn 'against their institutions' and '*mores*'; it could ruin that lifeworld which they found secure in moral terms. One more element in Sartre's philosophy which might be threatening to the freedom they were practicing as bourgeois was Marxist dialectic; they were dubious about whether it was safe to be marching along the Communists into the new phase of History for which Sartre was inviting them. Beauvoir clearly understood the dubiousness of attitude of those came to Sartre for ideological guidance and their half-hearted attachment with Existentialism and Sartre's apparent influence that could not be penetrated into that culture. She judges:

> In Sartre, the bourgeois recognized themselves without consenting to the self-transcendence he exemplified; he was speaking their language, and using it to tell them things they did want to hear. They came to him, and came back to him, because he was asking the questions that they were asking themselves; they ran because his answers shocked them.[14]

Sartre found himself 'a celebrity and a scandal' simultaneously, and this simultaneity loaded with a huge fame was absolutely unexpected for him and it did not in any way match with what he being a writer had ever dreamed of. Beauvoir reports about Sartre that he considered literature divine, sacred and eternal, and this eternity lied in its being alien and misunderstood in its facticity and in its transcendence of the epoch in which it was created to be properly understood and admired in the future. Sartre imaginatively aligned himself with great genii like Baudelaire, Stendhal or Kafka whose works did not reach more than a very small group of admirers in their lifetime; but the meanings they created were to transcend that facticity to become eternal in their impact when in the coming generations they found a hugely wider audience to appreciate them. Sartre's becoming a scandalous celebrity in a younger age robbed him of that fateful solitude necessarily belonged to a genius and which had to be transcended by the future generations interpreting the meanings with the due attention. This loss of eternity of meanings for Sartre, estimates Beauvoir, 'was

[13] *Ibid.*, 47.

[14] *Ibid.*

truly the death of God, who up till then had survived under the mask of words.'[15] This was completely catastrophic for Sartre, but for Beauvoir while in this regard she compared herself with him it was not so horrible, as she had never believed in the divinity or eternity of literature. She explained that for her 'God had died' when she was 14 and then nothing (even literature) 'had replaced him.' She appeared to be more intense existentialist than Sartre, as she experienced the meaning of the death of God in her teens while Sartre had to be matured enough to experience the same. Besides this for Sartre the absolute or the sacred was reincarnated in the form of literature whereas for Beauvoir 'the absolute existed only in the negative, like a horizon forever lost in view.' She confesses that she had a fantasy of becoming a legend like Emily Bronte or George Elliot, but this fantasy was absolutely mundane without even any traces of divinity, as she was 'firmly convinced' that once she died nothing would exist to embrace such fantasies. She wished to succeed as a writer in her lifetime, she 'wanted to be widely read,' 'to be esteemed, to be loved,' as she believed that once she closed her eyes all meanings would perish with the age she lived in.[16] If seen from existentialist point of view, Beauvoir appeared to be more contented both as a writer and as a social being; and she was less deceived than Sartre 'by the illusion of being,' for she 'had paid the price of this renunciation during' her adolescence. Being a true existentialist, she was more able than he to enjoy 'the transitory,' 'the immediate'—like 'the pleasures of the body, the feel of the weather, walks, friendships, gossips, learning, seeing.' Sartre was saturated by his fame as a scandalous celebrity and by his success as a writer, but she was able to be everlastingly unsaturated by success, and she could infinitely enlarge the horizon of her hopes as an ever prosperous writer. She explicitly declared that she might be 'satisfied' as a creative writer but never 'satiated.'[17] That was the genuine form of existentialist freedom or liberation that one could experience with that much richness and depth in such explicit terms.

For them, the most suitable practical social framework for exercising such a form of freedom was democracy to which they felt adherence, but the complementary part of that social structure was socialism to which they hitherto felt abhorrence due to their fear of being lost in the collectivity having deprived of their individuality. But in any case, they hitherto saw both democracy and socialism as humanity's only chance of giving rise to social justice and as a necessary condition of their own fulfillment.

In spite of this intellectual confusion of identifying themselves as half Marxist and half petite bourgeoisie, both Sartre and Beauvoir convincingly found certain notions of phenomenology and existentialism as absolutely meaningful for human lifeworld like 'the concepts of negativity, of interiority, of existence and of freedom elaborated in *Being and Nothingness*.' Like socialist-democrats, they knew the significance of the idea of *praxis* in human life, but they were not ready to abandon

[15] *Ibid.*, 48.

[16] *Ibid.*, 54.

[17] *Ibid.*, 55.

their commitment with the existentialist ethics or the morality of authenticity in life. This phenomenological-existentialist meaningfulness primordially defined the mold of their existence on the ground of which they later chose to be Marxist or petite bourgeoisie or both simultaneously. They tore the element of 'humanism from the clutches of the bourgeoisie' and sincerely tried to make it a value for the Marxists. In Beauvoir's words, it was an attempt 'to bridge the gap between the intellectual *petite bourgeoisie* and the Communist intellectuals.'[18]

1.2 Life-Experience and the Philosophical Meanings: Feminism[19]

As I mentioned above of Beauvoir's belief that her version of phenomenological existentialism is not a matter of intertextual study rather of reflection on experience while interacting with friends, people, ideas etc. in one's lifeworld; in her case, the most important life-experience in this respect is her life-long companionship with Sartre. Being a genuine phenomenologist, she is convinced with the view that the philosophical meanings whatsoever one comes across with should not be separated from one's life-experience. Like phenomenology, existentialism and Marxism her notion of feminism can also be traced back in the nexus of her relationship with Sartre. The pivotal theme of Beauvoir's magnum opus, *Le Deuxième Sexe* (*The Second Sex*), is the idea that women in relation to men have placed secondarily in the lifeworld since the ancient times and further that this secondary position of women in the social order is imposed by the force of the patriarchal atmosphere rather than the feminine characteristics. She argues that this situation is a result of men's chauvinistic attitude throughout History intimidating women so that they have failed to claim a position of human dignity as liberated and independent beings along with men. Although she wrote that book when she was a mature woman (the year of publication was 1949), the idea had been there in her mind since she was in her early twenties.

In *The Prime of Life*, she recalls her memories of those days when she was struggling to become a writer and she had to begin her career as an independent individual not only socially and economically but intellectually as well. She quarreled with her childhood friend, Herbaud, who accused her of having betrayed that notion of 'individualism' which had previously won her his esteem; and he then did not only condemn her for that betrayal but also broke off their childhood friendship. In

[18] Actually Beauvoir cites from Sartre's work, *Les Communistes at la paix* (1952). His exact words are: "Coming from the middle classes, we tried to bridge the gap between the intellectual *petite bourgeoisie* and the Communist intellectuals." See *Ibid.*, 15.

[19] In this part of the article, we shall take the "Introduction" to Simone de Beauvoir, *Le Deuxième Sexe* (*The Second Sex*), trans. H. M. Parshley (New York: Vintage, 1989) as a reference and guide, submitting its principal theses to our interpretation. We shall give the other references, if any, accordingly.

the mean, Sartre was also to show his anxiety that he felt about her. He told her that she not only 'used to be full of little ideas' which was jeopardizing for her as a budding writer; he also warned her that she under the yoke of that orientation might become a 'female introvert' possibly leading her to turning into 'a mere housewife' rather than a creative writer. Reacting to that anxious feeling of Sartre's and accusation of Herbaud's, she confesses that she was not 'a militant feminist,' as she 'had no theories concerning the rights and duties of women.' As during her adolescence, she 'had refused to be labeled "a child,"' so then during her youth she did not think of herself 'as "a woman."' She explains that she had been reluctant to have 'the notion of salvation' in her mind since it lost 'the belief in God' while she was only 14. This was the time when in her mind the idea of individualism was firmly rooted, making her believe that 'each individual was responsible for securing his own.' She furthers that being a woman if she had accepted 'a secondary status' in lifeworld as compared to that of man, she would have been a mere parasite degrading her own humanity. She was clearheaded that she was suffering from those problems because she 'happened to be a woman,' and she could control the situation if she was to attempt 'qua individual' (*not qua woman*) to resolve it.[20]

The phrase 'qua individual' needs here to be explained further referring to Beauvoir's theorizing concerning feminism. This phrase may afford some space to be occupied by the meaning of 'nominalism,' which she renounces, for she finds it disproportionate regarding her notion of feminism lopsidedly defined by existentialist phenomenology. In *The Second Sex*, she begins her feministic theorizing by putting to criticism certain nominalistic remarks by Dorothy parker: "I cannot be just to books which treat of woman as woman....My idea is that all of us, men as well as women, should be regarded as human beings." According to Beauvoir, it is an 'inadequate doctrine,' as the antifeminists can easily falsify it by showing that 'women simply *are not* men.' It is more than evident that 'humanity is divided into two classes of individuals whose clothes, faces, bodies, smiles, gaits, interests, and occupations are manifestly different' and this truth demonstrates itself to one in one's everydayness without investing one's intellect. Woman's repudiating her eternal femininity is to Beauvoir like a Jew's denying his Jewishness or a Negro's denying his Negritude that cannot liberate a woman or a Jew or Negro to surge up, rather an escape from reality. So women's being defiant regarding womanhood reflects that their sense of perpetual femininity is haunting to them and they want to get rid of it; and this to Beauvoir is in no way an appropriate attitude of women. In spite of this nominalistic abstractness, she directs herself to the existentialistic clearheadedness that in the facticity every human being finds himself or herself a 'concrete' existent 'always a singular, separate individual.'

Being a true existentialist, Beauvoir first define the problem of feminism in the nexus of facticity whereby she expounds how woman being a For-itself is necessarily related to the In-itself—the world and its past; then she attempts to afford a morality of freedom by virtue of which an autonomous subject can transcend the given pro-

[20] Beauvoir, *The Prime of Life*, 54.

jecting her existence beyond the facticity. Drawing upon Lévinas's idea of Otherness,[21] Beauvoir defines feminine as the Other of the masculine. The humanity is compartmentalized in the masculine, the male and the feminine, the female; and the former being the self-sufficient subject, the autonomous and the essential defines the latter as the object, the incidental and the inessential—the Other. This meaning of the Otherness of woman's being contains certain connotations of the secondariness, the inferiority and the humility, and so the meaningfulness of the Otherness absolutely remains one-sided in its effect which is to say that it is only woman that is the Other of man not the vice versa. The negligence of the element of relativity that is obvious in one's considering somebody as an Other contributes to the humility of the feminine, as for Beauvoir 'the other consciousness, the other ego, sets up a reciprocal claim.' For every native a foreigner is a stranger, an Other, but when a native is to travel abroad, he finds that the natives of the country he is traveling consider him a stranger, a foreigner, an Other; and so a native's experience of being regarded as an Other by the Others forces him to realize the reciprocity of the meaning of Otherness. Beauvoir tempts to let women be aware of their collective deprivation of the sensibility of this reciprocity in the meaning of Otherness; this sensibility is the key to understand that it is the chauvinism and the sovereignty of the masculine that he absolutely defines himself as the One, the subject, the essential forcing the feminine to submit to be the Other, the object, the inessential. Beauvoir's feminism tasks to convince women to renounce this submissive attitude to be the Other, the object, the inessential and to attempt to regain the status of being the One, the subject, the essential.

'Whence comes this submission' of the feminine? While seeking the answer to this question, Beauvoir compares women as a class of individuals with other such classes of the submissive individuals exemplified in the nexus of History and culture. Such classes include the American Negroes, the Jews, the Proletarians and the Colonized nations suppressed to be the Other by the American racist Whites, the Nazis, the Bourgeois and the Imperialists respectively. But the case of women is the worst among all. The Negroes said "We" as the subject, the One, the essential when they struggled for their constitutional rights in America. The Jews said "We" while convincing the whole world that they were subject to the extreme suppression by the Nazis and so they translated the word Nazi into an abusive term. The Proletarians said "We" while revolutionizing certain nations by eliminating the bourgeois regimes. And the Colonized nations said "We" when they finally dragged the Imperialist forces out of their homelands. 'But women do not say 'We," complains Beauvoir, 'men say "women," and women use the same word [as a term of objectification] in referring to themselves.' By not saying "We," women show that they are

[21] Lévinas thinks that the feminine represents an absolute caricature of the otherness (*altérité*) as the contrariness of the masculine, 'this contrariness being in no wise affected by any relation between it and its correlative and thus remaining absolutely other. Sex is not a certain specific difference … no more is the sexual difference a mere contradiction … Nor does this difference lie in the duality of two complementary terms imply a pre-existing whole … Otherness reaches its full flowering in the feminine, a term of the same rank as consciousness but of opposite meaning.' See Beauvoir, *The Second Sex*, n. 3 on p. xxii.

unable to 'authentically assume a subjective attitude.' They do not will to assert in order to regain the status of the One, the essential, the subject rather they are satisfied with gaining only what men are willing to grant; 'they have taken nothing, they have only received' from men. 'The reason of this,' explains Beauvoir,

> is that women lack concrete means for organizing themselves into a unit which can stand face to face with the correlative unit. They have no past, no history, no religion of their own; and they have no such solidarity of work and interest as that of the Proletariat. They are not even promiscuously herded together in the way that creates community feeling among the American Negroes, the ghetto Jews, the workers of Saint-Danis, or the factory hands of Renault. They live dispersed among the males, attached through residence, housework, economic condition, and the social standing to certain men—fathers or husbands—more firmly than they are to other women. If they belong to the bourgeoisie, they feel solidarity with men of that class, not with proletarian women; if they are white, their allegiance is to white men, not to Negro women. The proletariat can propose to massacre the ruling class, and a sufficiently fanatical Jew or Negro might dream of getting sole possession of the atomic bomb and making humanity wholly Jewish or black; but women cannot even dream of exterminating the males.

Why women, as compared to the Negroes, the Jews, or the proletarians, are unable to unite themselves against their oppressors, men. The nature of bond, according to Beauvoir, that unites women to men is unique and so transcending all other bonds between the oppressed and the oppressors. The women-men sexual divide is not an historical event, but rather 'a biological fact.' The masculine and the feminine 'stand opposed within a primordial *Mitsein*,' and the latter is unable to break with it. The man-woman espousing is the fundamental institutional act that webs the whole lifeworld as a unit and then keep it so intact, thereby splitting a social order 'along the line of sex is impossible.' This natural mutuality of man and woman genuinely defines the mutual Otherness between them that both are the One and the Other simultaneously 'in a totality of which the two components are necessary to one another.' Out of this reciprocity women, should have asserted to be a free individual—the one, the subject, the essential, but men distorting the meaning of man-woman mutuality parenthesized them as the object of fulfilling their sexual need and the desire of offspring. So women remain failure in safeguarding their social emancipation through man's dependence on them rather that dependence makes the male define the female as an object of satisfaction whose readiness for the coupling is determined not by her but by the male appetite. As a result of this fruitlessness of the reciprocity of the Otherness, 'the two sexes have never shared the world in equality.' The burden of this fruitlessness or failure is not only on the male chauvinism but rather equally on the female potentiality to act as an accomplice in the process of parenthesizing herself as the Other. If women had raised her voice against that suppression, they would have faced the loss of 'the material protection' provided by men. So in a bad faith, she is contented to be an inauthentic existent remaining incapable of showing the moral urge of transcending that facticity of being 'the creature of another's will,' though she may be frustrated to be a 'passive, lost and ruined' self 'deprived of every value.' Thus, concludes Beauvoir, woman has failed to lay claim to be the subject, the One preferring to play the role of the object, the Other because of her being short of 'definite resources' that leads her to feeling contentment and pleasure with 'the necessary bond that ties her to man regardless of reciprocity.'

Beauvoir condemns in this regard the process of history and tradition that has made woman deprive in absolute term of the urge of transcending the state of the secondary being. It has been the process of the millennia that men—'legislators, priests, philosophers, writers and scientists'—have firmly been struggling to establish that 'the subordinate position of woman is willed in heaven and advantageous on earth.' The religions, philosophies, sciences and arts all have been contributing to this menace characterized by the unjustifiable male domination and female subordination. The female consciousness as an outcome of this traditional process is effected to be an historical consciousness ascribed with the meanings of inferiority and humility. But for Beauvoir all these meanings are prejudiced[22] and biased attempting lopsidedly to convince woman to feel contented with the stagnant and static life. The key to rejection of this notion is the existentialist ethics—the view that man/woman is condemned to be free and he/she has to play his/her role in life by projecting freely himself/herself through the mode of transcendence. One may genuinely experience the real meaning of freedom through a 'continual reaching out toward other liberties.' And if in that process of the projecting and surging up of the for-itself there arises any 'degradation of existence' coming across with the in-itself and if so one finds threat to one's freedom owing to an existential downfall that 'spells frustration and oppression,' then 'it will be an absolute evil.' In order to assert the authenticity of one's existence, one has to transcend the stagnation of the facticity by engaging oneself in 'freely chosen projects.' If one undertakes the particular situation of woman as an individual in the perspective of the existentialist ethics, one may propose that she has to transcend her stabilizing and static role as the Other, the object, the inessential ascribed to her by men through history. The transcendence is possible if she in good faith freely engages herself in projecting life beyond these false meanings attributed to her by men and that have overshadowed the real meanings of her existence as the subject, the One, the essential.

2 Beauvoir's Concept of Woman in the Nexus of Existentialist Feminism

'What is a woman?'—This is the first major question Beauvoir poses while setting out the discourse of existentialist feminism. And an appropriate answer to this question will lead one, as she believes so, to seeking the task of resolving the pivotal and

[22] Beauvoir's argument is in opposition to that of Gadamer's. The latter while construing his hermeneutics of tradition, argues that the tradition is not a dead past rather a living continuity, a flow of 'effective-history' that not only encompasses the past but also the relevant present. So the functionality of human consciousness cannot in any way transcend the process of history and tradition, on the contrary it is continued through the very process. On Gadamer's theory of tradition see Hans-Georg Gadamer, *Wahrheit und Methode* (*Truth and Method*), trans. G. Barden and W. G. (New York: Crossroad, 1975) specifically Part II.

ultimate problem of her feminist discourse: 'why is woman the Other?' To the former question the first answer that she discusses in detail is biological in orientation. The answer is: Woman 'is a womb, an ovary; she is a female.'[23] The word female loaded with the biological significance has been used throughout history by man that symbolizes an outrageous attitude towards woman. But using that derogatory epithet for woman as an animal, man forgets his own animal traits that make him 'male.' In his own case, the word—male with all its biological connotations shows instead of derogation the meaning of pride. Being a phenomenological-existentialist, Beauvoir is not supposed to make an appeal to the naturalistic definition of woman, and so she seems to be obviously not interested in how much the biological science is informed about the nature of woman if she has to define the term in accord with her intellectual inclination. Her long survey of the biological data in this regard is undertaken only to show that this huge corpus of information ultimately remains insufficient to conceive of woman contentedly. Under the influence of Heidegger, Sartre and Merleau-Ponty, she directs herself in order to grasp the meaning of femininity to the biological, or physiological or bodily characteristics of woman. For, Heidegger conceives of man as a being-in-the-world and to be there in the world 'implies strictly that there exists a body which is at once a material thing in the world and a point of view toward this world.' In the face of this bodily presence in the world, man is not supposed to gauge existence in the nexus of his sexuality, as nothing requires in this respect that the human body has 'this or that particular structure.' Instead of a particular bodily structure, 'the real nature of man' in terms of his relation to himself and to the world is defined by the phenomenon of death. Heidegger admits the finiteness of Being as it is 'bound up with death,' but simultaneously he does not deny the 'unlimited' nature of human existence as a temporal development in the openness of future. The death makes man's life finite, but before that, he projects life through time creating behind him the infinite past and before him the unlimited future; and in this perpetual progress of human species, man and woman both take part as correlatives, and so this 'perpetuation of the species does not necessitate sexual differentiation.'[24] Besides this, she incorporates Merleau-Ponty's dictum that 'man is not a natural species: he is a historical idea.' In this Heidegger-Sartre-Merleau-Ponty perspective, human body is conceived in the nexus of situation; for them human body 'is the instrument of our grasp upon the world, a limiting factor for our projects.' As a matter of fact, since woman is bodily weaker than man, her grasp upon the world is more limited, and therefore she cannot invest the bodily efforts to accomplish the projects of life with such firmness and steadiness as man can show in seeking his tasks. These facts concerning the bodily traits of man and woman are undeniable, but Beauvoir thinks that the biological level is not significant for gauging one's capability of accomplishing life projects. The concept of human strength or weakness 'can be defined only with reference to

[23] Beauvoir, *The Second Sex*, 3.

[24] Beauvoir explains the Heideggerian dictum with reference to Sartre's *L' Être et le néant*. On this see *Ibid.*, 7.

existentialist, economic, and moral considerations.' That is to say, in the animal kingdom, it is a fact that the male more strongly asserts his individual existence than the female counterpart, but particularly 'in the human species individual "possibilities" depend upon the economic and social situations.'[25]

For Beauvoir, like biology, the model of psychoanalysis developed by Freud and Adler etc. is also not a sufficiently appropriate paradigm for defining woman. She expresses her interest in psychoanalysis, as it offers a perspective of conceiving human self on the substantial ground of sexuality; but her interest soon shows discontentedness, for the psychoanalysts assume sexuality to be irreducible or ultimate as a nexus of interpreting the self. She also has objections on the psychoanalysts' ensuring continuity and coherence in their choices of terms. The diction of psychoanalysis (terms like complex, tendency, unconscious etc.) reflects an attempt of interiorizing the self discarding it from the world whereas Beauvoir, being a phenomenologist, considers it the only field of human experiences. For her, human self asserts itself with all of its aims and projects in the nexus of the lifeworld it belongs to; and this is an experience of exteriorizing the self rather than interiorizing it as suggested in the psychoanalytic diction. According to Beauvoir, the most significant determinate challenge posed by psychoanalysis is its apparent incapacity to deal with man's volitional experience, since it directs the human self to be contented with 'normality' defined by its bodily or sexual existence. In this regard, a subject appears to be psychologically ill if it remains unable to show its normal growth in accord with its sexual tendencies; which may lead it to experiencing an inauthentic *normal* life rather than an authentic *moral* life guaranteed by freedom. For Beauvoir, sexuality is only one of the many traits of human existence and the psychoanalysts' attempt of reducing all of the realities to sexuality is not justifiable. In existentialist terms, this orientation of psychoanalysis leads a subject to experiencing one dimensional psychic reality ultimately but inauthentically determined by sexuality whereas the subject's existence is authentically defined by an absolute freedom that gives it a responsibility of projecting itself in the manifoldness of the lifeworld it belongs to. The case of defining woman in the perspective of psychoanalysis is more problematic. The psychoanalysts define man as a standard human being in relation to which one may interpret the female being as its other in that, when she acts like a human being, her actions are not her own; rather, she merely imitates the male. The psychoanalytic focus on virility as the ultimate source of human assertion makes the male a super-human and the female a sub-human,[26] as the latter is devoid of the virile power.

[25] *Ibid.*, 34–35.

[26] There are many places in Freud's account of femininity where his reader gets that impression that the discourse of femininity he is developing is lopsidedly oriented as subsidiary to that of masculinity. For instance while describing libido he explicitly says: "There is only one libido, which serves both the masculine and the feminine sexual functions. To itself we cannot assign any sex; if, following the conventional equation of activity and masculinity, we are inclined to describe it as masculine, we must not forget that it covers trends with a passive aim. Furthermore, it is our impression that more constraint has been applied to the libido when it is pressed into the service of

As we have seen above in case of two different perspectives for defining woman that Beauvoir, although does not reject them out right, shows her interest only partially in undertaking these frameworks as useful in this regard, the same is the case with the point of view of historical materialism as expounded by Marx and Engels. She again views this paradigm as well having limited scope for the task of defining woman aptly that she finds in the other two perspectives. What she appreciates about historical materialism is its contribution to bringing forth two truths regarding human world: first, humanity is a 'historical reality' rather than anything else like a biological species or spirit etc.; and second, 'human society is an antiphysis' in the sense that man defines his relation to nature antagonistically, as he arrogantly makes attempt to control it not merely subjectively in his ideas but rather objectively in his extraneous acts. Simultaneously, she exposes the limitations of historical materialism as a perspective for defining woman criticizing the Marxist emphasis on the abstracted connotation of man's being '*Homo œconomicus*' that can only project itself through the categories of proletarian and bourgeois which remain, Beauvoir believes, inadequate to aptly conceive of woman.

Throughout the discussion concerning the various perspectives—namely biology, psychoanalysis and historical materialism—what Beauvoir significantly discovers it is that in each paradigm man appears to be a primary being while woman the Other of man, the secondary being. The Otherness of woman is not merely a difference; rather, it shows a hierarchical relationship between two regions of being so that the male always a first order being and the female a second order existent. The reason behind this permanence of meaning of woman's inferiority shared among all of the perspectives used by Beauvoir for defining woman it is that every discourse of knowledge whether it is scientific or philosophical emerges and propagates in the sphere of culture and history already practiced by man with the heuristics of values. If one interprets this with reference to Gadamer's phenomenology, one may say that it is the effective-history that gives rise to consciousness whose progress is guaranteed by the prejudices of tradition the consciousness belongs to. Situated in ever developing sphere of tradition, consciousness adopts the prejudged meanings already available in the society and so in case of each contribution it makes to the propagation of knowledge one finds the meanings overshadowed by those prejudices. Biologist, psychoanalyst and historical materialist—all finds themselves situated within a living continuity of effective-history taking for granted all prejudged meanings (for instance, the male supremacy or the female inferiority) already prevailing in the society. Whatever contribution they make to the development of biology, psychoanalysis and historical materialism is coloured by the traditional prejudices. The point Beauvoir makes here it is that the Otherness or inferiority

the feminine function, and that—to speak teleologically—Nature takes less careful account of its [that function's] demands than in the case of masculinity. And the reason for this may lie—thinking once again teleologically—in the fact that the accomplishment of the aim of biology has been entrusted to the aggressiveness of men and has been made to some extent independent of women's consent." On this see Sigmund Freud, *The Complete Introductory Lectures on Psychoanalysis*, trans. James Strachey (New York: W. W. Norton, 1965), 576–599.

of woman is not naturally given or inherently found in the female sex; rather, it has been the process of historical unfolding that enabled man to treat women as a secondary being and women became convinced with that subordinate role of her in the making of human tradition.

The two words that Beauvoir finds most significant for phenomenological-existentialist interpretation of the male and the female statures in human tradition are *transcendence* and *immanence*. Men have always been free to transcend their natural or biological limitations in terms of involving themselves in the progressive life projects whereas women have never been so free to act their own in life projects; rather, they find themselves imprisoned in their immanence—the overall factual givenness of their being in the world.

If an anthropologist or ethnographer traces the human history back to its outset the most original or primitive form of human society, he may discover is that of the Nomads. The nomadic culture was defined by the man-woman balance in terms of their productive and reproductive contributions in the development of nomadic life form. Men's job was to face the terrors of the natural world outside home through engaging themselves in works like hunting, fishing etc. to earn life not only for themselves but for their female fellows as well. This struggle against the hostilities of natural world made man create new methods and procedures to attain their tasks more affectively and safely. He invented weapons and instruments like for instance club to negotiate the nature in a more productive way than doing the same with their empty hands. The induction of instrumentality in the mechanism of man's efforts to deal with the hardships he faced in his relationship to nature shows the nomadic man's ability to transcend the givenness of his facticity to make his life better through the projects he designed with the sensibility of improving the whole life-form he belonged to. This ability of transcending the facticity was missing in case of the nomadic women's both productive and reproductive contributions that she made to their social life. As far as her productive contribution to the society is concerned, she did many jobs like making pottery, weaving and gardening etc., but each of her works was of domestic nature devoid of any sense of struggle against the massive force of nature on account of which perhaps she did not call on for any inspiration to improve the situation. On the reproductive front, the situation of woman is no different. Although her role as a bearer of the human species was very generous, the process of pregnancy and childbirth caused unproductive elements in socio-economic life. On the one hand, 'the extravagant fertility of woman prevented her from active participation' in the generation of resources; on the other, if her fertility caused too many new born children it affects economy negatively, 'as the meager products from the soil and those' attained through man's efforts (hunting, fishing etc.) could not be sufficient to meet the challenge of their survival. Thereby, the nomads used to kill those newborn that they could not afford to feed; and it is not only infanticide which was common among the nomads they also left their newborn apathetically who later died from lack of due care. So the woman had a feel of uselessness and loss as regards her reproductive contribution to the nomadic society, for the social forces wasted her painful and creative efforts in order to perpetuate the human species on account of which found no sense of pride in concretizing that great task. That feel of uselessness and loss caused a huge anguish in woman's psyche and she experienced nothingness as an outcome of her jobs that she did for the society. The pregnancy and childbirth were simply found by her as biological functions rather than useful works that could make any positive contribution to the society. It was an experience of having no life projects in order to assert herself as an individual who could transcend her facticity defined by the burden of anguish; and the only way she found at that moment was to submit 'passively to her biological fate.' Her passive submission to the biological fate and the growing insignificance of her domestic jobs perpetuated for centuries and that reiteration of that passive behavior and anguish made woman imprison herself 'in repetition and immanence.'[27]

Beauvoir, while defining man's supremacy and woman's Otherness in terms of the meanings of transcendence and immanence, compares the male and the female attitudes rather than aptitudes in relation to their contribution in the making of human history. When man stood up to the hostilities of nature, it was not only to invent new methods to deal with the huge antagonistic force; instead, he created new lifeworld of his own in the process of dealing with the naturally given world. In this struggle against nature, 'he put his power to the test; he set up goals and opened up roads toward them; in brief, he found self-realization as an existent.' Once he created the lifeworld he then maintained it as well; and through the process of centuries, he has been perpetuating and continuously improving it. Thus, his efforts have never been restricted to 'conserve the given world;' he has broken through its frontiers, he has laid down the foundations of a new future.' This continuity of man's assurance to the perpetuation of life through his acts of existentially transcending the same life was to guarantee the project of value formation. Man's attempt of being face to face with the hostilities of nature, his creation of new instruments, his inventions of new methods, his shaping the future made him set himself up as a 'sovereign' existent who being a free individual had to have his own set of values. Therefore, his ability to transcend the given that he faced as an individual who was condemned to be free set his aptitude to challenge the present and to create the new future. This tendency of transcending the given is something that woman was deprived of not due to her aptitude as an individual existent rather her attitude determined by not only her own biological make up but the extraneous forces that suppressed her to be submitting to her biological fate. Beauvoir's being certain as regards woman's aptitude of transcending the given lies in her being an accomplice of man in all of his achievements as an individual existent. Man was like a mirror to woman, as she identified herself as an individual existent in comparison with man; like him, she felt an urge of transcending the given facticity going beyond the reiteration of life experience to attain the task of finding new future. According to Beauvoir, 'in her heart of hearts she finds confirmation of the masculine preten-

[27] *Ibid.*, 62–63.

sions. She joins the men in the festivals that celebrate the success and the victories of the males.'[28]

From the existentialist point of view, what Beauvoir tries to establish regarding woman's aptitude of being an existent it is that when she was to accept her biological fate to be a secondary contributor to the socio-economic life she was to do that in bad faith, as she was in fact an existent like man who could transcend that givenness of her being the Other of man. In the face of this facticity of being in bad faith, she is very much capable of showing her aptitude as an existent being free to engage in those life projects that could bless her with new frontiers in her future life. She always has the aptitude of getting rid of her bad faith by transcending the facticity to realize that she is a being-for-itself who can freely deliberate to develop her own life projects. This reminds us of Sartre's notion of the absolute individuality that he ascribes to man's being-for-itself while drawing parallels between the divine and the human freedom. According to him, man as an individual is to make of himself whatever he wills to be, which is to say, man's being is what he 'chooses to be.' Making this choice, man is helped neither by his inner essence, as he has none nor by any extraneous source like God, as he has already become what he has made of himself without the Divine help. Contrary to Gadamer's notion of effective history that gives rise to human consciousness and act, this explains the existentialist instance on the original of human act while man is 'condemned to be free' as an existent. It reflects that history cannot 'produce an act,' that is to say, fact cannot in any way be 'capable by itself of motivating any act whatsoever.'[29]

Beauvoir also refers to Hegel's definition of 'the master-slave relationship' in order to interpret the man-woman relationship as it has historically been unfolded. As regards Hegel's approach to the master-slave relation, the advantage of the master it is that he asserts 'the Spirit' against 'Life through the fact that he risks his own life' through that process of assertion whereas the slave although being aware of this risk remains unable to assert rather he is conquered by the master in the process of risking his life for conquering the world. Throughout the history, there has been a tension between man and woman and woman knows very well that she 'does not risk *her* life' despite the fact that she being an existent creates life herself perpetuating it through history. Thereby, she in accord with Hegel's master-slave model remains passive in that she being a consciousness does not assert but rather always remains dependent on the other consciousness (the male consciousness), which Hegel identifies as an 'essence' of 'animal life.' Beauvoir wants woman to ascend

---

[28] *Ibid.*, 64.

[29] Gadamer's notion of effective historical consciousness justifying the inevitable involvement of human self with tradition while producing act is simply antagonistic as compared to Sartre's notion of absolute individuality that Beauvoir takes for granted in construing woman as an existent. Sartre categorically says: "No factual state whatever it may be (the political and economic structure of society, the psychological "state," etc.) is capable by itself of motivating any act whatsoever. For an act is a projection of the for-itself toward what is not, and what is can in no way determine by itself what is not… Under no circumstances can the past in any way by itself produce an act." See Sartre, *Being and Nothingness*, 435–436.

from the animal level of existence where she leads her life depending on the male consciousness, as she believes that the female consciousness 'also aspires to and recognizes the values that are concretely attained by the male.' The positive element in that act of aspiration and recognition it is that 'women have never set up female values in opposition to male values' but men have always had that tendency of maintaining a culture of safeguarding exclusive rights and authorities peculiar for their rank in opposition to that of women. Beauvoir accuses men of being a 'prerogative' class of individuals who have in the process of history created a 'feminine domain of immanence' wherein women are imprisoned as slaves. She believes that this subordinating attitude of the feminine is not natural and necessary; women are like men the existents who are able to seek 'self-justification through transcendence.' And the contemporary movement of feminism demands that women should be 'recognized as existents by the same right as men' and should not be treated as beings that 'subordinate existence to life, the human being to its animality.'[30]

3 Beauvoir's Concept of Woman's Freedom in the Perspective of the Transcendence-Immanence Contrast

There may be several ways of defining Beauvoir's concept of woman's liberation, yet the most appropriate is one that will be considerate toward immanence-transcendence relation. Seen from the existentialist point of view, woman as far as her facticity is concerned is found as an existent being enclosed immanently within a particular set of meanings like weakness, passivity and dependence on man. This historical outcome of her facticity is something that woman has accepted in bad faith. The being of woman's consciousness does not only disclose that she is weak, passive or dependent with respect to the Others, it also shows negative attitude toward herself. She enjoys her reality of negating her own existence of a free individual subject directing her consciousness toward a life of dependence, un-freedom and passivity as an object. This attitude of immanence in which woman lies to herself, she deceives her own and she deliberates to find her consciousness having something as true which she can easily recognize to be false is what one may call bad faith. This is an experience of bad faith, as in this case woman attempts to hide truth of her being from her own in such a way that her consciousness is in complete possession of the truth she is hiding; the truth that she can transcend her facticity of un-freedom, dependence and passivity as an individual subject being absolutely free to make her own fate. This mutuality of truth and falsehood in the unity of woman's consciousness makes her bad faith a paradoxical existentialist experience for her. Her bad faith facilitates her consciousness to have ambiguity of meanings of her being as a subject and an object. She is unable to define her existence as a being

[30] Beauvoir, *The Second Sex*, 64–65.

for-itself that can deliberate to transcend her facticity of weakness and un-freedom, as her consciousness has been forced since the time unknown to admit that she is immanently enclosed as an object within the structure of a being-in-itself characterized by the fixed feminine traits of passivity and dependence. Beauvoir judges that owing to this ambiguity of meanings woman denies her being-for-itself as an individual subject in order to avoid the painful fact of her being condemned to be free. She plays with these ambiguities of meanings of for-itself and in-itself in bad faith when she deceptively realizes that she is a weak, passive and un-free being (in-itself) whereas she knows well that she is an independent, free subject (for-itself) which she may find painful to accept in her life-situation. This experience of self-consciousness of woman is certainly anguished; however, she may be able to release herself of this anguish if she in good faith accepts that in her facticity she is weak, fragile and un-free, yet she can transcend this for-itself of hers through her further projections in life as an independent individual subject. This life-projection of hers from immanence to transcendence requires her exercising freedom at two different levels in two different forms. On the one hand, woman should be "negatively"[31] free from immanence in the sense that she ought to release herself from the Others' interference that has been forcing her to be immanently enclosed within the structured feminine traits of weakness, passivity and dependence. On the other hand, she should be "positively" free to transcend through her life projections to be her own master, whose life depends on her own and not upon any external forces. This two-fold structure of liberty, namely the structure of freedom *from* (negative) immanence and freedom *to* (positive) transcendence affords us may be the most suitable paradigm for interpreting Beauvoir's concept of woman's liberation.

3.1 Freedom **from** *Immanence*

There are several dimensions of immanence that make woman enclose in her facticity yet the major aspect of that immanence is, according to Beauvoir, woman's economic dependence on man. Woman may be free from her obedience to man or she may be free to cast her vote like a male individual or she may have other socio-political forms of liberty but these liberties cannot guarantee her that she is truly free until and unless she emancipates from the economic dependence on male as a 'wife or as a courtesan.' Not a ballot in her hand or her being not obedient can emancipate

[31] We refer here to these meanings of negative and positive freedom as expounded by Isaiah Berlin in his famous essay, "Two Concepts of Liberty." In the negative concept of liberty Berlin conceives of freedom of a person who is free 'to the degree to which no man or body of men interferes' with his act of liberty. 'Political liberty in this sense is simply the area within which a man can act unobstructed by others.' In case of one's being positively free one becomes autonomous in the sense that one is 'a thinking, willing, active being, bearing responsibility for one's own choices and able to explain them by references to one's own ideas and purposes.' On this see Two Concepts of Liberty, in Isaiah Berlin, *Four Essays on Liberty* (Oxford: Oxford University Press, 1979), 118–172.

her truly in lifeworld but a gainful employment can make woman deny the alleged negative meanings attached with her existence by the society through traversing 'the distance that separated her from the male.'[32] Once she releases herself from being economically supported by male, the social order grounded upon that parasitic attitude of hers will be disintegrated; and there will be no requirement any more of a 'masculine mediator' between woman and her lifeworld. She can live like a genuine individual subject if she does not stand on others' economic support. In her facticity when she is not allowed to be economically active, she remains in 'the vain pursuit of her true being through narcissism, love, or religion.' However, when she actively participates an as economically productive individual, 'she regains her transcendence' by projecting her life through concrete assertions as subject; she asserts her responsibility as an individual, as she pursues her aims and earns her life through the money and rights she has as a productive self. This is one side of a picture; the whole picture of woman's having a job is not so simple rather it has a complex situation for woman; it has its own structure of immanence that may be worsening her facticity. In this regard, Beauvoir takes the case of woman as factory worker. As a factory worker, a woman obviously becomes economically independent yet simultaneously, she becomes a member of an oppressed class—the labor class. Her employer asks her to work in the factory for 40 h a week and the same load of work she has to take at home because she is also a housewife. So, the society and her husband both are exploiting woman in the sense that the former is using her productivity for its well-being while the latter is not assisting her in the household. Although women may experience a feel of being integrated in lifeworld as workers and also loving wives, they persist in the facticity of being unable 'to become in concrete fact the equals of the men' which is to say that they remain within 'the traditional feminine world.' It means that the gainful employment of woman cannot necessarily emancipate her from playing the traditional role as a suppressed being. And it is not the specific case of factory worker; the other professional women like the shop girl or the secretary etc. are having the same persistence of their being exploited in one way or the other.

One form of social interference in women's status as an individual like man and which, according to Beauvoir, woman should have freedom from is the custom concerning the male-female difference as regards their outlook—the physical appearance, the way they dress up, the way they look like etc. There is no such social issue with men that they have to take thought of how to dress up; he rarely thinks normally of what would the other opine or how would they judge of his outlook because he does not find himself an object to be looked at or judged as an object of beauty. In this regard, the case of woman is altogether different. She is always conscious of how she will look like if she dresses up this way rather than that way, as she knows traditionally that 'when she is looked at she is not considered

[32] The Beauvoirian text on which our interpretation of her concept of woman's liberation is based is Part VII of *The Second Sex*. So as far as Beauvoir's ideas are concerned all paraphrasing and citations in this part of the article to Simone de Beauvoir, *Le Deuxième Sexe* (*The Second Sex*), trans. H. M. Parshley (New York: Vintage, 1989), 679–715.

apart from her appearance; she is judged, respected, desired, by and through her toilette.' So woman, in order to look like woman in traditional feminine sense, requires a permanent budget for 'setting-hair, make-up materials, new dresses' which costs enormously as compared to what man requires to look like man through his outfit. Although this meaning of to be looking beautiful might be imposed extraneously upon woman, she does give care to her appearance because she 'wants to retain her womanliness for her own satisfaction.' In her childhood, she finds her mother and, in her adolescence, her elder sisters inculcating the desire for a home or an interior of her own. This sense of destiny drives her to entertain the 'narcissistic dreams'; to 'the male phallic pride,' she wants to be seen 'attractive.' The destiny of having home in the insecure masculine world symbolizes woman's own self as an interior wherein she finds the refuge out of the insecurities of life.

Woman's sexuality is a complex phenomenon, yet in its overall orientation, the most intensely she feels is the element of the male-female inequality in sexual experience and satisfaction. Beauvoir thinks that sexual experience is a unique level of the man-woman affinity where each party equally needs the other; so it should obviously manifest equality between the two. But even in that case woman has to experience inferiority as she finds herself as compared to man in all other domains of their existence. In this regard, Beauvoir takes the case of those women who are professionally successful and are up to certain extent enjoying an equal social status as compared to that of a successful man in the society. These are the women who can in no way be considered dependent on man economically or socially; the independent women like high professionals and intellectuals. Even the women of such a status when has 'access to the masculine world as does the male to the feminine world' in order to have the symmetrical sexual experience, what she faces intensely is the fact that 'the demands of the *other* are not symmetrical in the two symmetrical cases.' This situation, according to Beauvoir is less painful for those women who do not intellectualize issues deeply rather focus more on their physical appearance, for such women may easily fulfill the demands of their male partners by finding leisure for beauty care aiming to be more seductive. But for an intellectual woman like Beauvoir it is an experience of suffering, as it may be ridiculous for her to spend that much time in beauty care and in making herself seductive because it is demanded by a man. As she is a thinker, she cognizes her situation; she knows that she has to make herself attractive only for falling victim to the male phallic pride; it is for her like degrading herself into immanence of sexual desire that makes her suffer 'from an inferiority complex.' Instead of being relieved by sexual experience, an intellectual woman suffers from it, as it makes her feel awkward to please and seduce someone; and so 'she becomes vexed' with her being an object of sexual pleasure. According to Beauvoir, at that moment such woman begins to act for 'her revenge by playing the game with masculine weapons: she talks instead of listening she displays subtle thoughts, strange emotions; she contradicts the man instead of agreeing with him, she tries to get the best of him.' This attitude of a psychological complex may most of the time irritate men rather than conquer them, as there are men who have no intension of slaving women for their sexual desires; instead, they will to love an equal and through the way of loving them they experience their sexuality as a mutual process of pleasure for both of them. At this juncture of her argument, Beauvoir shows a hope that men have now begun 'to resign themselves to the new status of woman; and she, not feeling condemned in advance, has begun to feel more at ease.' At present, as compared to the past, the working women are 'less neglectful' of their 'femininity' without losing their 'sexual attractiveness.' This situation shows that the western part of the world is showing 'progress toward equilibrium' in terms of the man-woman sexuality; yet Beauvoir finds it 'incomplete.' As far as woman's 'erotic and affectional life' is concerned, she faces difficulties at different levels of her existence. The women who are not economically and socially independent rather living dependently as wives and courtesans 'are deeply frustrated' in their experience of sexuality yet they never challenge their husbands and masters to be emancipated. They remain silently so engaged in their everydayness that they never give thought to the fact that they have buried their will and desires. The case of an independent woman is a little different, but she is also facing difficulties in her sexual life. She is busy in her life as an active individual and she never compares herself with a dependent woman instead 'she considers herself at a disadvantage only in comparison with man.' This consideration of hers brings frustration for her when she being a woman acts taking man as a standard in expressing and satisfying her physical desires.

An independent working woman invests 'her energy' as a responsible professional and she 'knows how harsh is the struggle against the world's opposition;' in that process she 'needs—like the male—not only to satisfy her physical desires' but also to divert herself from the burden of her job by mutually agreeable sexual activities. Woman in this regard experiences the male-female inequality, as there are certain 'social circles' in which, according to Beauvoir, woman's free attempt to satisfy her sexual desires is not 'concretely recognized.' When a woman is to exercise her sexual freedom, she has to do this not without the fear that it might be a risky proposition for her, as it might be jeopardizing for her 'reputation' and 'career; at the least a burdensome hypocrisy is demanded of her.' As compared to woman, man is free to satisfy his sexual desires even if he has to do with sexuality as a matter of temporary interest; in that case, he can find institutions in society for sexual satisfaction. Beauvoir shows here her dissatisfaction on the unavailability of brothels[33] for the fulfillment of woman's physical desires even though she explicitly judges that 'woman does not obtain "appeasement" as mechanically as does the male.'

If woman acts like man in order to execute her sexual will, one more possibility that can be available for her 'is to pick up in the street a sexual partner' as a temporary arrangement. This attempt of a woman, according to Beauvoir, will prove to be jeopardizing for her in many ways. First, there is a risk of transmission of 'venereal disease' through the intercourse with an infected individual, 'because it is the man

[33] Beauvoir in this regard refers to a French novel titled *Le Numéro 17* that describes the proposal of a woman for the 'establishment of houses women could resort for "sexual appeasement" through the services of "taxi-boys." The novelist might not know that such an 'establishment...formerly existed in San Francisco; the customers were prostitutes, who were highly amused to pay instead of being paid.' *Ibid.*, 687.

who is responsible for taking precautions against infection.' Second, if man is not infected still there is a danger for woman to become pregnant. Third, and it is the most important one, man is physically stronger than woman; and in case of bringing some relatively physically strong stranger at her place woman is in permanent threat to be looted or treated violently.[34] If a woman is not interested in such temporary arrangements to satisfy her sexual desires, she can keep 'a permanent lover, as a man often takes a mistress.' But this arrangement is also not devoid of difficulties. First and foremost, this arrangement is possible for only those women who can financially afford it; a woman of ordinary means cannot have such a facility. Second, a young woman in her adolescence cannot take a man to be such a lover because in that age she is associated so profoundly with her lover that she remains unable to keep sex dissociated from the love sentiments. Therefore, only a relatively mature woman can successfully hold such a sexual relationship. Third, even if a woman is mature enough as well as granted the means, she, according to Beauvoir, never finds 'the purchase of a male a satisfactory solution' for her sexual problem, as she is more 'clear-sighted' as compared to man in defining her love relationship, and so she cannot be blinding herself as regards the fraudulence involved here 'but only at the cost of entertaining a more calculated bad faith.'

Sexual experience, for Beauvoir, is not significant for human beings (men and women both) in the sense that it brings them satisfaction rather it is important because by satisfying man's/woman's 'erotic desire' it in a way maintains 'dignity' for him/her as a human being. Man experiences himself to be a subject while deliberating to make his sexual partner enjoy her erotic experience, as in that case he feels to be like an 'imperious conqueror, or lavish donor—sometimes both at once.' On the other hand, woman also wills to bring her partner into subjection of her sexual pleasure by overwhelming him 'with her gifts'; and she remains convinced that she is giving that gift to her partner out of her generosity. That element of bounteousness toward her partner in sexual experience makes her feel freedom like a subject especially when she willfully chooses such a partner who remains thankful to her for her overwhelming sexual attitude.[35] But men generally, according to Beauvoir, do not like that feminine overwhelm in sexual experience in which they are 'chosen' by her 'as the means for satisfying her need in its generality: so chosen, they feel exploited.' As a matter of fact, men do not appreciate their female sexual partner if she tries to take the initiative in sexual experience; instead they like to arouse her erotic desires making her feel excitement at optimum. For in that case,

---

[34] Beauvoir here refers to an actual event reported to her by someone about two young women newly arrived at Paris. Both the girls were eager to "see life" in the Metropolis; and under the yoke of that feel they, 'after a look around at night, invited two attractive Montmartre characters to supper. In the morning they found themselves robbed, beaten up, and threatened with blackmail.' *Ibid.*

[35] In relation to woman's bounteous attitude toward her partner, Beauvoir mentions the character of a French novel, *Blé en herbe* who responding to her partner's caresses tells him: "I love only beggars and starved people." The author of the novel elaborates her attitude: "she made haste toward that obscure and narrow region where her pride could believe that the plaint is an avowal of distress and where beggars of her kind drink the illusion of liberality." *Ibid.*, 689.

only they can satisfy themselves being 'intent on conquering' woman as a passive object of their sexual fulfillment. This male dominion in sexual experience does not admit that women can satisfy herself as a subject rather she is always found to be 'the prey' of the male erotic desires. So, woman under the yoke of that social givenness 'is represented…as pure passivity, available, open, a utensil; she yields gently to the spell of sex feeling, she is fascinated by the male, who picks her like a fruit.' This social situation reflects 'a general refusal' to think of woman as a free individual which is a form of interference with woman's free act of satisfying her erotic desires; thus, woman is required to make herself free from this social negation. In this regard, Beauvoir warns women that in order to get rid of that social constraint they should not become a victim of 'masochism.' Woman's masochistic attitude toward her sexuality defines her readiness to feel dominated by her male partner with the hope of finding deep 'submissive pleasure' in his arms. This feel of having submissive pleasure may take the form of an addiction that may drive woman to have the painful climax of this adventure when she confronts frigidity in her sexuality, as in that case she may become habitual of the same feel leading her toward boredom.[36]

This social situation in which one either wants to win or avoid defeat in the battle of sexuality can be changed with pleasant outcome if according to Beauvoir 'the both partners recognize each other as equals' showing 'a little modesty and some generosity.' It is not difficult for a man to show the traits of modesty and generosity while falling in love with a woman. As he belongs to 'the male caste' having superior status in the society, he may have several reasons for this easiness in becoming affectionate to a woman belonging to the lower caste. First, his being introduced to the novelty of the feminine world brings him pleasure and joy in the sharing of feels of loving a woman and being loved by her. Second, since she belongs to a subordinated caste a woman may be excused by her lover if she mistakes or shows weaknesses in her personality. Her male partner can easily overlook if his beloved woman 'is not very intelligent, clear-sighted, or courageous,' as he may understand that she is such a person because she is a victim of not being allowed to develop herself as a free individual as he has the opportunity to realize his talent. He remains positively hopeful that his beloved is capable of molding herself in accord with his desires and dreams; 'she can be credited with any responsibilities, because she *is* nothing in particular.' As compared to man, a woman does not find it so much easy to hold a man in affectionate relationship on the ground of equality, as he is superior and powerful, 'he has made himself irrevocably' what he is. Since he is not like her nothing rather he is always something; therefore, there is no chance of his becoming in the future in accord with her dreams and desires. So love-relationship 'with him

[36] Consolidating her opinion Beauvoir describes the actual case history of a woman, 'who had been married several times since she was seventeen and had had several lovers, always with much satisfaction. After having successfully managed an enterprise in the course of which she had men under her direction, she complained of having become frigid. There was formerly a blissful submission that she no longer felt, because she had become accustomed to dominating over males, and so their prestige had vanished.' *Ibid.*, 691.

is impossible unless she approves his acts, his aims, his opinions.' When a woman meets a man in their shared lifeworld, what he offers is his accomplished self that has nothing like a promise to change that world by changing himself in accord with her dreams; and if he makes such a promise woman gets confused whether it will happen or not. So she has always to wait for such a man who can do that for her.

The societal interference, according to Beauvoir, with woman's existence in terms of the burden on her of the household in addition to the bearing, caring and bringing up of children also belongs to the area within which a woman can act unobstructed by man. That is to say, woman has every right to free herself *from* this burden in order to become equal with man.[37] But at present woman is unable to withstand such a forceful encroachment by the society on her private life. It is given in the social order that it should be woman's responsibility to do the housework and to take care of the children. The traditional meaning of the word, wife has the connotations of a loyal companion, 'a good housekeeper, a devoted mother' and other similar phrases. This multidimensional role of a woman is to set a 'task' for her that 'overwhelms' her whole personality. She has to lead several lives simultaneously; she as a wife has to 'assume' her husband's 'cares and participate in his successes,' as it is concerned with her own 'fate'; she as a mother has to give care to bringing up their children, as it is wholly and solely a responsibility of hers; and as an individual she is herself a person having her own desires and dreams which she wants to fulfill. In the overall givenness of this household world in the nexus of the traditional universe of the lifeworld, she intensely feels that 'the first place' both in the world and the universe 'is for man to occupy.' She wants to raise her voice against this unjust gender-based division of statures of human beings in the society, but she remains silent because 'she fears that in claiming it she would ruin her home.' Consequently, 'between the desire to assert herself and the desire for self-effacement she is torn and divided.'

One more space of the negative freedom where woman's biological nature and social customs both interfere with her liberty is the phenomenon of maternity. Beauvoir knows that as it is a natural givenness woman cannot 'perform' in this sphere 'in complete liberty,' yet she wants woman to exercise freedom as much as possible. She 'thanks' in this regard to the availability of certain 'contraceptive techniques' on account of which it has now become possible for woman to willfully avoid the unwanted pregnancy and 'decline maternity.' In the past when such medi-

[37] This is the reason why Beauvoir turned down the marriage proposal by Sartre assuring her that their marriage would remain a mere formality and it would not in any way affect their existentialist way of life. She explains her refusal: "Hitherto we had not even considered the possibility of submitting ourselves to the common customs and observances of our society, and in consequence the notion of getting married had simply not crossed our minds. It offended our principles. There were many points over which we hesitated, but our anarchism was as deep-dyed and aggressive as that of the old libertarians, and stirred us, as it had done them, to withstand any encroachment by society on our private affairs. We were against institutionalism, which seemed incompatible with freedom, and likewise opposed to the bourgeoisie, from which such a concept stemmed. We found it normal to behave in accordance with our convictions, and took the unmarried state for granted." On this see Beauvoir, *Prime of Life*, 65–66.

cal facilities were not available for woman, she had to take herself 'responsible for an unwanted child' that could 'ruin' her social and 'professional life.' This is the fate both of the married and unmarried mothers, whereas the latter might have confronted additional problems; she 'is a scandal to the community, and the illegitimate birth is a stain on the child; only rarely is it possible to become a mother without accepting the chains of marriage.' These are the problems a mother normally faces before the birth of her child, but thereafter she alone has to take the responsibility of the care and bringing up the baby that is enough for paralyzing the mother's 'social activity.' Although there are certain facilities like daycare centres and kindergarten, they are not perfectly convenient for her; 'she can go on working only if she abandons it to relatives, friends, or servants.'

3.2 Freedom to *Transcendence*

All of the aspects of the negative freedom that have been discussed above constitute the structure of immanence wherein woman has now found herself enclosed. If woman gets herself free from all the constraints society has traditionally imposed on her, what can she achieve most in that case? She can be economically independent, she can be able to rise and shine in the career she has chosen for herself, she can get less responsible in the household affairs etc. But is it the real meaning of freedom in the existentialist sense? One cannot be free in existentialist nexus if one is merely free *from* this or *from* that. The genuine freedom pertains to the notion of transcendence. In case of the feminine, the existentialist meaning of freedom can be realized when independent woman does not merely seek to shine in the sphere prescribed to her rather, she has to set herself daringly and bravely toward the goal created for her by herself. In such an act of 'setting out toward ends,' one can learn how to experience the risk of 'disappointments.' She has to develop 'a taste for adventure.' She does not only have to prove to herself that she can handle a job properly rather she has to educate herself how to 'passionately lose herself in her projects.' While progressing women are habitual of 'looking back to see how far they have come,' which shows their persistence to their immanence and 'interrupts' their life projects. This attitude may allow them to succeed in an 'honorable career,' but she remains unable to 'accomplish great things.' In order to do great things, woman has to transcend her facticity of being contented with the mediocrity as a professional, and she has to aim at the projections of herself that will lead her to finding the future space for the realization of her true self. Beauvoir says:

> What woman essentially lacks today for doing great things is forgetfulness of herself; but to forget oneself it is first of all necessary to be firmly assured that now and for the future one has found oneself.

Artists, writers and philosophers are the genuinely individual subjects who attempt 'to found the world anew on a human liberty.' Art, literature and philosophy are the most appropriate arenas for woman to experience themselves transcendent, as these are the spheres wherein the creator of meanings *ab initio* denies all constraints the world is to impose on him and thus he is able to assume clearheadedly 'the status of a being who has liberty.' One's finding oneself as a consciousness that emerges out of the process of history restricts one's ability to reflect on the world through traditional meanings. But Beauvoir wants woman to question those meanings and attempt to go beyond that knowledge limiting her existence within the mold of customs and conventions. Beauvoir here appears to be firmly antithetic to the Gadamerian hermeneutics. She thinks that since woman has lost herself with its true meaning within the conventional structure of education entrapping her in the inferiority complex in her relationship to the dominating male, it is therefore justifiable for her to transcend that historicity of her being. Art, literature and philosophy afford her the space to assert her freedom by going beyond what is factual to the world created anew by her. Beauvoir advices the woman who wants to become an artist, a writer or a philosopher that she 'must first emerge from [the world] into a sovereign solitude if' she has 'to regain a grasp upon it: what woman needs first of all is to undertake, in anguish and pride, her apprenticeship in abandonment and transcendence; that is, in liberty.'[38] A true piece of art is deeply embedded in the mass of its surrounding culture; and so a true artist must not only be cultured or civilized in normal sense but rather she must grasp culture 'through the free action' of a transcendent existent. That is to say, a true artist must be a transcendent self or a 'free spirit' that projects 'itself toward an empty heaven that it is to populate; but if a thousand persistent bonds hold it to earth, its surge is broken.' Such an experience of liberty as transcendence has an infinite impact on an artistic mind, through this richness of freedom of one's spirit one discovers how an individual can interiorize the whole external lifeworld as his own. Transcendence for Beauvoir is an experience of solitude that may take form of liberty from facticity to become an impetus for an artistic genius for creating a masterpiece. She gives example of Emily Brontë whose 'isolation' enabled her to write a great book, as in her solitude, namely – 'in the face of nature, death, and destiny, she had no other backing than her own resources.'[39] Still there are very few such creative writers from amongst women who were able to transcend the level of mediocrity to become an artistic genius. 'The constraints that surround her and the whole tradition that weighs her down prevent her from feeling responsible for the universe, and that is the deep-seated reason for her mediocrity.'

[38] In this regard, Beauvoir quotes Marie Bashkirtsev, who writes: "What I desire is liberty to go walking alone, to come and go, to sit on the benches in the Tuileries Gardens. Without that liberty you cannot become a true artist. You believe you can profit by what you see when you are accompanied by someone, when you must wait for your companion, your family! ... That is the liberty which is lacking and without which you cannot succeed seriously in being something. *Thought is shackled as a result of that stupid and continual constraint.... That is enough to make your wings droop*. It is one of the main reasons why there are no women artists." Beauvoir, *The Second Sex.*, 712.

[39] Beauvoir also gives example of Rosa Luxemburg. On account of her being ugly, Luxemburg 'was never tempted to wallow in the cult of her own image, to make herself object, prey, trap; from her youth, she was wholly spirit and liberty. *Ibid.*, 713.

According to Beauvoir, an artistic genius is one who is able to interiorize the whole universe as one's own, free enough to accuse oneself of 'its faults and to glory in its progress.' Such free individuals always find themselves 'in command to justify the universe by changing it, by thinking about it, by revealing it; they alone can recognize themselves in it and endeavor to make their mark upon it.' These transcendent spirits take the burden of responsibility for the whole universe and then they create new meanings for projecting its revival. This is something that has never been done yet by a woman. Woman has never been able to transcend the givenness of her historical existence characterized by passivity, weakness and dependence; she has never been able to believe in herself as a free subject who can 'authorize herself' 'to enact the fate of all humanity' in her personal existence. In the face of this absolute gloominess, Beauvoir shows that there is spark of hope for woman to revitalize her existence as a free individual. The primary issue in this regard it is that woman should no more think that the 'limitations' she has are not the essential trait of her existence; rather, it is only her historical 'situation' that she still remains unable to transcend as a being in liberty. She does not agree with the idea that the historically effected fact is established as 'an eternal truth.' Rather, historicity as facticity is only a situation that is already undergoing change; a subject can contribute to this changing situation if he or she attempts to project his or her life as a transcendent existence. Men and women both should transcend their 'gender differentiation' to behave like a human being who can unequivocally set his or her goals in the glory of liberty. Once this sexual difference is eliminated from this lifeworld and human beings experience the true meanings of freedom only then according to Beauvoir:

> will woman be able to identify her personal history, her problems, her doubts, her hopes, with those of humanity; then only will she be able to seek in her life and her works to reveal the whole of reality and not merely her personal self. As long as she still has to struggle to become a human being, she cannot become a creator.[40]

4 Conclusion

Through this study we have arrived at the task of Simone de Beauvoir's way of defining woman and her liberation in the nexus of phenomenological-existentialist feminism. This task necessitates to travel along all possible relevant ways in order to elaborate the two concepts and thus to have a feel of contentedness for the whole argument of ours.

[40] Beauvoir's dream of a free woman is inspired by Rimbaud's prophecy that she believes will be fulfilled in the future. In a letter to Pierre Demeny (May 15, 1871), Rimbaud writes: "There shall be poets! When woman's unmeasured bondage shall be broken, when she shall live for and through herself, man—hitherto detestable—having let her go, she too, will be poet! Woman will find the unknown! Will her ideational worlds be different from ours? She will come upon strange, unfathomable, repellent, delightful things; we shall take them, we shall comprehend them." *Ibid.*, 715.

This interpretation of Beauvoir's existentialist feminism regarding the concepts of woman and liberation is important for us, as we always found ourselves overwhelmed by her sincere attempt of redefining woman and her freedom through the mold of existentialism. Beauvoir deals with the problem of liberation as well as the problem of redefining woman in a more radical manner than any other feminist thinker. Once she makes us understand the problem of defining woman and she comprehensively defines the concept as an existentialist issue in the nexus of the transcendence-immanence contrast, the question of freedom is already raised as to whether or not the same nexus is proportional to comprehend. At that juncture, Berlin's concepts of negative and positive liberty afford us a heuristics to properly interpret the problem of woman's liberation as Beauvoir expounds in the nexus of transcendence-immanence contrast.

Our argumentation shows validly that Beauvoir's existentialist feminism is not a matter of intertextual study rather of reflection on experience while interacting with friends, people, ideas etc. in her lifeworld; the most important life-experience in this respect is her life-long companionship with Sartre. Being a genuine phenomenologist, she is convinced with the view that the philosophical meanings whatsoever one comes across with should not be separated from one's life-experience. We have thus tried in this study to trace her feminism back in the nexus of her relationship with Sartre. The pivotal theme of Beauvoir's magnum opus, *Le Deuxième Sexe* (*The Second Sex*), is the idea that women in relation to men have placed secondarily in the lifeworld since the ancient times and further that this secondary position of women in the social order is imposed by the force of the patriarchal atmosphere rather than the feminine characteristics.

This paper has construed a twofold argument. First, it has shown how Beauvoir defines woman in the existentialist mold and then how she expounds the possibility of freedom that can heuristically constitute woman as a true subject or individual projected so by her own.

Beauvoir condemns the process of history and tradition that has made woman deprive in absolute terms of the urge of transcending the state of the secondary being. It has been the process of the millennia that men—legislators, priests, philosophers, writers and scientists—have firmly been struggling to establish that the subordinate position of woman is willed in heaven and advantageous on earth. The religions, philosophies, sciences and arts all have been contributing to this menace characterized by the unjustifiable male domination and female subordination. The female consciousness as an outcome of this traditional process is effected to be an historical consciousness ascribed with the meanings of inferiority and humility. But for Beauvoir, all these meanings are prejudiced and biased attempting lopsidedly to convince woman to feel contented with the stagnant and static life.

Under the influence of Heidegger, Sartre and Merleau-Ponty, she directs herself in order to grasp the meaning of femininity to the nexus of phenomenology and existentialism. Heidegger conceives of man as a being-in-the-world and to be there in the world implies strictly that there exists a body which is at once a material thing in the world and a point of view toward this world. The real nature of man in terms of his relation to himself and to the world is defined by the phenomenon of death.

Heidegger admits the finiteness of Being as it is bound up with death, but simultaneously he does not deny the unlimited nature of human existence as a temporal development in the openness of future. The death makes man's life finite, but before that, he projects life through time creating behind him the infinite past and before him the unlimited future; and in this perpetual progress of human species, man and woman both take part as correlatives and so this perpetuation of the species does not necessitate sexual differentiation. Besides this, she incorporates Merleau-Ponty's dictum that 'man is not a natural species: he is a historical idea.' In this Heidegger-Sartre-Merleau-Ponty perspective, human body is conceived in the nexus of situation; for them human body is the instrument of our grasp upon the world, a limiting factor for our projects. As a matter of fact, since woman is bodily weaker than man, her grasp upon the world is more limited, and therefore she cannot invest the bodily efforts to accomplish the projects of life with such firmness and steadiness as man can show in seeking his tasks. These facts concerning the bodily traits of man and woman are undeniable, but Beauvoir thinks that the biological level is not significant for gauging one's capability of accomplishing life projects. The concept of human strength or weakness can be defined only with reference to existentialist, economic, and moral considerations. Beauvoir tends to define woman as an existent in existentialist connotation who being a free individual is to create her own life projects in order to control the structure of her immanence through transcending the persistence of the facticities. Beauvoir, while defining man's supremacy and woman's Otherness in terms of the meanings of transcendence and immanence, compares the male and the female attitudes rather than aptitudes in relation to their contribution in the making of human history. The tendency of transcending the given is something that woman was deprived of not due to her aptitude as an individual existent rather her attitude determined by not only her own biological make up but the extraneous forces that suppressed her to be submitting to her biological fate. Beauvoir's being certain regarding woman's aptitude of transcending the given lies in her being an accomplice of man in all of his achievements as an individual existent. Man was like a mirror to woman, as she identified herself as an individual existent in comparison with man; like him she felt an urge of transcending the given facticity going beyond the reiteration of life experience to attain the task of finding new future.

Brave New World: A Confinement Between Mythical and Behaviourist World-Views

Aydan Turanli

Progresses in computer science and biotechnology pave the way for the construction of developed robots and enhanced human beings. In the transitionary epoch we live in, we talk about not only the possibility of universal Turing Machine in the near future, but also the possibility of enhanced and super-intelligent human beings. These thoughts, in turn, lead a discussion about how we are going to interact with non-human but "intelligent" machines on the one hand, and enhanced post-human beings on the other hand.

As is well known, these issues are also thoroughly discussed in science-fiction books and films. Isaac Asimov's book *I Robot* and the film *Gattaca* are chief examples examining these issues. The first one primarily focuses on what kind of relations we will develop with non-human, but "intelligent" entities, the second one, on the other hand, concentrates on post-humans, perfect in health and appearance, with high IQ and long life-span, created through genetic engineering in the test tube. In the brave new world of the film *Gattaca*, there is the clash between biologically formed post-human beings, who are selected and reserved for superior jobs and naturally born human beings, who are assigned to carry out unqualified jobs.

These reflections on issues related to human, non-human and post-human of the future lead us to think about Aldous Huxley's *Brave New World* because it is also a science-fictive novel referring to the problems, which will be created in the future through the developments in biotechnology. The book is, on the one hand, critical of modern societies through allegorical associations and on the other hand, it draws our attention to a new world, which would be created through the developments in technology in the future. One of the problems considered in the book is related to the interaction between, human and post-human. The second problem is linked to

A. Turanli (✉)
Istanbul Technical University, Istanbul, Turkey
e-mail: turanliay@itu.edu.tr

C. A. Hornbuckle et al. (eds.), *Phenomenology of the Object and Human Positioning*, Analecta Husserliana 122, https://doi.org/10.1007/978-3-030-66437-4_8

the issue whether rational means are adequate to overcome clashes between different points of view, depending on different forms of life. In this article, I discuss these problems especially by focusing on the second problem within the context of *Brave New World* and *Island*.

As is well known, *Brave New World* is a futuristic novel, which describes a dystopian World State in the sixth century A.F (after Ford) in which genetic engineering and biotechnology are highly developed. Most of the human beings living in the state are born from human eggs; human eggs are treated by scientists and made clever, stupid, and average. There is a eugenically stratified caste-system of the society in which there are five divisions: Alphas are at the top, and then there are Betas, Gammas, Deltas and Epsilons. There is no family, there are no ancestors, and the primary motto of the World State is epitomized in three simple words: "Community, Identity and Stability." The first one implies that people should be at peace with one another and serve the state. The second one is that everyone in one social group should be exactly like everyone else and should not try to be different, and thanks to the conditioning systems they have, they cannot be different. The third one, on the other hand, requires citizens in the World State to be content and do not try to change society in any way. Because "History is bunk," there is no appeal to history, literature, art or even science. Hence, citizens of the World State are illiterate of the past cultures.

The principles of the State are inculcated into citizens by conditioning, which is a significant part of training. There is a wide-range of training methods, extending from sleep-teaching conditioning to even electro-shock conditioning.

On the other side of the coin, we see another part of the World State, in which there are savage Reservations isolated from the rest, surrounded by electric fences. Savages living there are at the bottom of the hierarchical caste system and there is no escape from a savage reservation. People, who are born there, have to die there. As they are different from genetically designed human beings, savages in the old world have ancestors. They keep their habits and customs such as marriage, religion, rituals, diseases, and wild animals.

The clash between the "civilized world" and the old world comes to the fore especially when John the Savage is taken to the civilized world and exhibited as if he is a cage-animal. The Savage's emotionality, beliefs, traditions, and rituals are in discord with the pragmatist, utilitarian and behaviorist way of living of the inhabitants of the New World. The conflict between the two different world-views reaches its peak when Mustapha Mond explained to John the Savage the reasons why the civilized world is designed in the way it is. This disagreement ended with a tragedy. Is this inevitable?

Along with other novels such as George Orwell's *1984*, *Brave New World* parodies, in Martin Heidegger's term "enframing" that is created in modern societies. Even the characters' names in the novel are inspired by 1900s world: "Ford," "Mond" have reference to actual industrialists of the time.

The New World is based on a form of life, in which human beings are insensitive, pragmatically driven and free of emotions. This reminds us of the material culture of modern societies in which one-dimensionality eliminates creativity, art, values, beliefs, sensitivities, and even science.

1 The New World vis-à-vis the Old World

In *Brave New World*, Huxley, on the one hand, satirizes the new economic and social world order established with the developments of technology especially after Fordism, on the other hand, he envisions a kind of a world we will have in the future with developments in sciences such as biology, genetic engineering and others. Hence, on the one hand, the novel is a critique of modernity and, on the other hand, a depiction of futuristic dystopia. Regarding the first depiction, the novel reminds us of Jürgen Habermas's analysis of the modern world in his early writings. As is well known, Habermas depicts the modern world as constituting a hierarchical order. On the one hand, there is technocracy, and on the other hand, there is the public, with no interest in participating issues concerning its own rights, which creates a split between technical and communicative action. Technocrats consider and solve technical problems as well as practical problems. Hence, Habermas states that in a modern society, communicative action is colonized and suppressed by technical action and science and technology become ideology (Habermas, 1970).

Huxley also underlines this hierarchy contained in a modern society,

> ...the significant facts about the governments of contemporary nations are these. There are a few rulers and many ruled.... The ruled, for the most part, quietly accept their subordinate position and even actual hardship and injustice.... the general rule is obedience (Huxley, 1941, 56).

In the New World portrayed in Huxley's book, we see the same kind of picture; the World State is under the control of the World Controllers, who constitute an exceedingly small part of society. An elite set of knowledgeable people, who control and implement technological advances and processes such as the Bokanovsky Process create meritocracy, in Jacques Ellul's terms. Technology becomes ideology because technology, specifically biotechnology and genetic engineering, plays a key role in eugenics.

Mustapha Mond in the dialogue with John the Savage says, "God isn't compatible with machinery and scientific medicine and universal happiness. You must make your choice. Our civilization has chosen machinery and medicine and happiness. That's why I have to keep these books locked up in the safe. They're smut" (Huxley, 1932, 207). Technology becomes ideology in another sense too: although there is no religion in the New World; "Ford," which is the turning point in technology is substituted for "God." People in the World keep saying "Ford forbid" instead of "God forbid."

However, we cannot say the same thing for science. Science, especially theoretical science, is banned to keep so-called "stability" and "universal happiness" safe.

The World Controller Mustapha Mond explains this in the conversation with John the Savage. The New World is created to secure happiness and stability. Anything that threatens stability is eliminated to secure the so-called "happiness." Emotions, sensitivity, new inventions, and discoveries are all threats to stability. Mustapha Mond says, "Besides, we have our stability to think of. We don't want to change. Every change is a menace to stability. That's another reason why we're so chary of applying new inventions. Every discovery in pure science is potentially subversive; even science sometimes be treated as a possible enemy. Yes, even science" (Huxley, 1932, 198). Therefore, science should be "carefully chained and muzzled" (Huxley, 1932, 198).

Not only science, but also art and literature are banned in the World State to keep stability safe and secure. Behaviorist conditioning of genetically engineered human beings requires elimination of all emotions; because art and literature stimulate feelings and sensitivity, they are banned. Mond accepts that Shakespeare's *Othello* is good; however, he underlines that the price they pay to eliminate instability requires banning it. He says that it is necessary to choose between happiness and high art, and they sacrificed high art (Huxley, 1932, 194). Tragedies need social instability, but the World State is stable and full of happiness; Mond continues, "People are happy; they get what they want, and they never want what they can't get. They're well off; they're safe; they're never ill; they're not afraid of death; .... *Liberty!*' He laughed. 'Expecting Deltas to know what liberty is! And now expecting them to understand *Othello*! My good boy!'" (Huxley, 1932, 193–194).

What happens when there is a confrontation between advocates of the New World and the Old World: namely confrontation between Mustapha Mond and John the Savage? Is it possible for them to convince one another by means of rational argumentation? The answer of this question will become clear after we discuss some political philosophers' views in the next section.

1.1 What Would Political Thinkers Say?

Influenced by Zen Buddhism, especially in his later period, Huxley thinks that arguments do not solve controversy. He says, "Argument and controversy are almost useless; in many cases, indeed they are positively harmful. But this, of course, is a thing that clever men with gift for syllogisms and sarcasm find it peculiarly hard to admit" (Huxley, 1946, 161).

Political philosophers, such as Habermas, are inclined to accept that any debate over different forms of life and views can be solved by means of a rational discussion. Hence, it requires argumentation; however, others such as Hannah Arendt, Michel Foucault and Chantal Mouffe do not believe that arguments by themselves can finalize a debate over world-views.

Inspired by Foucault, Friedrich Nietzsche, and Ludwig Wittgenstein; David Owen makes a distinction between ideological and perspectival captivity and identifies captivation by one world-view as "aspectival captivity." Owen explains "aspectival captivity" by correlating it with Wittgenstein's famous remark "A *picture* held us captive. And we could not get outside it..." (Wittgenstein, 1958, §115). According to him, background practices are one of the factors for being held captive by a picture because by acquiring a language we become minded (Owen, 2003, 84). Hence, they function as "nonphysical constraint on our capacity of self-government" (Owen, 2003, 88). Huxley also talks about "impersonal forces...pushing us all in the direction of *the Brave New Worldian nightmare*" (Huxley, 1958, 7).

According to Owen, we inherit a system of judgments and it directs our practical judgements in daily life and it helps us "make sense of (and hence to experience) ourselves as agents in the ways that matter to us" (Owen, 2003, 84). However, when there are changes in the circumstances we may experience, our world-picture or some aspect of it as problematic in the sense that we cannot "make sense of ourselves as agents in this way" (Owen, 2003, 84). Just at this point:

> ...we can call a given picture into question in order to assess its *value*, that is, its capacity to guide our judgments and activity such that we can make sense of ourselves as agents, such that we are intelligible to ourselves as agents, in the ways that matter to us. This is just where the issue of being held captive by a picture becomes pressing, since it suggests that in conditions where our picture is increasingly unable to guide our judgements such that we can make sense of ourselves as agents, we may find ourselves held captive by this picture and thus unable to revise or replace it in the ways required for us to make sense of ourselves as agents in ways that matter to us (Owen, 2003, 85).

In *Brave New World,* this is exemplified by individuals such as Bernard living in the New World and John the Savage, who comes from the Old World. Although Bernard is used to living in the form of life the New World presents, still unusually he feels himself imprisoned in that world. John the Savage, on the other hand, with his religiosity, commitment to old values and family life, cannot adapt himself to the New World. Being held captive by a picture of the Old World, and unable to revise and replace this picture within the New World, he cannot make sense of himself as an agent and feels trapped. In both cases, nonphysical constraints prevent people to act independently, hence they are in Owen's terms enslaved and entranced by a picture (Owen, 2003, 85).

In this case, Owen says, "...the way in which the exercise of our capacity for self-government *qua* agency is blocked by our captivity to a picture because the exercise of our capacity for self-government *qua* judging is obstructed by our captivation by this picture: we are enslaved because we are entranced" (Owen, 2003, 85). Freedom and self-government are possible by getting rid of enslaving pictures.

John Searle also underlines that norms entrenched in background practices create power and they function as directive type of speech acts, which he calls "standing directives" by means of which everyone may apply force to everyone else to make them conform to established norms. He points out that Foucault explicates this with the concept of "bio-power"; however, Foucault's view can be completed by pinpointing the agents of power exercisers (Searle, 2010, 155).

Owen characterizes assumptions implicit in background practices as constituting "nonphysical constraint on our capacity for self-government" (Owen, 2003, 88). Self-governance and autonomy are possible by freeing us from captivating pictures engrained in background practices. Captivating pictures suppress other ways of thinking, seeing and behaving; hence claim that they themselves are inevitable.

As in the case of the Mustapha Mond-John the Savage debate, each party is entranced with the picture they have in mind, which blocks sensible communication in dialogues over different forms of life. Then, the question whether a rational communication ends the debate, arises. Can we say that rational communication helps us see things differently and free us from the "despotic demand of the 'must' characteristic of our captivation by a given picture" (Owen, 2003, 88)?

In a debate, people captivated by a picture cannot see alternative ways of reflecting on the topic in question.

Owen contrasts two types of captivity: one is captivation by false beliefs and ideological captivity and the other is aspectival captivity (Owen, 2003, 88). Philosophers such as Habermas defend the first type. For example, Habermas's theory of communicative action takes into account all nonphysical captivity as captivation by false beliefs (Owen, 2003, 88).

Ideological captivity as different from aspectival captivity is related to "false consciousness" (Owen, 2003, 89). Hence, it is different from aspectival captivity with regard to the process it achieves. In the case of ideological captivity, self-reflection directs us from error to truth: while in the case of aspectival captivity, self-reflection leads us to perceive ourselves in such a way to evaluate the value of the picture in question (Owen, 2003, 90). In this sense, aspectival captivity is related to genealogy defined by Nietzsche and Foucault.

Genealogy allows subjects, who have been held captive by a picture, to free themselves from being trapped by revealing the possibility of other pictures and by showing that the picture in question has been otherwise. Genealogical practices aim at showing that anything could have been otherwise, contingent and historical in this sense. Histories of subjects have been otherwise and historical studies "free ourselves from ourselves" (Owen, 2003, 93). Owen underlines the difference: "whereas ideology-critique seeks to disclose a contradiction between our beliefs and our epistemic principles, genealogy aims to elucidate a disjuncture between our intelligibility to ourselves as agents, our form of subjectivity, and our cares and commitments" (Owen, 2003, 93).

When we analyze the debate between Mustapha Mond and John the Savage from the perspective of aspectival captivity, we can say that because of the Old World's form of life, John the Savage is under the "nonphysical constraint" of a picture; he is captivated by it. This captivation, on the other hand, blocks his ability to make sense of himself under different conditions of the New World. His conversations with Bernard Marx, Lenina and Mustapha Mond are not helpful to make sense of himself as an agent under the new conditions. Because he cannot re-describe the picture that helps set him free from the captivity, he cannot reorient himself in the New World, which results in a tragic situation.

We can reiterate the same question: is it possible to come to an agreement in a discussion about different world-views through fully-inclusive rational argumentation?

As stated, Owen correlates ideological captivity, which entails "false consciousness," with Habermas's view. As is well known, Habermas states that it is possible to come to an understanding through "universal pragmatics" within "ideal speech situation." In Habermas's case, freedom from captivity is possible through an enlightening self-reflexive process, which helps us realize truth and abandon falsity. Because it depends on universal principles, truth is not subjective here. According to Habermas, communicative action is based on rationality and issues concerning a debate between different points of view, especially in political matters, can be solved rationally. As Chantal Mouffe says, "deliberative democrats affirm that it is nevertheless possible to reach a consensus that would be deeper than a...consensus that could qualify as 'moral'" (Mouffe, 2000, 2). In this sense, they defend a form of "normative rationality" (Mouffe, 2000, 2).

Mouffe thinks that in Habermas's view "fundamental political questions belong to the same category as moral questions and they can be decided rationally" (Mouffe, 2000, 6).

As different from Habermas, Mouffe thinks that commitment to democratic institutions,

> ...is the constitution of an ensemble of practices that make the constitution of democratic citizens possible. This is not a matter of *rational justification* but of *availability* of democratic forms of individuality and subjectivity. By privileging rationality, both the deliberative and the aggregative perspectives leave aside a central element, which is the crucial role, played by passions and emotions in securing allegiance to democratic values (Mouffe, 2000, 10).

According to Mouffe, in Habermas's approach the "rational subject," is isolated not only from social, cultural, language and power relations, but also from the complete set of practices making individuality possible (Mouffe, 2000, 10). In this sense, in Habermas's account the question "what are the conditions of existence of the democratic subject?" is not asked.

As Hannah Arendt, Wittgenstein and Thomas S. Kuhn, Mouffe says that not only a rational debate, but also persuasion or conversion are operative in the closure of conflicts between two different world-views depending on different forms of life. This, according to Mouffe, implies acknowledging the limits of rational consensus (Mouffe, 2000, 12, 2005). Mouffe underlines the importance of practices, language-games, and power as constituting social relations. She thinks that deliberative democrats eliminate power and antagonism from politics by defining politics in terms of ethics. She, on the other hand, points out that social objectivity is constituted through acts of power. This implies that social objectivity is characterized with the political and has the sign of exclusion, which governs its constitution (Mouffe, 2000, 13–14). In this sense, according to Mouffe, any political order is the expression of hegemony. Because power relations are constitutive of the social, then the main question for democratic politics is how to constitute forms of power more compatible with democratic values. This, on the other hand, paves the way for abandoning the ideal of democratic society or "ideal speech situation," envisioned by Habermas as the actualization of perfect harmony and transparency (Mouffe, 2000, 14).

Mouffe defines the relation between power and legitimacy through two conditions: first, "if any power has been able to impose itself, it is because it has been recognized as legitimate in some quarters" (Mouffe, 2000, 14). The second is, "if legitimacy is not based on a aprioristic ground, it is because it is based on some form of successful power" (Mouffe, 2000 14). She also underlines that the deliberative approach excludes this by assuming rational argumentation where legitimacy is grounded on pure rationality (Mouffe, 2000, 14).

She identifies her alternative as "agonistic pluralism" and makes a distinction between "politics" and "the political." "The political" points to antagonism inherent in social and human relations; "politics," on the other hand, is the constellation of practices, institutions and discourses, which plays a part in forming an order to establish human coexistence. Because of the antagonistic nature of the political, politics is conflictual, and it tries to tame severe conflict implicit in human relations. Therefore, the question according to Mouffe, is not "how to arrive at a consensus without exclusion, since this would imply the eradication of the political" (Mouffe, 2000, 15). Inspired by Carl Schmitt's idea that every consensus is based on acts of exclusion, Mouffe underlines that it is not possible to come to an inclusive "rational" agreement in politics (Mouffe, 2005, 11–13). Contrary to a Habermasian view searching for uniform agreement, Mouffe defends the Schmittian friend-enemy distinction as a necessary starting point in politics. She defends the view that discord is necessary in the realm of public sphere. According to Mouffe, democracy necessitates the "we/they" distinction, because it is in accord with the appreciation of pluralism in modern democracy (Mouffe, 2005, 14). She thinks that every order is political and relies on a form of exclusion. Any fixed socially instituted established order is a "hegemonic practice." Hegemony of a view is contingent, and there are always possibilities that are suppressed and that can be reactivated (Mouffe, 2005, 18). The agonist view implies that every hegemonic order can be questioned by counter-hegemonic practices to infuse another form of hegemony. The struggle between contrary hegemonic projects, on the other hand, cannot be settled rationally (Mouffe, 2005, 21). As different from Habermas's deliberative model, Mouffe thinks that in democratic politics rational calculation or moral deliberation cannot be the only tool to convince others. Mouffe admits the role of passions as one of the important forces in the field of politics. Dynamics of the political can be understood in terms of not only reason, moderation and consensus, but also passions (Mouffe, 2005, 30). Consensus is necessary in politics; however, it must go together with dissent. Hence, the goal of democratic politics is to create a unity in conflict and diversity; it creates "us" by the determination of "them." In this sense, it does not try to overcome "us/them" opposition, which is an impossibility; on the contrary, it establishes this in a way, which is compatible with pluralist democracy; in a way that "them" in democratic politics should not be seen as an enemy, but as an adversary (Mouffe, 2000, 15). Mouffe says that the enemy is someone to be destroyed, while the adversary is a person, whose ideas are combatted, but her right to defend these ideas is not questioned: "This is the real meaning of liberal democratic tolerance, which does not entail condoning ideas that we oppose or being indifferent to standpoints that we disagree with, but treating those who defend them as legitimate opponents" (Mouffe, 2000, 15). "Adversary," in this sense, is a legitimate enemy (Mouffe, 2000, 15). "Legitimate" in the sense that we share "the ethico-political principles of liberal democracy: liberty and equality" (Mouffe, 2000, 15) with them. However, implementation of these principles and disagreement about them cannot be resolved merely through deliberation and rational discussion as Habermas suggests. Besides, ethico-political principles exist in many, different and conflicting interpretations, hence, such a consensus is bound to be "conflictual consensus" (Mouffe, 2000, 16).

In *Brave New World,* there are two worlds: the New World and the Old World, and these two worlds are actually in opposition with one another in an antagonistic way in Mouffe's sense. There is no tolerance towards John, who was brought from the Old World, he is treated as if he is a cage-animal and is exhibited accordingly. The New World's non-sensitive behaviorist people, who are completely devoid of passions and emotions cannot accept the Old World's John the Savage as a legitimate opponent but see him as an enemy in an antagonistic sense. Because the World State is a totalitarian system, it is not possible for each party to acknowledge one another as adversaries rather than enemies in an antagonistic sense. Although John the Savage has feelings, emotions, and passions, which in Mouffe's terms can be mobilized towards "conflictual consensus" and "democratic designs:" the New World's people for the sake of the motto "community, identity and stability" are "rational" enough not to give access to any "conflictual consensus," which results in the tragic death of John the Savage.

As stated above, *Brave New World* is a depiction of a totalitarian society with zero tolerance to alternative views; therefore, it is not possible to expect that society to value freedom and equality, which may flourish in a democratic society. There is a caste system within the New World, and a space is not opened for alternative ways of behaving and living, which cultivates freedom, and hence, the Old World is excluded from the system of the New World. This world is extremely far from satisfying the human needs for liberty and equality, because as mentioned above, it does not pave the way for discussion about alternative views, which may result in conflictual consensus. John the Savage does not have an alternative in that kind of suffocating form of life excluding people, who do not comply with their way of living. Huxley also expresses this in Foreword to *Brave New World* in 1946. He says,

> If I were now to rewrite the book, I would offer the Savage a third alternative. Between the utopian and the primitive horns of his dilemma would lie the possibility of sanity-a possibility already actualized, to some extent, in a community of exiles and refugees from the Brave New World, living within the borders of the Reservation. In this community economics would be decentralist and Henry-Georgian, politics Kropotkinesque and co-operative. Science and technology would be used as though, like the Sabbath, they had been made for man, not (as at present and still more so in the Brave New World) as though man were to be adapted and enslaved to them (Huxley, 1932, xliii).

Huxley also underlines that religion in that society will be "the unitive knowledge of the immanent Tao or Logos" (Huxley, 1932, xliii). Philosophy of life, on the other hand, is a kind of utilitarianist serving the Final End Principle in the sense that it should take into account the question "How will this thought or action contribute to, or interfere with, the achievement, by me and the greatest possible number of other individuals, of man's Final End" (Huxley, 1932, xliii). According to Huxley, that kind of living, which has philosophical completeness, may lead to sanity and aesthetic form of life.

Huxley does not see sanity in the use of science and technology in modernity and his dissatisfaction with modernity, reminds us of Herbert Marcuse's assertion that technological rationality implies irrationality, and therefore it is not sane. Huxley's last novel *Island* may be regarded as an effort to portray a sane society. As an alternative depiction of living, it might also be a proposal for John the Savage, who is confined between the New World's behavioristic form of life and the old World's mythical form of life. What does happen if we relocate John the Savage in the island of Pala? Can the island of Pala be the way out for John the Savage? Is it possible for John the Savage to feel at home in the island of Pala? What kind of conflicts will there be between John the Savage and the people living in the island of Pala? Before I discuss these, I want to focus on Huxley's *Island*.

2 Can the Island of "Pala" Save the Old World's John the Savage?

In contrast with the dystopian style of *Brave New World*, *Island* is a utopian novel. *Island* is Huxley's last novel, and it was published in 1962. As in the context of *Brave New World*, in this last novel, too, there is a confrontation of two different views depending on two different forms of life. Although there are some continuities between the island of Pala and the rest of the world, there are also significant differences. Huxley tries to demonstrate this through the eyes of an outsider, Will Farnaby, who is a journalist from England and is intentionally shipwrecked on Pala by his boss and owner of the newspaper—Will's dead wife's father—Joe Aldehyde, to make an oil deal in the island. The people of Pala have a policy of not bargaining about the island's natural resources; therefore, Joe proposes a huge amount of money to Will to make his way to the island and to create an opportunity to make a deal on the rich oil deposits on the island.

Pala is an unusual island and different from the rest of the world in several respects. The education system in Pala is quite different regarding the curriculum and priorities of disciplines. As an outsider, Farnaby realizes that with Psychology, Mendelism and Evolution, in capital letters, education in Pala, is mostly biological with the aim of creating an ecological awareness. The Secretary of Education Mr. Menon agrees and emphasizes that the primary emphasis is on the sciences of life, rather than physics and chemistry. This is for a reason: Menon states,

> You're irretrievably committed to applied physics and chemistry, with all their dismal consequences, military, political and social. But the underdeveloped countries aren't committed. They don't have to follow your example. They're still free to take the road *we've* taken-the road of applied biology, the road of fertility control and limited production and selective industrialization which fertility control makes possible, the road that leads towards

> happiness from the inside out, through health, through awareness, through a change in one's attitude towards the world (Huxley, 1962, 216).

As is seen, in contrast with the New World's, eugenically stratified structure in which sciences such as genetic engineering is used to produce bio-formed post humans in tubes; in the island of Pala, science and technology are used to produce harmonious ecosystem. This is achieved through the establishment of new vision of science and technology, which reminds us of Herbert Marcuse's proposal. Critical of a vision of science paving the way for technological rationality, Marcuse also says that our attitude towards nature should change, and we should develop fraternal relations with nature, which, in a way, is actualized in the island of Pala. In the island, animals are parts of people's lives and human beings living on the island develop friendly relations with animals such as lizards, birds and even snakes. The secretary of Education Menon underlines that the vision of science in the rest of the world should change to enable us to be attuned with nature, human dispositions and culture. This is possible by improving values internally—in human mind—"through awareness" and "through a change in one's attitude towards the world" (Huxley, 1962, 216). He also talks about applied biology as a possible savior of the future world (Huxley, 1962, 216–217). Therefore, because they think that it helps children to see the things in relation with one another, in schools, they start teaching ecology at the same time as they teach multiplication and division. He explains this: "[W]e always teach the science of relationship in conjunction with the ethics of relationship" (Huxley, 1962, 217). Besides, in Pala, educational environments, in which the children are raised, are not limited to schools only.

Ronald Lee Zigler emphasizes that there are similarities between John Dewey's view of education and that of Huxley. According to Zigler, education in the island of Pala integrates the strengths of Western and Eastern Cultures: pragmatic and practical values of scientific research, on the one hand, and subjective experience of Zen Buddhism, on the other (Zigler, 2002, 158). Besides, school is not the only educational force in service of children; there are many forces in a society affecting children's education. In Pala, *Mutual Adoption Club* is a place in which children practice being a son or daughter of other families, who are called "deputy" mothers and fathers, which also help them express themselves in extended family relations. This is also a part of education. Zigler states that, for both Dewey and Huxley, human society is not a completed whole, but it is an open-ended process, in which human beings actualize their latent potential for problem solving and growth (Zigler, 2002, 158–159). Zigler states, "…both Dewey and Huxley were mindful of what may be termed the 'educational ecosystem' in which we live and develop" (Zigler, 2002, 159). In this sense, the educational ecosystem of the island of Pala is in contrast with the New World's educational system depending on conditioning.

How would John the Savage be taken in the island of Pala? Before we discuss this, it is better to discuss Farnaby's transformation in Pala.

Farnaby discovers the form of life of Pala, not only by reading the Old Raja's *Notes on What's What, and What It Might be Reasonable to do about What's What*, but also by visiting educational institutions and other places. Especially in a close relationship with Susila during his recuperation period, he experiences an individual enlightenment and gives up helping Joe Aldehyde in the exploitation of Pala's natural oil resources. Unfortunately, it is too little too late. Murugan, who is the son of the Rani, who were both educated outside Palanese society and, who is very much under the influence of western modernization; will be successor of his mother to become a new Raja in Pala soon. He is also open to oil bargaining with other countries. After being rejected by Farnaby in an oil deal, Murugan forms the United Kingdom of Rendang and Pala through a military *coup d'état* and announces himself as the Monarch to set up Western type of modernization and to get into future oil bargains with companies from other countries.

As is seen in the novel, Farnaby experiences a kind of transformation in the island of Pala, which is similar to the transformation of the protagonist Jake Sully in the film of *Avatar*. Is this also possible for John the Savage? How could the people of Pala take him? Obviously, the people of Pala are friendly, cooperative and have respect for alternative ways of living. Therefore, he would not be doomed to the tragic destiny he had in the New World, and he would be taken well. However, he would live conflicts even in the island. For him to appropriate himself to the conditions of the island, he should go through various channels of training like Farnaby.

This confrontation between different forms of life and the debate over them is characteristic of Huxley's novels such as *Brave New World* and *Island*. In *Brave New World*, this conflict, in a Mouffean sense, takes an antagonistic form and cannot pave the way for democratic design and ends up with tragedy. In *Island*, on the other hand, in an optimistic way, the outsider Farnaby is dragged into Pala's form of life, appreciates it, becomes a part of it and gives up his earlier agenda by complying with the people of Pala. Unfortunately, in *Island* too, the individual transformation of Farnaby cannot save the island from drowning into the corrupt oil bargains with the rest of the world. Hence, in a manner similar to *Brave New World*, this utopian novel too, ends up with a dystopia.

Whether *Island* is a utopian book is also discussed by interpreters and some consider it as dystopia, because of the pessimistic end of the novel. Alex MacDonald, on the other hand, interprets Huxley's *Island* from an existentialist point of view and thinks that it is not a pessimistic, but an optimistic book, because it demonstrates not only the possibility of individual enlightenment, but also the possibility of utopia in the social sense (MacDonald, 2001, 103). He thinks that the end of *Island* in a way demonstrates to us that "the idea and possibility of utopia does not end but remains to be worked for" (MacDonald, 2001, 111). It does not show us the impossibility of utopia "but that we have to keep choosing it, over and over again, and something like Pala, or something different from but just as utopian as Pala, can happen again. Our future is not deterministically programmed but depends to some considerable extent upon the choices we make today" (MacDonald, 2001, 111). Hence, MacDonald is optimistic about the future implementation of utopia because he thinks that the future of the world is up to our choices. He underlines that we are always "at that balancing point, at that moment of existential choosing, and the meaning of lives depend on what we choose to believe and to do" (MacDonald, 2001, 112). In line with Huxley, he also states that utopia is always relative and not absolute. MacDonald does not use the concept "utopia" in its commonly used sense, as a completed end-point, or ultimate end, on the contrary, he considers it as an open-ended process depending on our continual choices.

Huxley's understanding of utopia is also different from standard meaning of the concept. He does not consider utopia as an impossible idea or an absolute ideal, but quotes Aristotle, who says that "In framing an ideal we may assume what we wish, but should avoid impossibilities" (Huxley, 1962, i). Hence, when we envision a new alternative life, it should be imagined within real possibilities. Whether Huxley's island of Pala can be a possibility is another question, however, it gives us insights to develop new, alternative forms of life devoid of the materialistic interests, and greediness, which modern technological life presents to us. Imagining alternative lives will help us free ourselves from perspectival captivity and the confinement between two views portrayed in *Brave New World.*

We can turn back to the question of what happens to John the Savage when he is re-located onto the island of Pala. Huxley has something to say here:

> Brought up among primitives, the Savage (in this hypothetical version of the book) would not be transported to Utopia until he had had an opportunity of learning something at first hand about the nature of society composed of freely co-operating individuals devoted to the pursuit of sanity. Thus altered, *Brave New World* would possess an artistic and (if it is permissible to use so large a word in connection with a work of fiction) philosophical completeness, which in its present form it evidently lacks (Huxley, 1932, xliii-xliv).

Huxley underlines that John the Savage would not have been received well even in the island of Pala, unless he had been educated about the nature of a self-governing society beforehand.

I think that tragic destiny of John the Savage can be avoided by putting him onto the island of Pala. This is because he is not going to be taken as an antagonistic enemy in the Mouffean sense in this friendly environment and it is possible for him, like Farnaby, to get trained and educated through many channels to embrace a society, which promotes ecological science, co-operation and participatory involvement. Then, in this self-governing, co-operative and sane society he gets flourished and free himself from aspectival captivity.

Hence, not only rational means, emotions and sensitivities, but also education and engagement in practices from inside is necessary to free ourselves from a perspectival captivity. As discussed earlier, this is not possible by a rational debate only, as is assumed by philosophers such as Habermas; a Kuhnian type of gestalt shift, or a Kierkegaardian type of leap of faith or an Arendtian type of enlarged mentality is also needed. Wittgenstein underlines that discussion or debate over different forms of life cannot be solved through rational means only, because "At the end of reasons comes *persuasion*" (Wittgenstein, 1969, § 612).

3 Concluding Remarks

In this article, I discuss Aldous Huxley's novel *Brave New World* to demonstrate that, where there is a conflict between two world-views or forms of life such as the New World and the Old World in the novel, this conflict cannot be solved by rational argumentation only. From this perspective, I primarily appeal to the views of political thinkers such as David Owen and Chantal Mouffe to explicate that besides rational means, there are other ways of helping our transformation of vision. Inspired by Owen, I say that the conflict between the defenders of the Old World view and the New World view is not dissolved because each party is entrapped by its own view, which reminds us of "aspectival captivity." Influenced by Mouffe's ideas, I say that the New World's people can appreciate the values of John the Savage, only if their emotions and passions are also mobilized towards "conflictual consensus" and "democratic designs". Devoid of emotions, however, they are unable to do so, hence the communication collapses and John the Savage cannot be seen as adversary as suggested by Mouffe's "agonistic pluralism."

In the article, I also discuss that whether it would be possible for John the Savage to be well regarded by the people of Pala, which is another imaginary utopian island in Huxley's novel *Island*. I state that engagement in the island of Pala's practices and educational training of John the Savage are necessary for him to adapt himself even to that kind of life. Hence, not only rational argumentation and emotions, but also educational training is necessary for overcoming "aspectival captivity;" which, in return, allow for accessing to another form of life. Only under the fulfilment of complex conditions, can transformation of vision be accomplished. In that case, as Wittgenstein says, "Light dawns gradually over the whole" (Wittgenstein, 1969, §141).

References

Habermas, J. (1970). *Toward a rational society* (J. J. Shapiro, Trans.). Boston: Polity Press.

Huxley, A. (1932/2007). *Brave new world*. London: Vintage Books.

Huxley, A. (1941). *Ends and means*. London: Chatto & Windus.

Huxley, A. (1946/1947). *The perennial philosophy*. London: Oxford University Press.

Huxley, A. (1958/2004). *Brave new world revisited*. London: Vintage Books.

Huxley, A. (1962/2005). *Island*. London: Random House.

MacDonald, A. (2001). Choosing utopia: An existential reading of Aldous Huxley's *Island*. *Utopian Studies, 12*(2), 103–113.

Mouffe, C. (2000). *Deliberative democracy or agonistic pluralism* (Political science series) (Vol. 72). Wien: Institut für Höhere Studien (IHS). https://www.ihs.ac.at/publications/pol/pw_72.pdf.

Mouffe, C. (2005/2009). *On the political*. London: Routledge.

Owen, D. (2003). Genealogy as perspicuous representation. In H. J. Cressida (Ed.), *The grammar of politics: Wittgenstein and political philosophy* (pp. 82–97). Ithaca: Cornell University Press.

Searle, J. (2010). *Making the social world: The structure of human civilization*. Oxford: Oxford University Press.

Wittgenstein, L. (1958/1968). *Philosophical investigations* (G. E. M. Anscombe, Trans.). New York: The MacMillan Company.

Wittgenstein, L. (1969). In G. E. M. Anscombe & G. H. Von Wright (Eds.), *On certainty* (D. Paul & G. E. M. Anscombe, Trans.). Oxford: Basil Blackwell.

Zigler, R. L. (2002). Revisiting Huxley's *Island*: Advancing pragmatic ideals through a vision of utopia. *Journal of Beliefs and Values, 23*(2), 157–168.

The Thing/Beast/Human Relationships in the Autobiographies of Camara Laye and Wole Soyinka

Tony E. Afejuku

The Guinean Camara Laye's *The African Child* and the Nigerian Wole Soyinka's *Ake: The Years of Childhood* are two major, two pre-eminent African autobiographies that present vivid pictures of childhood that can compel one to call them thinly disguised fictions, thinly veiled novels, that is.[1] But this is not what I am setting out to prove or defend here. My purpose here is to examine how *The African Child* and *Ake: The Years of Childhood* respectively widen our consciousness of man's, of human being's, that is, practical handling of cosmic, existential situations as revealed in his relationships with other beings—other things or beasts—in creation. How does man's intuitive creativity and sensibility elucidate the myth and otherwise of human destiny in the African world as he goes about perceiving, feeling, thinking in his attempt to understand his experiences in the cosmological order? The essay seeks to answer this question mainly through an exploration of the writers' engagement with the subject of the relationship between the living and the living-dead as well as spirit-beings which the autobiographies excellently focus on.

Each writer's approach underlines the essential differences and divergences of the European frame of mind from the African one. Laye and Soyinka do not in any way lead us to gain, for example, the impression that there is "no other universe except the human universe" as the French existentialist Jean-Paul Sartre argued.[2] Laye's and Soyinka's universe is the cosmic universe as known by each one's

[1] The respective editions of the texts studied are as follows: Camara Laye, *The African Child,* trans. James Kirkup (London: Fontana, 1955); Wole Soyinka, *Ake: The Years of Childhood* (London: Rex Collings, 1981).

[2] Jean-Paul Sartre, "Existentialism Is a Humanism." *Philosophy and Choice: Selected Readings from Around the World,* ed. Christensen (Mount View, California: Mayfield Publishing Company, 1999), p. 255.

T. E. Afejuku (✉)
University of Benin, Benin City, Nigeria
e-mail: tonyafejuku@yahoo.com

C. A. Hornbuckle et al. (eds.), *Phenomenology of the Object and Human Positioning*, Analecta Husserliana 122, https://doi.org/10.1007/978-3-030-66437-4_9

community which acknowledges and recognizes the existence of beings—spirits, things or beasts—that exist side by side human beings in their community which is also the community of the other-worldly. Thus, both human beings and other-worldly beings relate with one another in the African belief that life is really nothing and meaningless until it is lived by both in shared relationships. Indeed, in the African belief system the human community is never devoid of the community of other-worldly beings, through whom human beings gain deep understanding and knowledge of existence. In fact, in a typical African (traditional) community (even of the modern times), according to John M'biti, "Spiritual Beings, Spirits, the Living-Dead" and the living-being mingle and inter-mingle.[3] The import of this quotation can easily pass for an African myth, but it is a living myth well nourished by the people who cherish it in their African worldview which accepts and accommodates the denizens of its "un-demarcated world," to borrow Soyinka's words in his seminal essay, "Drama and the African World-View,"[4] in the endeavour to dissect its "heart of cosmic mysteries," to use Soyinka's words again.[5] On this point, Soyinka and M'biti are clearly and classically of the same mind because they are positivist African existentialists. It is necessary here to quote M'biti in full:

> The departed of up to five generations are in a different category from that of ordinary spirits.... They are in the state of personal immortality and their process of dying is not yet complete. We have called them the living...these are the spirits with which African people are most concerned.... They are still part of their human families.... The living dead are still 'people'.... They return to their human families from time to time, and share meals with them, however symbolically.[6]

These are the words of a true, positivist African phenomenologist who is recording for posterity the cosmic belief of his people, a cosmic belief that should always remain in their consciousness. The value the quotation espouses underscores the respect for ancestors which enriches the culture and quality of life or of existence of African peoples. Indeed, the exchanges between the living and the dead help the living to record properly history, and to chronicle meaningfully different times and epochs for posterity. The exchanges also promote intra- and inter-family unity that solidifies families, communities and societies at large. The living also benefit immensely from the spiritual and moral authority which the "living dead" offer them in the forms of warnings and admonitions time after time. Such mystical, cosmic warnings and admonitions help the living to improve their spiritual and moral conduct.

Camara Laye and Wole Soyinka make it clear that they are interested in the mysteries of their people's cosmic world whose inter-reactions between man and things of nature are perfectly underlined. In several instances in the autobiographies,

---

[3] John M'biti, *African Religion and Philosophy* (London: Heinemann, 1970), p. 47.

[4] Wole Soyinka, "Drama and the African World-View." *Exile and Tradition: Studies in African and Caribbean Literature, ed. Rowland Smith (London: Longman and Dalhousie UP, 1976),* p. 189.

[5] *Ibid.*

[6] M'biti, p. 47.

especially in *Ake*, our attention shifts from the individual selves of the narrators, the autobiographers, and focuses on the relationship between the things/beasts/beings who leave their abode and enter that of man, who equally enters theirs without qualms because of the seeming lack of demarcation between both worlds—which actually are the same or seem to be the same as the narrators want us to understand what they are presenting. It does not matter whether the reader believes or does not believe what is happening before his very eyes.

In *The African Child*, the scene where the narrator's father turns trinkets into gold, the reader sees the sensations that are associated with his formula, which include and involve his magical invocations, and transformation from an ordinary being to a special, supernatural being—who moves in and out of the world of man and that of the other-worldly—when the goldsmith is at work. His magical communication and relationship with beings of the other-worldly, which his invocations underline, constitute an experience of a world which is always present to Laye's father (and to Laye himself—as the recorder of the everyday experience which he wants us to see as a unique experience which he took to be ever present in his childhood years, at least). If or when we "look to the pure immanence of experience," in line with Edmund Husserl's ideas, wherein he tends to posit that we should avoid interpreting a given experience, we might not need to examine "the sensations ... perceptions or pictures of the world" Laye recalls, to borrow Claire Colebrook's words (as applied to Husserl).[7] But not wanting to see the sensations and pictures or perceptions Laye recalls through the act and art of memory, his memory, that is, will be un-African. The myth of an African experience, such as the one referred to above (and any symbol it generates) is always a subject and an object of inquiry in African Literature, African Philosophy and different kinds of African cultural studies, including African ethics and hermeneutics. I think Wole Soyinka's "Drama and the African World-View" makes the point persuasively well even though not precisely in the manner of my discourse.[8]

In the African worldview, the everyday experience is not independent of the existing world whose existence is tied to the existence of man through the experience of man who understands his environment better through his relationship with this environment and the numerous things and beings in and outside it. From an African perspective, this is not an assumption—as Husserl may wish us to accept it; it is an actuality. In any case, we are aware that Husserl, the acknowledged father of Western phenomenology, to quote Colebrook, argued that "Each region within the world has its own specific way of being known".[9] Everything outside a specific region's specific ontology in the forms of misconceptions, assumptions and preconceptions must be bracketed off. Thus, any misconception, assumption and preconception about African life and worldview which Laye and Soyinka present in

---

[7] Claire Colebrook, "Edmund Husserl (1859–1938)." *The Continuum Encyclopedia: Modern Criticism and Theory,* ed. Julian Wofreys (New York: The Continuum International Publishing Group Inc., 2003). p. 74.

[8] Soyinka, "Drama and the African World-View", pp. 173–190.

[9] Colebrook, "Edmund Husserl", p. 78.

their autobiographies relating to the kernel of our discourse should be adjusted out of the consciousness of those who conceive of them as such. In other words, an incult Eurocentric consciousness relating to a reading of an African experience and ontology generally should be declaimed against.

The opening pages of Laye's and Soyinka's autobiographies are pertinent here. The first page of *The African Child* draws our attention to the autobiographer's unique experience as a child of 5/6 years. His first moment of awareness and of discovery centres on his relationship with a snake, a poisonous beast the innocent child is playing a perilous sport with. Just about the moment when the deadly reptile is set to sink its fangs into his hand with which he tantalizes the beast with a reed, the young child is saved by one of his father's apprentices who hurls him away to the bewilderment of Laye's emotionally shaken mother who slaps him severely. Suddenly the child becomes aware that the snake in question is a harmful one which he mistook for the harmless one who everyone in his parents' household is familiar with, and who his father in particular relates with properly (as his blood relation) because it is the guiding spirit of his lineage and their protector. The good snake is a living-dead, an ancestor who is in perpetual contact and communication with the living. Clearly, the child suddenly is made aware of the nature of good and evil, and of the nature of the relationship that must exist between man and the good and evil things (beings) in creation. The narrator speaks thus:

> There were good spirits, and there were evil ones; and more evil than good, it seemed to me—And how was I to know that this snake was harmless? It looked the same as any other snake.[10]

This is one of the significant quotations in which Laye revisits a perplexing experience. By remembering his perplexity, he tries to abide within himself. In fact, the recollection is metaphorical and mystical, and is akin to that of a modest phenomenologist who cherishes the reverie of childhood or the poetic imagination (of the uncollective unconscious).

Commenting on this aspect of the autobiography, and specifically from the perspective of the child's perplexity, David Carroll says as follows: "From this point, the child is self-conscious and frequently perplexed by the world in which he finds himself, he never regains the security and complete happiness of the first two paragraphs of the novel".[11]

Anyone who inclines to agree with Carroll fully will miss the point of this autobiography. Laye, obviously narrating his experience as he does, and as underlined by the aspect recalled above, is saying that man is nothing else but that which he propels of himself and which he looks forward to in existence with or without a sense of security or with or without a feeling of complete happiness but in obedience to some force. How the individual carries himself (even in a communal society such as Laye's) with other beings—human or not—ultimately will put him in

---

[10] Laye, *The African Child,* p. 16.

[11] David Carrol, "*Camara Laye's The African Child:* A Reply", *African Literature Today* (1971) p. 130.

possession and will give him a good measure of existential happiness in this world so long as he acts in obedience to the force that propels him. From an African perspective, obedience to messages from the mythical or supernatural beings such as the little black snake that guides and guards his father, and by extension Laye himself, cannot but lead the obedient one to his existential destination.

Laye's eventual departing Guinea to France is to be seen—in fact, must be seen—in this light. His father faithfully obeyed and followed the guiding light of the little black snake, an action that made it possible for him to train Laye properly and encourage him in the direction of his existential direction and destiny. And Laye himself followed—as a child up to the point of his departure for France—the dictates of his father as prescribed by the little black snake of joy and happiness in his father's homestead.

This perspective offers this autobiography its existential value. Laye, the adult recalls his childhood to underscore this essence of his autobiography, and to let us know what it means to follow one's destiny in existence from the prism of one's regional ontology—which his African ontology depicts. As Laye really makes us to understand it, the force—in the form of the little black snake—that linked him to his destiny, away from his father's gold-smith profession, the profession of the lineage, is one that is greater than him. The autobiographer in his design links the adult Laye's present with his past, a past that enables him to look forward to the future in his present. Will his relationship with the character called the Little Black Snake continue to the future? Will the Little Black Snake keep on moving and motivating him anywhere he is or in which he finds himself? Although the book's suspenseful ending leaves much to be desired, we can affirm confidently that the adult Laye succeeded in his endeavours in Europe—a success which his writing of this autobiography attests. The Little Black Snake which we can liken to a biblical angel did not waive his guidance of Laye even in very distant Paris as Laye's father prophesied in line with "the wheels of destiny"[12] that guaranteed the autobiographer's existential destination.

In *Ake*, Soyinka's introduces us straightaway to his childhood world of *egungun* (ancestral masquerade), apparitions of the dead and *oro* (tree daemons) and other spirits that populate his Yoruba (African) people's world. Without qualms Soyinka tells us that his mother's brother, Sanya was an *oro* who came to live as a human being. This explains the reason why he was always "at home in the woods"[13] when he and Soyinka's mother (and others) would go to the woods to hunt snails—as the autobiographer reveals in the words of his mother, the owner of the Uncle Sanya stories in the autobiography. Indeed, the *oro* phenomenon runs from pages 5 to 11 of the text. The good nature of the *oro* (as well as their bad nature) is related with great gusto that underlines the existential, cosmological reality of the text and the characters of interest from the stand-point of the relationship between things/beasts and human beings in creation. It is desirable here to note that the autobiography

[12] Laye, *The African Child*, p. 158.

[13] Soyinka, *Ake*, p. 5.

contains passages which teem with pictures of things and beings in existential relationships, for example:

> Well, you know your Uncle Sanya. He was angry. For one thing the best snails are on the other side of that stream. So he continued to complain that those *oro* were just being selfish and he was going to show them who he was. Well, he did. About a week later he led us back. And he was right you know. We gathered a full basket and a half of the biggest snails you ever saw. Well, by this time we had all forgotten about the warning, there was plenty of moonlight and anyway, I've told you Sanya is an *oro* himself…
>
> 'But why? He looks normal like you and us.'
>
> 'You won't understand yet.' Any way, he is *oro*. So with him we felt quite safe. Until suddenly this sort of light, like a ball of fire began to glow in the distance. Even while it was still far we kept hearing voices, as if a lot of people around us were grumbling the same words together. They were saying something like, "You stubborn, stiff-necked children, we've warned you and warned you but you just won't listen…."[14]

In the above passage, there are no sharp distancing indices between fiction and reality, between other-worldly things/beings and authentic creations/humans; and exaggerations are not relied on by the narrator to drive home the import of the other-worldly forms and children who strayed into the woods hunting for choice snails. But the other-worldly ***oro*** had ceaselessly warned Sanya and the other kids against invading their abode. In fact, it was not only the other-worldly ***oro*** who had warned them again and again not to enter the woodland.

Earlier on, Uncle Sanya and the other children, including Soyinka's mother, had been warned by their highly mystical uncle, the Reverend J.J. Ransom-Kuti, who possessed invisible powers, not to venture into the woods again when the daemons pursued them to their house. In any case, in the context of our discourse, Uncle Sanya's stubbornness was a rebellious and courageous attempt to conquer his existential environment. Of course, Uncle Sanya's existential environment included that of his fellow-human beings and also that of his fellow-*oro*, both with whom he maintained a relationship. From the stand-point of a Euro-centric existentialist Uncle Sanya may be called a perfect existentialist who wanted to be in firm control of the two existential worlds he inhabited. Soyinka's streak as a rebel against constituted authority and establishment in Nigeria was probably partly acquired from the mysterious Uncle Sanya whom he had admired (as he admired the equally mysterious Reverend J.J. Ransom-Kuti) in his years of childhood.

In several profluent passages which are significant in the context of the subject of attention Soyinka convincingly relates Uncle Sanya's ***oro*** world. Clearly, the following passages, like the previous one, underline the existential implications of Uncle Sanya and of Soyinka's autobiography, or at least the fact that the narrative has existential, cosmic implications presented in the form of direct self-narration and pure report of actions, scenes and quoted spoken words that buttress the narrative's phenomenological mode:

> Even Sanya had however stopped venturing into the woods at night, accepting the fact that it was far too risky; daytime and early dusk carried little danger as most wood spirits only

[14] *Ibid.*, pp. 6–7.

> came out at night. Mother told us that all this occasion she and Sanya had been picking mushrooms, separated by only a few clumps of bushes. She could hear his movements quite clearly, indeed, they took the precaution of staying very close together. Suddenly, she said, she heard Sanya's voice talking animatedly with someone. After listening for sometime she called out his name but he did not respond. There was no voice apart from his, yet he appeared to be chatting in friendly, exalted tones with some other person. So she peeped through the bushes and there was uncle Sanya seated on the ground chattering away to no one that she could see. She tried to penetrate the surrounding bushes with her gaze but the woods remained empty except for the two of them. And her eyes came to rest on his basket.
>
> It was something she had observed before, she said. It was the same, no matter how many of the children in the household went to gather snails, berries or whatever. Sanya would spend most of the time playing and climbing rocks and trees. And yet, whenever they prepared to return home, his basket was always fuller than the others. This time was no different. She came closer, startling our Uncle who snapped off his chatter and pretended to be hunting snails in the undergrowth.[15]

Furthermore,

> Mother said that she was frightened. The basket was filled to the brim, impossibly bursting. She was also discouraged, so she picked up her near empty basket and insisted that they returned home at once. She led the way but after some distance, when she looked back, Sanya appeared to be trying to follow her but was being prevented, as if he was being pulled back by invisible hands. From time to time he would snatch forward his arm and snare.
>
> "Leave me alone. Can't you see I have to go home? I said I have to go."
>
> She broke into a run and Sanya did the same. They ran all the way home.[16]

These passages from the early pages of ***Ake*** are brilliant compositions relating to Soyinka's surroundings which reveal his ecstasy of objects and places. But each of the quoted passages is a verbatim report of seemingly actual accounts related to Soyinka by his mother. Soyinka relates his mother's childhood experiences here to underscore his own phenomenological experience of a lived reality which resonated from his memory—from the nooks and crannies of his phenomenological childhood. The environment, the vegetation, the plants, the rocks, the animals, mushrooms and spirits and images of his childhood which they generate are juxtaposed with his mother's with whom he seemingly shared the same phenomenological reality. His creative imagination in the passages (and elsewhere in the first two chapters of the autobiography), are characteristically those of the phenomenologist of homescapes and landscapes. Soyinka's phenomenological imagination in the passages testifies to his unique visual and verbal, auditory and tactile perceptions, felicitous feeling of consciousness by displaying anthropomorphic and spiritual essences of his and his mother's diurnal childhood landscapes of discovery, which were hunting places for survival, places where courage was tested, and, very importantly, places where natural wealth existed in plentiful supply. Gaston Bachelard's view of the phenomenologically poetic image which reflects the coming into being of perception and imagination that are not radically opposed modes of intentionality, which

[15] *Ibid*, pp. 9–10.

[16] *Ibid*, p. 10.

he focuses on, among other things, in his *The Poetics of Space,* is reflected in Soyinka's quoted passages. Soyinka's autobiographical recollections about images of his childhood landscapes reflected a great deal in his mother's story are themselves linguistic descriptions of his childhood perception and imagination. Phenomenologically speaking, they induce the reader and critic to harvest an ability for creative spectacle.[17]

Interpretation and language constitute the central themes of Soyinka's phenomenological experience/reality which the passages reflect in Soyinka's language of spectacular simplicity which provides social, cultural, mythological and historical evidence of his people's existence right from time immemorial. Obviously, to restate, to re-underline, and drive home the point, the quoted extracts alternate very rapidly between autobiographical report and reflection, creating the impression that Soyinka's mother's ruminations have irresistibly led to Soyinka's conceptual expression of the phenomenological perception—his mother's and his—which the cited extracts profile. All the extracts, more or less, contain "narrated monologue"[18] sentences which Soyinka, as reporter and reflector, utilizes for rendering phenomenological consciousness and solicitude.

When the children eventually got home, after their escape from the ***oro***, we are told thus: "That evening Uncle Sanya took ill."[19] It was after the *oro* were properly appeased that Sanya became sound again in health. The story as related entirely from pages 10 to 11 is too true to be believed. But it is a story regarding the belief of Soyinka's people regarding the relationship between mankind and his environment and other beings in creation. As far as the *oro* incidents and other similar ones are concerned, Soyinka does not present any illusion of reality, but reality itself. As already said, this kind of stories are believed—and widely so—by the people. But the wide belief of these *oro* and other stories among Soyinka's people of Western Nigeria does not vitiate the term "mythology" we may ascribe to them. We must, however, recognize each one as a cosmological, existential mythology which is not disagreeable to experts in African ontology—or, better stated, phenomenology. The ideas contained here need not make me a ruffler of phenomenological or ontological feathers relating to this discourse. The adult Soyinka who recalls the *oro* stories believes them as he believed them when he was a young child growing in awareness in the world of adults and that of the *oro* and other spirits who collaborated with mankind.

---

[17] Soyinka's *Ake*, from this perspective, certainly makes really interesting when compared with Gaston Bachelard's *The Poetics of Space,* trans. Maria Jolas (New York: Penguin, 1958) as I have tried to argue here.

[18] "Narrated Monologue" is significant as "a reported discourse that has the syntactic forms of indirect discourse…., a vision of reality that is not the narrator's own, but that of a fictional (or "real") character." See Dorrit Cohn, "*From* Transparent Minds: Narrative Modes for Presenting Consciousness in Fiction," *Theory of the Novel: A Historical Approach,* ed. Michael McKeon (Baltimore and London: The John Hopkins UP, 2000), p. 500.

[19] Soyinka, *Ake*, p. 10.

Yet we must note thus: even though the belief of Soyinka's people may have had nothing to do with the actual existence of the *oro* what is significant is that each *oro* story belongs to a body of beliefs with their own distinct, independent existence—verifiable or not verifiable—in the symbolic thought of the people. Such stories and their symbolic value constitute part of the people's existential civilization. This, in fact, is what constitutes Soyinka's artistic creations evidenced in his many works varying from drama, poetry, fiction, the essay and autobiography. The reality of his childhood world and experience generally conditions and enriches his creations as myths that can only be understood "mythically," to borrow Jean Rudhart's expression used by Maximilien Laroche.[20] If it is true that Uncle Sanya was *oro* (as Soyinka wishes us to believe), then in him we see the theme of, to use Jacques Roumain's words, "Life as an eternal return," an expression that Laroche also employed in his essay,[21] as a human being or as an invisible spirit or whatever—which Uncle Sanya's transformation or metamorphosis in the text suggests.

Now it needs to be emphasized that the little black snake in Laye's narrative and the *oro* in Soyinka's are not to be considered as two of a kind—for they are not. While Laye's little black snake belongs to the African religion of ancestor worship, Soyinka's *oro* does not, although it underscores the African belief in the existence of other-worldly beings, which are in bond, relationship, communication and collaboration with and between mankind and these beings in creation. Clearly, both exist in a boundary-less proximity. I don't think that this point needs to be focused on or explained beyond this point. My argument on the common ground between mankind and these other beings is affirmative enough.

The two autobiographies contain other pertinent recreations regarding the subject of this discourse. But it is needless to over-labour the point relating to "The Thing/Beast/Human Relationships in the Autobiographies of Camara Laye and Wole Soyinka." Yet I must anticipate Eurocentric critics and scholars who may insist that the recreations in the narratives are superstitions. My counter-insistence is that the communal realities of Laye's Malinke and of Soyinka's Yoruba are not different from the reality of the serpent who lured Adam to eat the forbidden fruit he ate in the garden of Eden, an action that was part of Adam's existential destiny. If the scriptural story of the relationship between a serpent and man is not disagreeable to many a Western critic as superstition or myth, why must we question the African worldview the autobiographies present? As an African scholar rightly stated, "Every society builds up its own beliefs, its own myths and its own legends."[22] If the mystical, cosmological or existential experiences relating to Laye's and Soyinka's African societies are not applauded as similar scriptural experiences are applauded then, in reality, we must insist that the scholarship that promotes this kind of discrimination be bracketed off of our phenomenological palates in order not to make Anna-Teresa

[20] This borrowed expression is quoted in Maximilien Laroche's "The Myth of the Zombie," *Exile and Tradition,* ed. Rowland Smith, p. 52.

[21] *Ibid.*, p. 57.

[22] Peter Igobnekwu Okeh, "Two Ways of Explaining Africa:" *Exile and Tradition,* ed. Rowland Smith, p. 76.

Tymieniecka (a famous life-long student of phenomenology) turn in her grave. Already, she has joined Edmund Husserl as a phenomenological ancestor whose spirit we will be invoking from time to time in our eternal relationship with her through the subject of phenomenology.

Let me note right away that Husserl's investigations relating to the structures of human existence and the meaning of death are apposite here. Indeed, in his epistemological phenomenology, he can be said to anticipate his post-existence as a phenomenological ancestor going by my reading of his meditations in his *Cartesian Meditations: An Introduction to Phenomenology*.[23]

Furthermore, from the perspective of the reflective phenomenologist which we can also rightly claim Husserl to be, his reflective or perceptive consciousness reaches an essence of his post-existential mode, his phenomenological ancestorship, but this is a pre-supposition which we deduce from his obscurantist *Logical Investigations* where we see his idea of *epoche* in which he dwells on the notion of phenomenological consciousness that demands a return of things including all beings, to their real states. Thus Husserl, as a human who has died, has returned to his pre-existential state which enables or allows us to perceive him as an ancestor. I hope I am neither ruffling feathers nor splitting hairs.[24]

But by way of rounding off this discourse, I must give the final words to Camara Laye: "These unbelievable things, I saw with my own eyes, and I still see them as I saw them before. Are there not everywhere things that one cannot explain?"[25]

We must accept that the things that Laye and Soyinka recreate in their autobiographies are marvellous examples of supernatural forces that man experiences in his cosmological existence.

References

Bachelard, G. (1958). *The poetics of space* (M. Jolas, Trans.). New York: Penguin.

Carrol, D. (1971). Camara Laye's *The African Child:* A reply. *African Literature Today, 5*, 129–136.

Cohn, D. (2000). *From* transparent minds: Transparent modes for presenting consciousness in fiction. In M. McKeon (Ed.), *Theory of the novel: A historical approach* (pp. 493–514). Baltimore: The John Hopkins University Press.

Colebrook, C. (2002). Edmund Husserl (1859–1938). In The continuum encyclopedia: Modern criticism and theory & J. Wolfreys (Eds.), (pp. 73–81). New York: The Continuum International Publishing Group.

Husserl, E. (1900–1901). *Logical investigations* (J. N. Findlay, Trans.). London: Routledge and Kegan Paul.

---

[23] Edmund Husserl, *Cartesian Meditations: An Introduction to Phenomenology*, trans. Dorion Cairns (The Hague: Martinus Nijhoff, 1967).

[24] I should like to note here that hardly can all readers of Edmund Husserl's *Logical investigations*, trans. John Niemeyer Findlay (London: Routledge and Kegan Paul, 1900–1901) agree on the perspectives the work contains or does not contain.

[25] Laye, *The African Child,* p. 16.

Husserl, E. (1967). *Cartesian meditations: An introduction to phenomenology* (D. Cairns, Trans.). The Hague: Martins Nijhoff.
Laroche, M. (1976). The myth of the Zombi. In R. Smith (Ed.), *Exile and tradition: Studies in African and Caribbean literature* (pp. 62–73). London: Longman and Dalhousie UP.
Laye, C. (1955). *The African child* (J. Kirkup, Trans.). London: Fontana.
M'biti, J. (1970). *African religions and philosophy*. London: Heinemann.
Okey, P. I. (1976). Two ways of explaining Africa. In R. Smith (Ed.), *Exile and tradition: Studies in African and Caribbean literature* (pp. 74–84). London: Longman and Dalhousie UP.
Sartre, J.-P. (1999). Existentialism is a humanism. In R. Christensen (Ed.), *Philosophy and choice: Selected readings from around the world* (pp. 242–255). Mountain View, CA: Mayfield Publishing Company.
Soyinka, W. (1976). Drama and the African world-view. In R. Smith (Ed.), *Exile and tradition: Studies in African and Caribbean literature* (pp. 173–190). London: Longman and Dalhousie UP.
Soyinka, W. (1981). *Ake: The years of childhood*. London: Rex Collings.

Part IV
Art and the Question of Humanity

The Force of Things Unknown

Christopher S. Schreiner

1 Cosmic Alterity

Philosophical approaches to alterity relations—encounters that compel or force us to commingle or communicate with otherness—have typically focused on personal contexts, such as the nuanced and frankly intimate ethical challenges of the face-to-face, dialogical relation studied by Mikhail Bakhtin, Martin Buber and Emmanuel Levinas. Less commonly treated by philosophers is the hermeneutic challenge posed by impersonal metaphysical or cosmic forces, such as the oceanic storms and maelstroms in Edgar Allan Poe's *The Narrative of Arthur Gordon Pym of Nantucket* and "Descent into a Maelstrom." It seems less problematic for philosophical critics to explicate the verbal and facial appeals of the Other than to understand the language of the skies, tropical tsunami, and earthquakes. Philosophers don't do that, or rather, they no longer do it. The cosmic challenge has long been delegated to the earth and atmospheric sciences, who use the technical language of seismology and meteorology, for example, to "read" and predict forces of nature, storm patterns, avalanches. It is no longer feasible to accomplish what Seneca did in his fascinating study, *Natural Questions*, where he melds philosophical wisdom and conjecture with naturalistic observations in "a special concoction of physicomoral investigation" (Williams, 2012, 77).

With regard to the problem of "force," it is unwise for philosophers and literary critics to rely entirely on science and thereby forego the hermeneutic challenges posed by this phenomenon. In his paper "The Distinction between Mental and Physical Phenomena," Franz Brentano (2002) describes the reduction of force by science to a physical phenomenon of magnitude and reciprocal effects of pressure extended in space and flowing in time (47). Force is identified in terms of its

C. S. Schreiner (✉)
University of Guam, Guam, USA
e-mail: cschreiner@triton.uog.edu

C. A. Hornbuckle et al. (eds.), *Phenomenology of the Object and Human Positioning*, Analecta Husserliana 122, https://doi.org/10.1007/978-3-030-66437-4_10

calculable impact within a material nexus, scientifically measured by potential forcefulness—the real crater it will make, or tidal wave, and the ecological devastation these will cause. But such empirical measures of force (quality and intensity; extension and spatial location) are not necessarily pertinent for psychophysical experience conceived as *intentional*, as acts of meaning bestowal. There are, Brentano argues, mental phenomena which "appear without extension and without spatial location" (40). Literary writers often deftly render such phenomena, their experience having taught them that the phenomena comprise embodied, emotionally fraught acts and not merely mental acts. Consider the following passage written by Thomas Traherne (1952):

> Another time in a lowering and sad evening, being alone in the field, when all things were dead and quiet, a certain want and horror fell upon me, beyond imagination. The unprofitableness and silence of the place dissatisfied me; its wideness terrified me; from the utmost ends of the earth fears surrounded me. (149)

The narrator expresses a lugubrious mood by personifying the evening as "lowering and sad," that is, encumbered with melancholy. The field he occupies of silent and inanimate things becomes a space of nocturnal observation for the writer's morbid consciousness, which narcissistically describes his *own feelings*; and rather than shelter him, the space is widened to "the utmost ends of the earth" by the encompassing reach of his "fears," which are likewise personified, as is "wideness" itself. It is as if there is no place to hide from his own fearful discontent, his inability to *enliven* existence. The rhetorical effect of these multiple personifications is to reroute agency from the narrator's self-determination (*Selbstbesinnung*) to that of external phenomena, amplifying his sense of incapacity or infirmity ("a certain want"). Infirmity and unprofitability are synonyms in this context for being incapable of achieving a joyful outcome. His morose affects work against him, thwarting meaningful action. As we will see, Gilles Deleuze's Spinoza (1990) also theorized about the potential forcefulness of affects and our capacity to be empowered (made joyful) or disempowered (made sad) by them. In Traherne's passage, it is as if the narrator's multiple fears populate a sort of teeming amphitheater in which the writer is besieged or terrorized by his agoraphobia, among other fears. The meaning of this passage relies on the convergence of the "sad evening," typically a temporal phenomenon, with the "silent place," a spatial phenomenon, within the encompassing milieu of fear. Sadness and fear, two volatile affects, are diffused atmospherically through a perception of "wideness" which pertains to silence as well as space. The wide silence of a wide space become the milieu of a melancholy mood that becomes global in scope and all enveloping. The critic is challenged to clarify this nocturnal commotion of tormented feelings. The phenomenological outcome is that a private world (*Eigenwelt*) is amplified by a terror-stricken mood into *Weltschmerz*, a world consciousness of weariness and pain that was popularized by the Romantic Movement in literature and philosophy.

Let's briefly consider another example of unmeasurable phenomena which "appear without extension and without spatial location." Such phenomena appear as the mystical sublimities, the "chasms and sunless abysses, depths below depths" recorded by Thomas De Quincey's (1975) opium eater, for whom "the sense of space, and in the end, the sense of time, were both powerfully affected" by his opium ingestion: "Buildings, landscapes, etc. were exhibited in proportions so vast as the bodily eye is not fitted to receive. Space swelled, and was amplified to an extent of unutterable infinity" (103). Brentano would say here that hallucinatory objects and events scorned by science have intentional structures characterized, in a manner of speaking, by their "inexistence" (*Inexistenz*) within various modes of presentation—"joy, sorrow, fear, hope, courage, anger, love, hate...astonishment, admiration, contempt, etc." (36). Even so, according to Brentano, "Knowledge, joy, and desire really exist" (43). They exist in their dynamic potential, as do guilt and paranoia, each depending on our capacity to be affected, to be moved by what befalls or captivates us in the world. Our attentiveness intensifies the colors of the rainbow, their *Abgehobenheit* or saliency, which in turn awaken our feelings of sublimity. The incipient panic and claustrophobia that Raskolnikov feels after murdering the pawnbroker and her daughter in *Crime and Punishment* are real feelings for him in the concreteness by which they compel him to abandon the squalid rooms and dark stairwell that are the external correlatives of a twisted psyche tormented by ideological intoxication. His guilty conscience, something elusive and impalp able to the methods of science, is likewise real to Raskolnikov as its affects of shame and regret compel him to confess and seek salvation.

2 The World as Affective Sphere

According to phenomenology, location is not about mapping but retracing attentional patterns within the inner and outer horizons of lived body awareness that constitute the field of consciousness: "My world includes zones or regions of intimacy, familiarity, and strangeness" (Natanson, 1968, 89). When Mark Fisher (2016) says that weird fiction "de-naturalizes all worlds, by exposing their instability, their openness to the outside," he means that weird phenomena, experienced as eerie and ominous, even numinous, by the characters in a story, destabilize their normative assumptions about reality—for example, about gender, appetite, and identity, as we will see in a moment. The preconceptions of readers, in turn, are "othered" or altered as they bracket their lives and temporarily bestow reality status on the story that absorbs their attention. Serious readers who make a habit of conferring reality status on fictional narratives take existential and ethical truths from them of lasting value for their "real" lives.

Inspired by his teacher, Brentano, Husserl (1970) described our consciousness of the world as a "total sphere of affection":

> The consciousness of the world, then, is in constant motion; we are conscious of the world always in terms of being conscious (intuitive, nonintuitive, determined, undetermined, etc.) and also in the alteration of affection and action, in such a way that there is always a total sphere of affection and such that the affecting objects are now thematic, now unthematic; here we also find ourselves, we who always and inevitably belong to the affective sphere,

always functioning as subjects of acts but only occasionally being thematically objective as the object of preoccupation with ourselves. (109)

In this passage from *The Crisis*, Husserl underscores the pervasiveness of affective tension or tonality as consciousness fluctuates in an attentional arc, thematizing phenomena within a continuum of feeling and action guided by reason. We attend to things through modes of affection, or affects, which codetermine with reason the particular salience of an experience and the relative intensity with which it compels us into action. This phenomenology of world consciousness echoes the continuum theory of the ancient Stoics, who described human existence in the cosmos as mediated by the *pneuma*, through which traverses a force of *tonike kinesis* or tensional motion (Sambursky, 1987, 30). Granted, the pneuma (in Latin, *spiritus*) is a sort of material spirit, whereas Husserl's intentionality is noetic; but both have an inner relation to *kinesis*, kinetic bodies. One can argue that Husserl's myopic concern with the *Lebenswelt* at the end of his career, limits his investigation to world consciousness and prohibits inquiry into cosmic consciousness, but there is an integral rationality giving structure and purpose to both forms of consciousness. Also, the intentional relation between transcendental ego and world on one hand, and between the *hegimonikon* and *kosmos* on the other hand, creates an ambient, affective force field, a field of affects inseparable from constitutive acts of meaning.

As a consequence of being deemed unmeasurable by science, intentional existence and its affects, the affective imagination of force as a *subjective disturbance*, have come under the scrutiny of psychology, philosophy, and literature, which conduct exegeses. To be sure, Jacques Derrida (1978) asserts that "one would seek in vain a concept in phenomenology which would permit the conceptualization of intensity or force. The conceptualization not only of direction but of power, not only the *in* but the *tension* of intentionality" (27). But it can be argued that this contentious observation by Derrida, which was penned early in his career, evinces an ignorance of the theory of affects, the *tonike kinesis* that dovetails without fanfare between Stoic physics, Spinoza's expressionism, and Husserlian phenomenology.

3 Waiting for Daylight with Arthur Gordon Pym

In Edgar Allan Poe's only novel, *The Narrative of Arthur Gordon Pym* (1975), the shipwrecked narrator reports, "In this frightful situation we lay until the day broke so as to show us more fully the horrors which surrounded us" (123). As is to be expected, Poe's narrator *waits* for daylight to *see*, to distinguish something—the natural forces that besiege him. But the meaning of what confronts him does not cohere in the unity of a *distinction*, and this is what we seek to do—to distinguish (Luhmann, 2002, 121). To "see" something will not mean to await a weather report delivered by a third party; the frightfulness of the situation manifests itself as an affect at once compelling and deranging, and is co-constitutive of the line of sight. In his desolate mood, his enveloping *Stimmung*, the human adrift relies on analogy to produce his *own weather*, in which sublimity occurs as one among other affective outcomes. A horizon both horrible and wonderful stimulates the affective imagination of Pym the narrator, who thinks he sees a gigantic waterfall in the sky: "I can liken it to nothing but a limitless cataract, rolling silently into the sea from some immense and far-distant rampart in the heaven" (238). As Poe's survivors drift south toward Antarctica, off the cartographic grid, it seems that their spectacular disorientation is tolerable to the narrator, Pym, due to the "novelty and wonder" of so many "unusual phenomena," recorded in his journal as follows:

> March 1. Many unusual phenomena now indicated that we were entering upon a region of novelty and wonder. A high range of light gray vapor appeared constantly in the southern horizon, flaring up occasionally in lofty streaks, now darting from east to west, now from west to east, and again presenting a level and uniform summit—in short, having all the wild variations of the Aurora Borealis. *The average height of the vapor, as apparent from our station, was about twenty-five degrees*. The temperature of the sea seemed to be increasing momentarily, and there was a perceptible alteration in its color. (236, emphasis added)

Given the desperate situation in which he finds himself, it is noteworthy that Pym estimates the height of the atmospheric vapor. As is typical of narrators in Poe's writings who get a sort of affective "lift" and serotonin boost from instrumental thinking, Pym's enthusiasm for scientific discovery overcomes his circumstantial anxiety and quickens his concrete ability to respond to events. This development dovetails with Deleuze's (1990) explication of the tonic power of affect as conceived by Spinoza, that is, when a state of affairs or exigency "agree with my nature" and enable action (239). Readers of Poe (2003) will recall that the protagonist of "The Pit and the Pendulum" alleviates his sense of doom by measuring, step by step, the dimensions of the dark torture chamber in which he awakens, thereby gaining data that might later facilitate his escape from the Inquisition: "There were in all, then, a hundred paces; and, admitting, two paces to the yard, I presumed the dungeon to be fifty yards in circuit" (216–217). Consciousness of terror gives way to embodied inquiry as the body copes with a situation that causes cognitive dissonance. With the captive's affection for ratiocination, the body is forced to "think," demonstrating a hitherto unrecognized capability (Deleuze, 1990, 257). Likewise, the force of things unknown stimulates an affection that agrees with Pym's nature, producing from within his being a "joyful passion" that quickens his awareness and increases his chances for survival (Deleuze, 1990, 239). Rather than shut down and become immobilized in panic mode, Pym evinces resourcefulness at sea. In short, the "force" of things unknown is as much an affective power intrinsic to his psychomotor capability as it is oceanic turbulence. Notice the narrator's recurrent use of analogy—"*I can liken it to*..." In this regard, Kant (1987) argues that "the broad ocean agitated by storms cannot be called sublime," but rather, "The imagination... is very mighty when it creates another nature out of the material that actual nature gives it" (182).

In Poe's novel, we witness a compulsion to visualize natural forces from the perspective of the embodied spectator and victims, but identification is deferred; this deferral, which *meanwhile* resorts to the imagination until confirmation of empirical data, is integral to explication of force and alterity, which relies on intuitive synthesis, as Jacques Derrida (1978) writes, "between meaning and literality," echoing Kant's assertion that imagination "schematizes without a concept" (7). "To say that force is the origin of the phenomenon is to say nothing," argues Derrida. "By its very articulation force becomes a phenomenon…Force is the other of language without which language would not be what it is" (26–27). From these observations we can infer that words and discourse sublimate force into communication; we know that so long as opponents speak to each other, they appear to renounce force. To be sure, specialized discourses sublimate force and affect in different ways, depending on the degree of threat posed to the integrity of the discourse. Challenged by the enigma of force, philosophy resorts to metaphor, "the analogical displacement of Being"; but rather than accede to the affective imaginary as does literature, philosophy inexorably seeks clarity and a grammatical subject: "Diaphanousness is the supreme value; as is univocity" (26–27). In other words, for the sake of clarity and conceptual rigor philosophy resists the creative and often emotion-fraught invention that literature employs to envision strange forces. But it does not follow that philosophy is devoid of affect, or can avoid affect, any more than humans can avoid being in a mood. Nietzsche and Heidegger saw that the affective core of philosophy is logic, or technical rationality, rhetorically *forceful* in its institutional hegemony, but *flat* in its rhetorical delivery. In this regard, Cary Wolfe (2010) quotes Heidegger to the effect that there is a sort of "sublimated violence" in the discourse of philosophy (71). This is the outcome of suppressing its expressive range and diversity. The rational drive of philosophical thought, which ironically took it speculative intensity from wonder, its inaugural passion, ends up suppressing emotional discourse over the years, and excludes women from institutional membership, at least until recently. Edgar Allan Poe, a stylistic virtuoso, finds a secret grandeur in logical thinking; in his stories a *fondness for ratiocination* not only stimulates depressed or disadvantaged protagonists, but, according to Deleuze's Spinoza, provides an affective boost that serves to further the "power of action" (Deleuze, 1990, 239).

Literature and art teach us that there is no greater threat, or saving power, than the affective imagination. The psychoanalyst Andre Green (1999) argues that force is integral to affect, that "imagination enhances affect, and affects are the strongest stimulation to imagination," and that affects such as fear and wonder have a "potentially disorganizing power…that eventually takes hold of vast areas of the mind" (285–286). On the other hand, Deleuze's reading of Spinoza finds in affective life an enabling force that "increases or aids our power of action" (Deleuze, 1990, 239). Deleuze asks: "Of what is a body capable? Of what affections, passive as well as active? How far does its power extend?" (256).

It is regrettable that there is a stubborn "human-all-too-human" resistance in philosophy to identifying forces as inhuman or posthuman. Nietzsche argues that an anthropomorphic agent or agency is conjured and projected, *dissimulated*, to make human sense of primal force; causality and human agency are merely *presumptions* that reduce the polymorphic dynamism of being in a play of differential forces, "that of the perceptual and the sense-giving forces on the part of the subject, together with the quantum of natural forces that he initially encounters" (Allison, 1985 xxiii).

Husserlian identification, as Theunissen (1986) observes, involves "integration in the world order" (29). However, the incomparability of the transcendental ego—the otherness of the Other—is compromised by this compulsory membership. We are world bodies; the world is a prolongation of our bodies and laws of thought, and everything we recognize fits in the world as a rational continuum. "The lived world," says Pierre Thévenaz (1962), "is therefore never foreign to reason for Husserl; it is only that reason remains too often latent and requires much effort and many ruptures to make it appear" (53). Husserl insists that the "removal of the alien system of appearances" is achieved "through the identification of the ego in the Other with my ego" (Theunissen, 1986, 215). As such, in reducing otherness to the same, to an integral rationality, Husserl's philosophy is another chapter in the Anthropocene.

Yet there are philosophers who struggle to extend a cosmic line of inquiry beyond the human sphere. We have already mentioned the ancient Stoics, who endorse a continuum theory in which bodies are linked to the cosmos via the tensional motion of the *pneuma*. Granted, theirs is a human cosmos, like Husserl's, and logical, like Husserl's; but the Stoics insist one always live in harmony with both reason and elemental nature (earth, water, air, and fire), whereas Husserl's transcendental epoché unnaturally brackets empirical nature and the "natural attitude" to foreground an intentional manifold of passive and active rays (or acts) of sense bestowal (or constitution). Hence the good Stoic is sympathetic, sensitive to the correspondences between her microcosm and the cosmic macrocosm (Seneca, 2010, 3). In arguing for a return to cosmology, Stephen Toulman (1985) recognizes that even Kant "knew firsthand what it means *an die Grenze der Vernunft anzurennen*—that is, to run up against, and to overreach, the inherent boundaries of rational speculation"; but Kant had too much faith in Newtonian physics to press beyond his own self-imposed critical limits (5). Merleau-Ponty (1968) quotes Bergson to the effect that our bodies extend to the stars: "For if our body is the matter upon which our consciousness applies itself, it is coextensive with our consciousness. It includes everything that we perceive, it extends unto the stars" (Merleau-Ponty, 57). This cosmic dimension inhibits or challenges anthropomorphism and raises the possibility of transversal or transspecies developments. Gilles Deleuze (1991) also picks up on this tendency in Bergson's philosophy: "Bergson is *not* one of those philosophers who ascribes a properly human wisdom and equilibrium to philosophy. To open us up to the inhuman and the superhuman (durations which are inferior or superior to our own), to go beyond the human condition. This is the meaning of philosophy" (28).

4 Capitalizing Oceanic Alterity (The Gold Standard Destabilized)

The cosmic dimension glimpsed by Nietzsche and Bergson is ignored at our peril when we reduce the grandeur of cosmic forces by measurement and calculation. Poe, Melville, Ambrose Bierce, and Arthur Machen have produced writings in which inhuman and superhuman forces are encountered by humans who often fatally misinterpret the scope and gravity of their encounters. Such a reductive misprision comprises the tragedy of Melville's *Moby-Dick* (1992). It is out of Poe's phantasmal horizon in *The Narrative of Arthur Gordon Pym* that Melville conjures the force of a giant whale in *Moby Dick*, published 6 years after Poe's novel of shipwreck and disorientation. The doomed men on Melville's whaling ship, the Pequod, presume creaturely force when envisioning Moby Dick, but the alterity of the tumultuous encounters with the whale exceeds collective memory; the forceful impression left by the giant whale is something *das Ungeheuere*—monstrous and unsettling (Otto, 1958, 40). Ahab's hatred of the whale, expressed in words and body language, is the forceful affect that arouses the crew's spirit of revenge, and the specter of the monstrous force is fitted or downsized to the scale of the ship's world—hence to perception and action—when Captain Ahab posts a reward of gold for the capture or destruction of the whale. Gold is a known thing with stable value; Moby-Dick is an unpredictable force that engulfs the nautical efforts of little men. The arrogant economizing that capitalizes the oceanic threat with gold, is a valuation that only serves to underestimate the destructive force of the creature, exposing the ship to its doom. The whale is misinterpreted as conquerable, as a commodity equivalent to gold; the fatal outcome of Melville's novel is the result of a misprision, an economic and industrial feasibility study gone awry. "All events in the organic world are a subduing," says Nietzsche, "a becoming master, and all subduing and becoming master involves a fresh interpretation, an adaption through which any previous 'meaning' and 'purpose' are necessarily obscured or even obliterated" (Allison, 1985 xxiii). Not a fresh interpretation, but an older meaning and purpose for Melville's whalers is what prevails, who seek to merchandise a living being for its lamp oil, a practice rooted in the slave trade and then resource exploitation for profit. This practice is now at the forefront of posthuman research, for example in Cora Diamond's essay "Injustice and Animals," which receives attention in Cary Wolfe's (2010) book *What is Posthumanism?* But for the men under Ahab, capitalization of whales is less an injustice than a kind of reckless envisioning sanctioned by human laws in an unhuman sea. The sea is neither pre- nor posthuman for Melville, but a cosmic abyss at once splendid and catastrophic. The final confrontation with the giant whale—a behemoth embodying oceanic and cosmic forces—results in mutually assured destruction for all but the lonely survivor, Ishmael. Here the restraint of rational philosophy in resisting affective fabulation could have served an advisory caution to the men of the Pequod. They were too literary, too emotional in their envisioning of the whale, and should have been more rational, assessing its true nature, its size, its ferocity, its unpredictability and inhumanity as a force. But to do so they would have had to check their arrogance, minimize their own power and significance amid the cosmic horizon first depicted so sublimely by Poe.

The men of the Pequod should have taken to heart the counsel of Algernon Blackwood, who wrote, “There are forces here that could kill a herd of elephants in a second as easily as you or I could squash a fly. Our only chance is to keep perfectly still. Our insignificance perhaps may save us.”

While capitalization of a monstrous force characterizes the mercantile compulsion of the mid-nineteenth century in the throes of the industrial revolution, a few decades later literature represents forces that resist capitalization and test the limits of common sense and the popular imagination. In the literary genre of so-called *weird tales* spanning the later nineteenth and early twentieth-century—whose influence reaches into the films and literature of today by such creators as John Carpenter and Stephen King—forces are described reminiscent of Nietzsche’s fusion of being and becoming. In stories by the likes of Ambrose Bierce, Arthur Machen, and Algernon Blackwood, there are struggles to distinguish and epitomize indiscernible forces, forces that mask or camouflage themselves, whose scale or color or insatiability challenge human preconceptions. This inability to make distinctions is further clouded by the tendency of weird fiction to evoke remote, nocturnal, and uncanny atmospheres that engulf perceiving agents, as it were *foregrounding the natural background* in a colossal or obscure way to suggest *supernatural* forces. These stories give the cosmic element its due scope and respect in a manner opposite of that demonstrated by the whalers in Melville. Three factors seem to prevail in the weird stories by Bierce and Machen, the first being the reach of science and medical research into the late Victorian popular imagination, the second being a recourse to Gestalt theory by the writers mentioned, the third factor being the sexualization of force. In these stories the affective imagination is severely stimulated and challenged in an initial encounter with unknown, unhuman forces but subsequently checked by rational inquiry that falsifies its assessment for the sake of preserving a comfort zone. In both stories there is initially the adumbration of the force as a subversive power, volatile concreteness of encounter without direct visualization; the concreteness is unstable and metamorphic—powerful but *indeterminate* salience.

5 The Force of the Indeterminate

In Ambrose Bierce’s story “The Damned Thing” (1970), a brutal force displaces vegetation, rips shrubs, kills a hermit hunting it, without revealing its own contours or color. The body of the hunter is shredded and bludgeoned. In the rural cabin of the deceased, a coroner’s inquest is held to determine cause of death. An inquest is a corporeal mode of exegesis in which the cause of death is interpreted through careful scrutiny of the available evidence. The exegesis is thwarted to some extent by the dim lighting and poor quality of the book, but the deeper problem is the inscrutable cause of death. The other men in the room are nameless jurors except for a journalist, William Harker, who will be called as witness to the fate of the deceased. He came to study the hermit hunter, the mountain man, and saw him suffer a gruesome death when they went hunting together. The narrative unfolds via three "texts": the journal of the deceased hermit read silently by the coroner, the body of the deceased himself, and thirdly, the narrative carried in the coat of William Harker, which describes his report of the violent death of the hermit Morgan. The corpse is described as follows:

> The coroner rose from his seat and stood beside the dead man. Lifting an edge of the sheet he pulled it away, exposing the entire body, altogether naked and showing in the candle-light a clay-like yellow. It had, however, broad maculations of bluish black, obviously caused by extravasated blood from contusions. The chest and sides looked as if they had been beaten with a bludgeon. There were dreadful lacerations; the skin was torn in strips and shreds. (102)

What caused the savage death of the hunter lying on the table? The jury will hurry to dispel ambiguity; for lack of better ideas, the inquest presumes that Morgan was killed by a mountain lion, hence defaulting to animal analogy. Yet William Harker's report says otherwise. On the day he witnessed the death of the hunter, he encountered a force of unrecognizable color in the agitation of shrubs and undergrowth:

> I observed the wild oats near the place of the disturbance moving in the most *inexplicable* way, I can hardly describe it. It seemed as if stirred by a streak of wind, which not only bent it, but pressed it down—crushed it so that it did not rise; and this movement was slowly prolonging itself directly toward us…the *apparently causeless movement* of the herbage and the slow, undeviating approach of the line of disturbance were distinctly disquieting (101). [emphasis added].

Harker's report exposes two problems: absence of agency, hence lack of a perpetrator; and a breakdown of natural laws. However, the testimony by Harker is disqualified when he is accused of being mad. Neither the coroner nor the jurors are credentialed psychologists and lack the expertise to diagnose the only surviving witness, yet a discriminatory fusion of medical and legal systems, analyzed in detail by Foucault in his lectures on the "abnormal," prevails to condemn his testimony. Yet they find him mad, and as I said, falsely identify the perpetrator as a mountain lion. Harker's testimony does not help the matter when it refers to the indiscernible force making "a loud savage cry—a scream like that of a wild animal" (101). His use of analogy is in this case inescapable but misleading, for the "thing" is not an animal. To put it philosophically, the jurors (all elderly men) lack Harker's enthusiasm and imagination for alterity relations. Their willful ignorance and fear prevents them from pursuing the mystery that confronts them in the death of the man on the table. Their verdict: Morgan was killed by a mountain lion. Their collective affective disposition, an arguable form of hermeneutic nihilism, leads them to *go with what they know* rather than explicate the unknown.

6 The Force of Pan's Libido: Hypersexual Alterity

As in Ambrose Bierce's "The Damned Thing," in Arthur Machen's story "The Great God Pan" (1894), we encounter a forceful menace via misunderstanding; our viewfinder and Gestalt go awry. In both narratives, perceptual and hermeneutic conventions thwart intelligent responses to alien encounters. Arthur Machen's primordial force eviscerates human flesh without accurate confirmation of the perpetrator by visual witnesses; identification is deferred until mounting evidence provokes an *inference of agency* and subsequent confrontation with the alien perpetrator. No one knows precisely what they are looking at when they look at a posthuman libido unconstrained by social norms. Uncanny events of a sexual nature begin to disturb the late Victorian surface of life after the operation on the brain of a young woman named Mary, apparently a waif. The physician's intent is to unlock latent transcendental forces of human nature, but a sexual predator is born, is released into the world. Although Mary is herself indisposed by the brain surgery, her daughter Helen, born 9 months later, becomes a polymorphic, predatory nymph, savage, insatiable.

Arthur Machen's tale of the libidinal uncanny was preceded by a less horrific, more glamorous and mythical emergence in Vernon Lee's "Dionea" (1890). The latter veers from our focus on unstable visualization; the sexually charismatic goddess Dionea, who emerges from the sea, is disturbing in her aloof, exotic charms, her hint of sorcery and magnetic attraction, but above all in the obscurity of her purpose. From the start, the narrator describes Dionea with the language of the fine arts, as if she has stepped out of a painting by Da Vinci or Botticelli. Yet her primordial relation to the sea is never forgotten and forms a powerful backdrop to all of her behaviors.

In their respective fictional settings, the influence both women, Dionea and Helen, have on men is spellbinding and turbulent. In "The Great God Pan," men with whom the nymph Helen becomes intimate end up as wasted souls, traumatized—as if electrocuted. There are intimations of Dionysian frenzy in each case, each encounter. Even when Helen was younger, there were ambiguous "sightings" of Helen Vaughn by shaken, almost speechless individuals such as Trevor, a 7 year-old boy who reports seeing Helen "playing on the grass with a 'strange, naked man'" (193). Trevor is permanently traumatized by this spectacle in the woods, experiencing "paroxysms of fright" (194). There is speculation he was molested, but the alleged pervert, Helen, refutes the charge. What is disturbing to Trevor's father and physicians is the powerful, deranging affect aroused by Helen in young people. Presumably promiscuous encounters in the forest between Helen and her young neighbor, Rachel M., transform Rachel into someone "different from herself" who appears "languid and dreamy" (195). She tells a "wild story" of her obscure adventures with Helen (195). The wild context in which these events take place make it problematic to entirely blame human agency: "the swaying leaves and the quivering shadows on the grass" are the elemental milieu whose behavior as it were—like a charged atmosphere, electric and tingling—cannot be separated from the pagan perversions of Helen H (196).

As Helen matures, so she often *disappears* after intimate encounters and then returns under a *different guise*, mask, name, and persona: Mrs. Beaumont/Mrs. Herbert, and so on. The force depicted by Arthur Machen is gendered as a woman both metamorphic and insatiable; hence, the gender of woman is an *unstable* force, and, to echo Judith Butler's coinage, "The Great God, Pan" is most fundamentally about *gender trouble*. But the trouble starts with science, which is a male domain in the late Victorian era; much like *Dr. Jekyll and Mr. Hyde*, and before it, *Frankenstein*, experimental science represents a transgression of nature for the sake of the advancement of knowledge and power. In "The Great God Pan," nature bites back and re-establishes dominion in the exotic, elusive, fatal sublimity of a pagan dominatrix.

The force remains mysterious. The libidinal awakening that issues from the brain surgery on Mary result in multiple effects at first traceable not to a single persona but to a weird space in the forest, a sort of new dimension or force field: "neither man nor beast, neither the living nor the dead, but all things mingled, the form of all things but devoid of all form" (188). Attributing a subjective agency to such a mysterious field effect will be troublesome. For this reason, ownership (e.g., "patent rights") of the outcome of this experiment, decreed by Dr. Raymond in his haughty remark, "I think her life is mine, to use as I see fit," is stymied or made absurd by the events that ensue. How is ownership, crucial to scientists as a means to discipline and distribute knowledge as power, established? Identify the causality of the force, identify it as a thing, and name it—personal pronoun, a grammatical subject: Mary or Helen. But this process seems frustrated when, late in the story, Villiers says to Clarke, "*She* is the mystery" (206). In other words, having finally identified the perpetrator, the mystery is still intensely compelling and unresolved, as if not Helen the person but *the feminine gender* is the mystery that draws these Victorian gentlemen into nocturnal speculations, and, finally, the investigation that leads to a fatal confrontation with the polymorphic perpetrator, Mrs. Beaumont/Mrs. Herbert/ aka Mary's daughter, Helen Vaughan.

The monster named Helen created under surgery, this experiment from which has emerged unbound, posthuman libido, is most of all a haunting affect whose circulation, "its capacity to seep into other domains and inhabit them and finally to transform itself" is as much apparent in the worrisome, somewhat paranoid conversations and gossip of the local British culture, as it is in the ruined husks of men who encountered Mrs. Beaumont in the flesh (Green, 1999, 285). The Protean obscurity of her libidinal force is its "dynamic attribute"; the verb to *affect*, Andre Green reminds us, means to pretend, to adopt disguises (285). This dynamic attribute, this force which according to Andre Green "is less susceptible to control than any other source of information" bedevils the investigatory logic of Victorian gentleman who are minor versions of Sherlock Holmes (286). They do not recognize the force as dynamic, nor conceive of it in positive terms, only as a threat. If we employ the term "posthuman" to describe the dynamic force embodied in Helen a.k.a. Mrs. Beaumont, it is because her unleashed libido is denied by Victorian men as an *integral element of female identity*, and of the human body in general. In this regard, given an abject disgust and horror with the human body as such, "unhuman" is arguably a more accurate term (Trigg, 2014, 5). But the affective *otherness* depicted in "The Great God Pan" as libidinal monstrosity *disturbs* while also *liberating* certain individuals from Victorian norms. Such is the case of Rachel, Helen's neighbor, described earlier in these pages as becoming "different from herself" after forest encounters with Helen. Sadly, her transformation is fatal.

7 Polymorphic Alterity

In 1982, the film release of John Carpenter's *The Thing* brought to the screen a polymorphic creature of interstellar origin whose alterity manifests itself as violently mimetic as humans and animals are absorbed and simulated at an Antarctic research station. The protean nature of the alien force, which appears as a grotesque xenomorph in various stages of assimilating life forms (hence never stabilizing as a fixed form or identity), is not in the service of seduction or libidinal frenzy, as in "The Great God Pan," but creaturely self-preservation on a cosmic scale. The John Campbell (1938) novella, *Who Goes There?* carries forward from Machen's story the idea of a shape-shifting monster: "It is unhuman; it has powers of imitation beyond any conception of man" (Campbell, 2009, 65). Although Machen's Helen is not strictly mimetic, both monsters are polymorphic, and both pose a fatal challenge to the complacency of phallocentric anthropomorphism. These authors are aware that by having the monstrous Other be polymorphic, they destabilize the entire comfort zone—i.e., lifeworld—of male identity. Here it is worth repeating Mark Fisher's observation that weird fiction "de-naturalizes all worlds, by exposing their instability, their openness to the outside" (29). While the monsters point to an unnatural space beyond the human sphere, the men who are confronted this alien spectacle demonstrate a stark lack of flexibility when coping with their sudden exposure.

It is noteworthy that John Carpenter's cinematic rendition of John Campbell's novella follows the latter by excluding women from the research station. The total repulsion and dread expressed by the exclusively male characters as they witness the alien creature assimilate their colleagues, turns to paranoia as they suspect each other of becoming alien. The self-assured identity politics of this smug scientific enclave is shattered by otherness from within; as Rimbaud famously said, *Je suis l'autre*. Only Blair, the resident biologist, admires the alien's invulnerability and genetic perfection, the biological supremacy of a design for self-preservation and cosmic hegemony. His awe harks back to the initial Victorian self-assurance of Machen's medical researchers who are ultimately chagrined by the gendered mayhem they set loose. The male characters in "The Great God Pan" describe Helen as "a girl of the most wonderful and most strange beauty" (198), but their aesthetic wonder turns to horror when they hear reports of Helen's insatiable appetite. As we said, Helen's feral, voracious sexuality, which threatens to breach the norms of Victorian gender politics and sexuality, is dissimilar to the asexual, ungendered process by which John Carpenter's monster assimilates and simulates life forms. But where *The Thing* and the "Great God Pan" overlap is in their portrayal of monstrous appetites. As a visceral, highly suggestive metaphor, appetite encompasses pressing psychosexual issues that are inescapable for late Victorian and early modern self-understanding.

The reader must decide for herself which is most disturbing to the characters in these stories, the shape-shifting force of the monsters, or their insatiable appetites. What is the common threat? We saw how the Victorian gentlemen are haunted by the fear of personal dispossession and cultural ruin instilled in them by Helen's libidinal excesses. Theirs is a paranoid vision of a new capacity attributable to women, forces hitherto undiscovered in the *female body*: "such forces cannot be named, cannot be spoken, cannot be imagined except under a veil and symbol" (Machen, 2009, 225). Likewise, John Carpenter's "Thing" is always someone or something else's body, be it a dog or lab technician, so that, as we mentioned, the scientists and technicians in the film begin to suspect that their co-workers, not to mention themselves, are already infected and dispossessed. The deepest threat posed by the polymorphic "Thing" is to the certainty of concepts of selfhood based on a fixed identity, whether gendered or anthropomorphic. This is not unlike the threat posed by Machen's Helen.

8 *"Was ist der Mensch?"*

In the context of the present study, this issue raised by Deleuze's Spinoza pertains to the hermeneutics of alien encounters, the often violent and confused resistance of humans to seemingly unnatural forms of otherness that challenge anthropomorphic and phallocentric identities. The story "The Great God Pan" and the film, *The Thing*, both raise issues bearing on the uncanny, abject possibilities of our own bodies when we encounter alien bodies animated by forces unknown. In this respect, Dylan Trigg (2014), who studies the horror of the body in films like *The Thing*, is on target when he insists that films such as *The Thing* bring a new challenge to phenomenology insofar as unhuman or posthuman bodies, artifacts, and geographies are increasingly brought to our attention for investigation (9). It is metaphysically unsettling for all parties concerned, from Arthur Gordon Pym to the whalers on Melville's Pequod, from Victorian ladies and gentleman to modern Antarctic researchers, to consider that, as Gilles Deleuze puts it in a paraphrase of Spinoza, *we still do not know what our bodies can do*. In our digital era characterized by its distractions and addictions, the lived body, liveliness (*Leiblichkeit*), focus, optimal attention or perceptual salience, are forces unknown. Gender and sexuality remain frontiers of conflict and discovery. Kant's epic question, "*Was ist der Mensch?*" is still pertinent. Echoed by Husserl (Cairns, 1976), this same question leads the phenomenologist to profound disillusionment: "Man has been taken, like the table, as something *there*" (45). In short, as something given, a static thing without force, a known object: humanity deracinated from its transcendental and affective dynamism. To the Stoics or, arguably, Deleuze's Spinoza, it would seem in our digital era that the cosmic continuum has been discontinued, reducing the range of affective sympathy and expression to the sphere of the human-all-too-human: the Anthropocene. The internet is a universe unto itself of hearsay refractory to reason, producing its own storms, controversies, and fabricated evidence. The sheer number of people who ridicule the scientific evidence of global warming suggests something more astonishing, that many do not live in continuity with nature. They do not feel climatic change, meteorological perturbation; nor do they panic over the rising seas that displace island populations, much less comprehend the evidence that proves it; nor do they feel the impoverished suffering of millions who live elsewhere than the upscale suburbs. Husserl's last writings in *The Crisis*, focused on describing the structures of the *Lebenswelt*, sought to reestablish the rational continuum according to his phenomenological *logos*. Nowhere in his phenomenological treatises does he mention the encroaching barbarism of Hitler's Reich, which was instituting a diabolical form of reason that the Frankfurt School saw as immanent to the Enlightenment in its regressive turn via the uncritical psychosocial need for authority. Husserl's intense and exhausting recovery effort was, he repeatedly claimed, *spiritual*, for the sake of establishing and safeguarding *meaning* in a world increasingly dominated by experimental science in the service of military armament design and production. This crusade of Husserl's seems to at best abstractly dovetail with that of the Stoics, for whom "the earth is a living creature, and the whole world is a living creature with a soul" (Seneca, 3). In the final analysis, Husserl evokes world consciousness, and Seneca, cosmic consciousness. But since both forms of consciousness are governed by integral rationality, neither are particularly prepared for alterity relations that challenge the claims of reason and, as Deleuze puts it, our understanding of "what our bodies can do."

What would it mean to become a celestial body? What would one see and feel? And what, Wolfgang Hildesheimer (1986) wondered, does it feel like to become a nightingale? The trouble is that transspecies migration has until recently been frowned upon; we are only now getting used to transgender migration. Such questions will remain absurd until our interrelated concepts of self and world break with their anthropomorphic preconceptions, and we practice imagining and performing posthuman and polymorphic modes of *Leiblichkeit*.

References

Allison, D. (Ed.). (1985). *The new Nietzsche: Contemporary styles of interpretation*. Cambridge: MIT Press.

Bierce, A. (1970). The damned thing. In *The complete short stories of Ambrose Bierce* (pp. 97–105). Lincoln: University of Nebraska Press.

Brentano, F. (2002). The distinction between mental and physical phenomena. In D. Moran & T. Mooney (Eds.), *The phenomenology reader* (pp. 35–50). London: Routledge.

Cairns, D. (1976). *Conversations with Husserl and Fink*. The Hague: Martinus Nijoff.

Campbell, J. W. (2009). *Who goes there?* Rocket Ride Books.

De Quincey, T. (1975). *Confessions of an English opium eater*. New York: Penguin Books.

Deleuze, G. (1990). *Expressionism in philosophy: Spinoza*. (M. Joughin, Trans.). New York: Zone Books.
Deleuze, G. (1991). *Bergsonism*. (H. Tomlinson & B. Habberjam, Trans.). New York: Zone Books.
Derrida, J. (1978). Force and signification. In *Writing and difference* (A. Bass, Trans.) (pp. 3–30). Chicago: University of Chicago Press.
Fisher, M. (2016). *The weird and the eerie*. London: Repeater Books.
Green, A. (1999). *The fabric of the affect in the psychoanalytic discourse* (A. Sheridan, Trans.). New York: Routledge.
Hildesheimer, W. (1986). Why I transformed myself into a nightingale. In D. Halpern (Ed.), *The art of the tale* (pp. 347–350). New York: Penguin Books.
Husserl, E. (1970). *The crisis of European sciences and transcendental phenomenology* (D. Carr, Trans.). Evanston: Northwestern University Press.
Kant, I. (1987). *Critique of judgment* (W. Pluhar, Trans.). Indianapolis: Hackett Publishing.
Luhmann, N. (2002). *Theories of distinction* (J. O'Neil & E. Schreiber, Trans.). Stanford, CA: Stanford University Press.
Machen, A. (2009). The great god Pan. In R. Lockhurst (Ed.), *Late Victorian Gothic tales* (pp. 183–233). New York: Oxford University Press.
Melville, H. (1992). *Moby-Dick or the whale*. New York: Penguin Books.
Merleau-Ponty, M. (1968). *The visible and the invisible*. (A. Lingis, Trans.). Evanston: Northwestern University Press.
Natanson, M. (1968). Phenomenology and the theory of literature. In *Literature, philosophy, and the social sciences* (pp. 86–100). The Hague: Martinus Nijhoff.
Otto, R. (1958). *The idea of the holy* (J. H. Harvey, Trans.). New York: Oxford University Press.
Poe, E. A. (1975). *The narrative of Arthur Gordon Pym of Nantucket*. New York: Penguin.
Poe, E. A. (2003). The pit and the pendulum. In D. Galloway (Ed.), *The fall of the house of Usher and other writings* (pp. 212–227). New York: Penguin.
Sambursky, S. (1987). *Physics of the stoics*. Princeton: Princeton University Press.
Seneca, L. A. (2010). *Natural questions* (H. M. Hine, Trans.). Chicago: University of Chicago Press.
Theunissen, M. (1986). *The other: Studies in the social ontology of Husserl, Heidegger, Sartre, and Buber* (C. Macann, Trans.). Cambridge: MIT Press.
Thévenaz, P. (1962). *What is phenomenology?* Chicago: Quadrangle Books
Toulman, S. (1985). *The return to cosmology*. Chicago: University of Chicago Press.
Traherne, T. (1952). Centuries of meditations. In H. Read (Ed.), *English prose style* (pp. 149–150). New York: Pantheon Books.
Trigg, D. (2014). *The thing: A phenomenology of horror*. Winchester, UK: Zero Books.
Williams, G. D. (2012). *The cosmic viewpoint*. Oxford: Oxford University Press.
Wolfe, C. (2010). *What is posthumanism?* Minneapolis: University of Minnesota Press.

Paul Klee's *Ad Parnassum* and the Reworking of Consciousness

Bruce Ross

The tumult in modern art at the beginning of the twentieth century centrally incorporated redefinitions of painterly treatment of form and color and, moreover, subject matter. It could easily be stated that a new form of consciousness was being expressed, most likely because of the undermining of the established empires by World War I, resulting in the phenomenological experience of an almost inhuman transformation. More than a simple reaction to long standing points of focus, including structural arrangements, in cityscape, landscape, or portraiture, some deep phenomenological connection to the nature of art was being uncovered. Explorations of primitive art, metaphysical understanding of time and space, and spiritual concepts like the aura were supporting Pointillism, Cubism, and "pure painting." The result was a repudiation of mental or painterly stasis as the nexus of art, with perhaps a reconsideration, in some cases, of the aesthetic idea of the sublime, as in Paul Klee's *Ad Parnassum* (1932) (Partsch, 2003, 65). The importance of these directions in Klee's art, and specifically, this painting, is examined in Hajo Düchting's study of Klee's understanding of music: "Klee saw polyphonic painting as superior to all other arts, because he felt that spatial and temporal dimensions can be given visual expression in a two-dimensional representation as two intersecting planes. His masterpiece *Ad Parnassum* constitutes the great synthesis of this endeavor" (Düchting, 2002, 78).

This paper examines Klee's use of structure, color, language as language, North African visits, and symbolism of the sun and moon as a project to reorient consciousness and the nature of painting itself as it impinges upon *Ad Parnassum* to reveal an unstated relation to prehistoric art, the shamanic trance state, pre-Socratic issues of flux and stasis, a neo-Platonic interest in pulsing energy, and Orphism's

B. Ross (✉)
Independent Scholar, Bangor, Maine, USA

C. A. Hornbuckle et al. (eds.), *Phenomenology of the Object and Human Positioning*, Analecta Husserliana 122, https://doi.org/10.1007/978-3-030-66437-4_11

understanding of natural form, all to the emphasis in Klee of a lyrical use of color and form to evoke an affect, in this painting, of metaphysical sublime.

Part of this revolution was making painting a vehicle of its own aesthetic intentions as opposed to being a subject to pre-modern orientations to so-called objective reality. Structure as structure and color as color became a central focus. Klee's early *Dogmatic Composition* (1918) gives us a re-thought cityscape dominated by vertical patches of color and a density of x's and some asterisks within squares and rectangles, and a few arches, all in a richer color hue to designate buildings, a red roof at the lower left, a pale yellow moon or sun in the upper right gives a spatial orientation. Humor is introduced with a bright red heart, the name Maria in capital letters, and an abstract drawing of a woman below the name. The later *Individualized altimetry of layers* (1930), with its horizontal arrangement of different blocks of color in darker and lighter hues and its humorous title referring to measuring elevation, here to reflect an internal emotion, prefigures as "pure painting" modern abstract expressionist art.

Klee was a member of The Blue Four group (1923) and probably learned from group member Wassily Kandinsky abstract form technique and from Lyonel Feininger abstract architectural form. Klee was also probably influenced by Robert Delaunay's approaches to Cubism which revolutionized form, color, and subject. Klee would learn to create abstract structures and to illuminate such structures through color. Düchting notes: "Klee used the musical term 'polyphonic' (many-voiced) in titles to draw attention to the simultaneous sounds created by the various pictorial elements and stylistic devices in his painting" (Düchting, 2002, 48). Düchting noted Klee used the term in relation to Delaunay's paintings and quotes Klee on one of these paintings: "Polyphonic painting is superior to music in so far as the temporal element has more of a spatial quality. The sense of simultaneity emerges in an enriched form. With his choice of an over-sized horizontal format, Delaunay endeavored to accentuate the temporal dimension of the picture in the manner of a fugue" (Düchting, 2002, 27–28).

The proposed metaphoric relation of music and painting, the subject of Düchting's study, gives some purchase to one of the aims of Cubism. Not unrelated, Edmund Husserl somewhere noted the way to look at a plant was to look at it from all directions, a kind of polyphony of natural entities. So, out of a breaking down of structure, there may be an illumination of a given painterly subject, as in the metaphoric music structure equivalent, supported by the emotional effect of color and form. Certainly there is an evident rhythm to be seen in Cubist paintings, including Klee's. Many of Klee's paintings have apparent connections to representational subjects. For example, in *Mountain Village* (1934), there is verticality offered by the painting's orientation and hints of the mountain forest in the upper left quarter with light green openings beyond the many shapes and bright colors of the village itself, both producing a kind of rhythm. *Revolution of the Viaducts* (1937) presents a cartoonish rhythmic procession of brightly colored viaducts. *Park near Lucerne* (1938) is dreamlike with abstract black plants, trees, and, perhaps, paths, each surrounded by different pale colors, possibly a kind of aura. These objects are seemingly floating in a pale blue backdrop of sky or perhaps consciousness itself. *Highways and*

Byways (1929) is a construction of a countryside, an abstract landscape of delicate colors and various small rectangular shapes organized projecting a subtle rhythmic quality through form and color. Approaching modern abstract art in its forthright use of bright color (here crayon on burlap) and rhythmically grouped, variously sized, squares and rectangles, *Glass Façade* (1940), with its façade whimsically backed by slight sections of the chartreuse element, perhaps the true self, behind the façade, has an upbeat feel.

Klee developed several distinct approaches to color and form, usually extensions of other painters' work. He had his own theory of complementary colors and their emotional qualities as color. *Blossoming* (1934) is constructed entirely of squares and rectangles. The larger ones in more somber hues surround a section of brightly colored small forms, creating an expression of burgeoning life. *Harmony of rectangles with red, yellow, blue, white and black* (1923) also offers various shaped squares and rectangles on a black background. All the colors are in dull hues and overall produce a somber painting. Such patterns of brightly colored checkerboard and simple geometric patterns occur in prehistoric cave art, such as the colored squares at Lascaux or the engraved grid at Grotte de Bara-Bahau and in Southwest pictographs of Native North America suggesting a tantalizing commonality of discovery. *The Day in the Forest* (1935) prefigures *Park near Lucerne* with stark black trees and some of their shadows lodged near boulders of pastel colors or overhanging an open area in pale orange, perhaps a huge boulder, pond, or open earth, and perhaps a lake in pale blue at the painting's top, the rhythm here created by color. The approach of blocks of non-representational color on representational form here is indebted to *The Blue Rider* (1911) approaches of Kandinsky and Franz Marc, such as Marc's blue horse paintings. Klee was also influenced by Pointillism which he renamed "divisionist painting." Düchting describes the process: "A coloured ground is applied either in a cloud pattern or divided into rectangles and covered by a grid of dense dots of contrasting colours, giving the image depth and transparency. In these pictures the contrast between the coloured ground and the pattern of dots results in an intense luminosity" (Düchting, 2002, 68–69).

In *Palace Garden* (1931), the abstract spires, roofs, and supporting structures are highlighted in such clouds and rectangles of pale pink and dark brown dots within the overall mosaic-like pattern of dots. The interaction of these elements produce a sparkling effect, as if patches of light flicker in, perhaps, a night scene here. Düchting cites Klee's own understanding of this process: "this work was created by placing a highly dynamic, individual element based on colour on a tonally-dynamic structural rhythm" (Düchting, 2002, 70). In *Tendril* (1932), the pastel pink and blue cloudlike patches of dots provide the dynamic background for the static abstract black or blue forms reminiscent of rock art geometric shapes and a bright sun made up of minute yellow dots, almost pulsing energy to the tendril below it. This kind of construction is the central building block of *Ad Parnassum* and produces the mesmerizing effect in the painting. Painted 3 years after this work, *Light and Sharpness* (1935) is a compressed version of the key aspects of *Ad Parnassum*, as if Klee wanted to point to the centrality of "divisionist painting" in his masterwork.

Though not directly related to *Ad Parnassum*, there are hints at the direction that perhaps led to this work in Klee's frequent incorporation of language in the broadest sense in his art work. Klee states this influence simply: "The genesis of writing is a very good analogy for movement. Also the work of art is primarily genesis, it is never experienced as a product" (*Taking a Line for a Walk*, 2014, 5). The authors for the brochure to the Paul Klee Center 2014 exhibit "Taking a Line for a Walk" accordingly note, "For him writing and drawing were rooted in one, because both writing and drawing are based on the line that arises out of a point setting itself in motion" (*Taking a Line for a Walk*, 2014, 5). Motion was the primary issue for Klee, whether the relation of form and color, the kinetic effect of pointillism, or color as color whether in the mind or in a work of art. The watercolor patchwork quilt-like *Once emerged from the gray of night* (1918) is topped by a handwritten poem, probably Klee's own. The many small brightly colored squares are filled with letters and proto-letters, simple geometric forms such as triangles, and other shapes. *Legend of the Nile* (1937) has a pictograph-like stick man in a boat at its center. Protolanguage forms and entoptic like abstract images float around him. A fish confuses the spatial/temporal focus, but centering the painting is an under grid of squares and rectangles. *Hot Pursuit* (1939) is another imagined rock art scene with stick figure combatants in a rose cloud surrounded by simplistic picture language. Likewise, in *She bellows, we play* (1928), childlike animal figures wrestle while another cries out to their right hovering in pale pink and white cloud patches as if in a shaman's other world visit. Similarly, in *Dances caused by fear* (1938), stick figures with geometric shapes for torsos dance surrounded by precise groupings of red dots not unlike those found in cave art. In *Death and Fire* (1940), white-faced death is surrounded by pale yellow and bright orange color fields, the symbolic fire, which introduces movement into the scene. Like others of his period, Klee theorized on the art revolution of his period. The abstract geometric forms of his *Theory of Modern Art* are used unchanged in his art work, such as the active line that centered *Tendril*, squares, crosshatch forms, triangles, and circles as empty intermediate lines, and rectangles, triangles, and circles as full, surface colored active lines (Klee, 1985, 72, 77, 78).

Klee's interest in architectural form including the geometric structures that underline them as well as picture and geometric language elements were solidified in his trips to North Africa. The Arabic calligraphy introduced forms that suggest entoptic images and show up as discrete shapes in many of his paintings. What was called "architectural rhythm" in a brochure for an exhibition of work related to this influence at the Bern Paul Klee Center is commented on:

> Paul Klee was so impressed by Moorish architecture, that he developed what he called his own "picture architecture." He was fascinated by the pictorial aspects of the architecture, the narrow lanes and striking silhouettes, but even more, also by the geometric-cubic structure of the Moorish building and city constructions. In numerous, at times wholly abstract water colors fascinating parallels between city architecture and picture structure are apparent. (*Paul Klee Carpet of Memory*, 2009, 2)

This discovery sets itself in the center of his painting aesthetics that also incorporated a new appreciation of color as color, as the brochure notes: "The importance of the 1914 journey to Tunisia for Paul Klee lay in the release from the structured drawing practice as the one and only means possible for the creation of a picture. The picture was now entirely structured from the color, the experience of nature leading to pure abstraction" (*Paul Klee Carpet of Memory*, 2009, 4). Another brochure from the Paul Klee Center on his travel in Tunisia cites Klee's journal on these breakthroughs, which led, as this brochure suggests, to "developing a colorful structure consisting of geometrical planes" (Klee, Macke, & Moilliet, 2014, 12): "My head is full of the impression of last night's walk … Art/nature/Self. Went to work at once and painted in watercolor in the Arab quarter. Began the synthesis of urban architecture and pictorial architecture" (Klee et al., 2014, 11). Suggesting the structural procedure to be explored in *Ad Parnassum*, he used "a rectangular grid he brings mountains, buildings and plants into a structured field of view" (Klee et al., 2014, 13). He embedded in this approach "geometrical color fields as a basic structure for his watercolors," the pointillism-like focus in *Ad Parnasssum* (Klee et al., 2014, 15). His well-known statement from his journal in Tunisia expresses this focus on color: "Color possesses me. I don't have to pursue it. It will possess me always, I know it. That is the meaning of this happy hour: Color and I are one" (Klee et al., 2014, 15). The intensity of Klee's assertion suggests a shaman's heightened state. The more recognizable representational structure in *Ad Parnassum* is suggested in Klee's watercolor created during a 4-week visit to Egypt, *Three Pyramids* (1929).

A key to *Ad Parnassum* can be found in Klee's engagement with what might be termed mountains and pyramids of the mind. The authors of the Paul Klee Center brochure for the exhibit *Paul Klee Carpet of Memory* explain the possible attachment of Klee to these forms:

> Since his earliest youth, Klee had been using mountains as the subject for his pictures, where he immortalized the Swiss Alps with absolute precision in his sketch book. Later the mountains became more abstract and adopted mostly a simple triangular shape, not infrequently reminiscent of the Egyptian pyramids. This ambiguity was clearly intended and occupied Klee following the journey to Tunisia. The connection between mountain and pyramid was the link between nature and art, landscape and history. The pyramid shaped mountain appears constantly as a motive of longing, as the enticing object stands in the far distance. (*Paul Klee Carpet of Memory*, 2009, 3)

What makes *Ad Parnassum* singularly enticing is its close up exploration of the mystery of that longing. The early *Cacodemonic* (1916) is an experiment in geometric blocks of color that is dominated by triangular shapes, particularly two rising into the watercolor's top, suggesting two mountains surrounded by dark sky or two triangular rooftops in an abstract representation of color, setting up the singular construct of triangular forms reaching into the heavens, like Swiss mountains or Egyptian pyramids, a convincing image of longing. *Lonely* (1928) emphasizes the emotional nature of this longing, really a mystery, through a grid of variously colored slender lateral lines from darker at the bottom to lighter at the top. The brightest color occurs where lines cross a distant pyramid with a minute doorway and above the pyramid as a multi-colored sky. *Double Tent* (1923) uses the lateral grid in a similar manner but with a darker abstract treatment. The two tents are the mountain/pyramid/triangular form Klee was obsessed with. Here the brightest colors are contained in this form, with two inverted triangles touching the apexes of the two central triangles. These inverted triangles descend from the presumed dark sky chromatically moving from dark to bright grids as they approach the central triangles, a suggestion of something coming out of mystery and bringing some sort of illumination to the central triangles. Klee is definitely depending on color for *Ad Parnassum* to create a sense of rhythm metaphorically similar to music. He also clearly understood the effect of gravity in a representation of the world, as suggested in a drawing of the force of gravity on a mountain/pyramid/triangle's waterfall, as presented in his *Theory of Modern Art* (Klee, 1985, 92).

The second major element and the brightest element in *Ad Parnassum* is the orange sun. Most reproductions of the painting depict a solid orange sun, the only solid colored form in *Ad Parnassum*, thus making it dramatically relational in contrast to the mesmerizing pattern of pointillism-like multi-colored dots. In 2013, I viewed the original painting in the Bern Kunstmuseum. The room in which it was inconspicuously mounted near a doorway was darkened. The lack of strong artificial lighting took the luster out of the painting. What stood out is the concentric circles of pointillism-like small dots awash in the pale orange/yellow color of the sun. As such these dots, more like miniature ears of corn lined up in each circle, were in extreme contrast to the presiding vertical dots in the rest of the painting. Most reproductions somehow heighten the orange color and many do not have a trace of this circular pattern. This is a special image, almost a symbol of perhaps, some Neo-platonic pulsing godhead, as it maintains itself in the almost turquoise blue sky. In *Ad marginem* (1930), 2 years earlier, an orange sun pulses energy-like aura at the center while a collection of human and natural forms circle at the painting's four sides. Düchting notes that Klee called *Ad Parnassum*'s sun a "warm circle" (Düchting, 2002, 76). Nonetheless, Klee was as concerned with sun and moon images as he was with mountain and pyramid images. He was probably even more than simply concerned. On his visit to Tunisia, he notes in his journal, "The evening is deep inside me … I myself am the moonrise of the South" (Klee et al., 2014, 12). He thus expresses the meaningful relation of aesthetic consciousness to the moon and to the sun. In the watercolor *Moonrise* (1915), painted in St. Germain, Tunisia pastel buildings made up of colored squares, rectangles, and one yellow triangle darken at their tops and are surmounted by a full white moon. A later oil *Colorful Lightning* (1927) highlights an orange full moon to the right of the beginning of a line of entoptic-like lightning bolts, stasis juxtaposed to dramatic flux, highlighting this exchange by the totally black-purple background. In the playful punning *The Bavarian Don Giovanni* (1919), this reworked Don Juan climbs ladders and pyramid-like mountains to his conquests whose names are spelled out in some of the structures. Yet above this scene of merriment are the motionless crescent moon and a star. Suggesting the later structuring of *Ad Parnassum*, *Red Balloon* (1922) places the sun-like solidity of a red balloon above bright color field patches in the abstract houses. In *Around the Fish* (1926), abstract expressionist geometric and human forms circle the almost realistic fish on a platter. In the circle but centered at the top are a crescent moon and a probable full moon, the symbol of change on the Earth, from emptiness of the new moon to completion in the full moon. The two major elements of *Ad Parnassum*, a triangular structure and the sun in the upper right hand corner obsessed Klee, and 5 years after his masterwork he created *Castle and Sun* (1928) which was made up of colored Cubist geometric forms, including many triangles, and a representational luminous sun with a kind of aura.

Is there some rigorous symbolism, such as the mountain and pyramid, sun and moon, color itself, surrealist depictions, point of focus, geometric structures, and so forth in Klee? Is *Double Tent* a metaphysical meeting of two worlds or dimensions, for example, or merely an illustration for color theory? In *With the Eagle* (1918), is the eagle placed above with outstretched wings atop an arch the conventional symbol of power? That the arch encloses a gigantic blue eye suggests another recognizable symbol. It has been suggested that this is an allegory of war, with the artist as the eagle about to fly above it (Partsch, 2003, 42). Klee, who served in World War I, has noted somewhere in his commentary on The Blue Rider aesthetic that the more terrible the world, the more abstract the art, the more peaceful the world, the more secular the art. Is the eye Klee's eye, which was also blue? This eye is placed like the sun in the center of the painting's universe, the only unmoving thing. Klee has used the sun as a dominant image, as in *Ad Parnassum.* He has also placed a dominant image in the center of a circle of other objects as in *The Goldfish* (1925) where the huge seemingly motionless goldfish, like Aristotle's unmoved mover, is surrounded by, perhaps, fleeing fish, like the planets of the solar system in their revolutions. In *Moonshine* (1919), the moon and a kind of aura of glowing white, placed in the upper right, like the sun in *Ad Parnassum*, presumably passes this glow to the similarly glowing white bare tree.

In *Figurative Leaves* (1938), the bright orange sun is in the upper left and providing life to the plant depicted below. On the first leaf, and perhaps on the others, are human forms surrounded by an orange aura as are the leaves themselves. Klee is expressing here the values of Orphism's lyricism of color treatment as opposed to Cubism's architectonic structure. There are elements of theosophy in these possible symbols. Kandinsky attended Rudolf Steiner's lectures and wrote "On the Aura of Humankind." In the essay, he suggested four elemental bodies: physical, astral, mental, and Buddhic. Klee's *Theory of Modern Art* includes a drawing of the density of the various parts of a human body, such as ligaments, muscles, and the brain, and the density of earthly elements, incorporating air, fire, water, trees, and earth. Superimposed on the diagram are the touching triangles that appear in *Double Tent* (Klee, 1985, 90). These symbolic elements, including geometric structures, such as the pyramid in *Ad Parnassum* may be supported by an uncovering of the basic functioning of the mind in different states. In *Ad Parnassum* the relation of the seeming stasis of the sun, through its solid circle of orange, the brightest element of solid color, to the flickering effect of incandescent white dots in various color fields in the rest of the painting, expresses some essential truth of the universe and the human psyche in relation to it. To put it another way, breakthroughs in modern art in Klee's period, were exploring the idea of flux, what Vedanta would term *prana*, the life-sustaining force in the universe as expressed in the poem of the Indian mystic poet Kabir (1440–1518):

> Between the consciousness and the unconscious, the mind has put
> up a swing:
> all earth creatures, even the supernovas, sway between these
> two trees,
> and it never winds down. …
> Everything is swinging: heaven, earth, water, fire… (Mitchell, 1993, 70)

As expression, this breakthrough was in the earliest art. The geometric forms, including pyramids, squiggles, and other shapes in relation to bright color, have occurred as what David Lewis-Williams terms entoptic art (Lewis-Williams & Dowson, 1988, 1–45). These are internal geometric shapes perceived in a heightened state that turn into recognizable images from a given culture. Abstract forms recognizable in Klee's art, and that of others of his period, as stark geometric form or groups of color within a geometric frame, appears in prehistoric art work. The possibly oldest work, 75,000 years old, is a cross-hatched pattern carving in a block of red ocher found in Blombos cave at the tip of South Africa (National Geographic, 2014, 41). A similar form occurs as a 5000 year old pictograph created with hematite, manganese, gypsum, and other minerals for color in Big Bend State Park (Roberts, 2014). I have frequently seen this cross-hatched pattern and other geometric entoptic forms in North American rock art, such as the Titus Canyon petroglyphs in Death Valley National Park (Whitley, 2001, 86). Patchwork-like brilliantly colored pictographs occur at Lascaux Cave, and I have seen similar prehistoric pictographs in the American Southwest.

The dehumanizing trauma of World War I upon painters produced the need for new modes of expression, and some of this relates to an interest in primitive art exploration. Some of the results of this exploration led to heightened states of awareness, as in Klee's exclamations on his possession by color and his identity with the moonrise. A number of his subsequent paintings are un-representational abstract images.

World War II and following worldwide inhuman acts perhaps provoked contemporary new directions in art, such as hyperrealism, high definition photography, and the like. Marshall McLuhan could have predicted Jean-François Lyotard's "postmodern condition" when he expressed how culture works as "the medium is the message." In the current period, the prominence of computers and new communication devices have reduced the creativity of human beings, in a sense conceivably provoking a post-human world.

A brochure from the Paul Klee Center noted: "Paul Klee saw control and intuition as the two cornerstones of artistic work" (*Taking a Line for a Walk*, 2014, 3). It could be said that the formal structures of Cubism and the like, however connected ultimately to the psyche, was the control Klee sought, such as the pyramid in *Ad Parnassum*, and the color approaches of Cubism and the like, was the intuition Klee sought, such as the pattern of flickering and complementary color relation in *Ad Parnassum*. What makes this painting compelling is its impressive use, ultimately, of both control and intuition to produce a work of depth and dynamic movement to evoke the inner dimension of time and space within its aesthetic affect.

References

Düchting, H. (2002). *Paul Klee, painting music*. Munich: Prestel.

Klee, P. (1985). *Théorie de l'art moderne*. Saint-Amand: Denöl.

Lewis-Williams, D., & Dowson, T. A. (1988). The signs of all times: Entoptic phenomenon in upper Paleolithic art. *Current Anthropology, 29*(20), 1–45.

Mitchell, S. (Ed.). (1993). *The enlightened heart, an anthology of sacred poetry*. New York: Harper.

National Geographic. (2014). Blombos cave photograph. *National Geographic* Magazine, *41*.

Partsch, S. (2003). *Paul Klee*. Köln: Taschen.

Paul Klee Carpet of memory. (2009). Bern: Zentrum Paul Klee.

Roberts, T. (2014, December 26). Rock as canvas, native American rock art in the big bend.

Taking a line for a walk. (2014). Bern: Zentrum Paul Klee.

Walter, C. (2015). The first artists. *National Geographic*, 32–52.

Whitley, D. S. (2001). *A guide to rock art sites, Southern California and southern Nevada*. Missoula: Mountain Press.

Transhuman and Posthuman in Popular Culture on the Basis of Miura Kentarō's *Berserk*

Natalia Kućma

The purpose of my paper is to concentrate on transhuman in popular culture on an example of the most popular manga *Berserk* by Kentarō Miura. I base my considerations on comics because popular art often allows to visualize future, and thus to emphasize and point out in an exaggerated way many problems and challenges, making it possible for us to see them clearly and analyze them. I read *Berserk* from the point of view of the philosophy of culture. Firstly, I will summarize the plot of this manga. Secondly, I will compare the two main characters. My aim is to show how thanks to the mediation of technologies human subjectivity is changing. On the basis of Rosi Braidotti's ideas, I want to illustrate how the concept of posthuman and transhuman opens human to other transitions: transgender, life beyond the species and so on. Thirdly, I will look at ethical problems which this piece of art signalizes.

Berserk, by Miura Kentarō, has been published since 1989, and it is still ongoing. This story is very complex. I will summarize only the crucial events. The main character is Guts—a strong mercenary, who one day joins The Band of the Hawk—a mercenary army under the leadership of Griffith, called The White Hawk. There, Guts finds a friend and a lover—the only woman in this army—Casca. The dream of Griffith, with whom the main character establishes a close relation and whom everyone firmly trusts, is to have his own kingdom. And when on a day of an eclipse he gets a chance from Fate, he decides to sacrifice (which means to condemn to death) the whole Band of the Hawk. During the eclipse, demon monsters appear, kill his friends, and rape Casca, as does Griffith. Guts loses his eye and arm, but, with Casca, they are the only survivors. From this moment on, Guts travels across the world to find Griffith and fights countless demons, which want to kill and devour him because he remains an incomplete sacrifice from the eclipse. A craftsman gives

N. Kućma (✉)
Jagiellonian University, Krakow, Poland
e-mail: nmkucma@gmail.com

C. A. Hornbuckle et al. (eds.), *Phenomenology of the Object and Human Positioning*, Analecta Husserliana 122, https://doi.org/10.1007/978-3-030-66437-4_12

him an incredibly huge sword and a new metal prosthetic hand (with a little cannon inside). A few years later, Guts receives a cursed armor of berserk.

At the moment, we can divide the time in *Berserk* into four parts:

- The Golden Age: it is an age of humans and war, the gold time for winning mercenaries;
- The Eclipse: the liminal moment, when human world and the world of inhuman start to entangle (with Griffith's decision the new evil is born);
- The Time After The Eclipse: the monsters boldly appear in the human world and destroy it while the followers of a very important religious group are waiting for the coming of the savior called the Hawk of Light;
- The Time of the Hawk of Light: the end of the war.

In one of the few interviews Miura (2002) that he gave so far, Miura Kentarō admits that, probably due to their pop cultural adaptations, such as *Excalibur* (1981), Arthurian legends and the stories of the Knights of the Round Table, though rooted in the European culture, are much closer to him and his generation than Japanese accounts of the brave samurais. That is why I intend to read *Berserk* mainly through its connection to European culture. It is important, in so far as the first assumption upon which this reading is based, to see the opposition of nature and culture, as understood differently in Europe and Japan.

Transhumanism and posthumanism in popular culture are most of all associated with science-fiction, when people of the future have reached an incredible level of technological development, leading them either to a glorious rise or to a disastrous fall. While choosing *Berserk*, stylized as a medieval story, as my primary source, I was led by the assumption that technology should be broadly conceived as a tool enlarging human possibilities. That is why I include magic, treated as an ability gained by enlarging one's knowledge, in the scope of technology. For, as Arthur C. Clarke rightly points out, "at some level, science and magic are indistinguishable" (1973). The philosophers whose ideas I address in this paper, writing about technology, probably thought about contemporary times, while the story which I am referring to is stylized on the European Middle Ages. However, in my opinion, this transfer into seemingly inadequate times, can show that technology has always had in it the revolutionary potential to blur the differences, interestingly described in the works of Rosi Braidotti. And because this manga is a product of our times, like every work of popular culture, it presents contemporary struggles. As John Fiske says (1989, 139), it is like that, because popular art is never disinterested, which means it does not distance itself from everyday life, but is involved in it.

Max More in his *A Letter to Mother Nature* More (2009) enumerates seven amendments that mankind should carry out in order to perfect its natural, original condition. To fight the process of ageing, to broaden the scope of the senses, to control emotions... all of them focus on exceeding biological limitations resulting from evolution. To effectively conquer various imperfections, we cannot wait for the coming centuries of regular evolution; our task is to effect evolution on our own terms, using available tools offered by technology. Biology, determinism, nature, and even silent God are thus juxtaposed with human being and its products: culture and technology. When Nature-Fate appears as an enemy (death stealing friends, body refusing to obey will, emotions and reason blinding one another), a human must escape into technology. And with the turn to technology, not only is their relation to nature changing, but also other relations, attitudes and values are reformulated. The struggle with Nature-Fate as depicted above is the main axis of *Berserk* and its two main characters follow the paths that can strengthen them as men (both physically and mentally) and free them from human imperfections, disease and death. The result of this struggle is transcending what is human and heading in the direction of transhuman and posthuman.

I understand "transhuman" as a subject between "human" and "posthuman," as a being exceeding the possibilities of man, but still bearing the resemblance to its original. Transhuman is thus an enhanced human, whereas posthuman is an entirely new and peculiar subject. N. Katherine Hayles defines the posthuman subject as "an amalgam, a collection of heterogeneous components, a material-informational entity whose boundaries undergo continuous construction and reconstruction" (1999, 3). It is an open-ended subject uniting human and all that could not have been included in its old, limited constitution. Rosi Braidotti calls posthuman "transversal assemblages" (2013a) that, having opened on transcending the human, open also to other transitions. Because of the interaction between human and technical the boundary between genders, races, species becomes blurred. The Technological Other is the one who operates in the domain of an egalitarian obliteration of differences. I want to look at another possible "trans-" which occurs in this story, and that comes into being thanks to technology.

Each kind of transition is examined on the basis of the example of the two main characters, Guts and Griffith. Guts, as a warrior, reminds us of the old truth that technologies develop due to the needs of the army. By constantly exercising his body, he smoothly wields a ridiculously large sword. When during the Eclipse he loses an eye and his left arm, it is replaced with an artificial one, which is designed so that it can both handle the sword and be a little gun. During a fight, when he is already wounded, he receives a cursed berserk armor, which removes pain, and, in this way, enables its wearer to perform deeds unthought-of in a regular human, for whom pain is an impediment and a warning. For "[a] human without pain realizes unbelievable strength, reflexes, and perception" (Miura, 2008, 219). The armour heals broken bones, but also wakes up in a warrior wild instincts and recklessness.

Let's look at the transformation that Guts undergoes after putting on armor. Guts feels something wild and murderous awakening in him. Something that wants to get outside and take over (Illustrations 1, 2). This page contains three pictures of his eye, which evolves from a recognizable, human shape into an abstract vortex, whereas the next page presents a furious wolf with an eye resembling a lightning or a zigzag. According to old Nordic beliefs, berserkers possess the powers of a wolf. In Nordic culture, a wolf personifies evil, chaos, demonic powers. Guts, wearing this armor, will be more and more receptive to madness, which is represented as a demonic *lupus*. The armor protects only from external injuries, calling and waking inner demons. It is very important that it is like a living organism, eating up into bones and feeding on the emotions of fear, hate, resentment, anger. Thus, technology uses existing emotions, fueling and reshaping them again. The armour will become a catalyst for these feelings or even their embodiment. Along with the development of the story, Guts will be more and more sensitive to its influence. Not only will it weaken his body when not worn (making his hands tremble or distorting his vision), but it will also poison his mind with terrible visions (e.g. with the idea to rape Casca just as Griffith did). The armor, wholly dependent upon its wearer, will try to make its wearer completely depend upon itself, so that he is not able to function without it. All the fears of technophobes centering on the idea that technology possesses the power to beguile and control us are best illustrated by the panels presenting a demonic wolf, trammeled by shackles in Guts's mind, who tries to free itself and threatens that dominating its master is only a matter of time. In this way, Miura Kentarō illustrates common fears connected with technology, underlining its destructive influence upon man, and his/her moral and emotional standing. And, according to the author, there is a remedy for the dangers connected with technology: one that is seemingly a step backwards since it makes a human look back into the past—in Gut's case at Casca, mad after the catastrophe of the Eclipse.

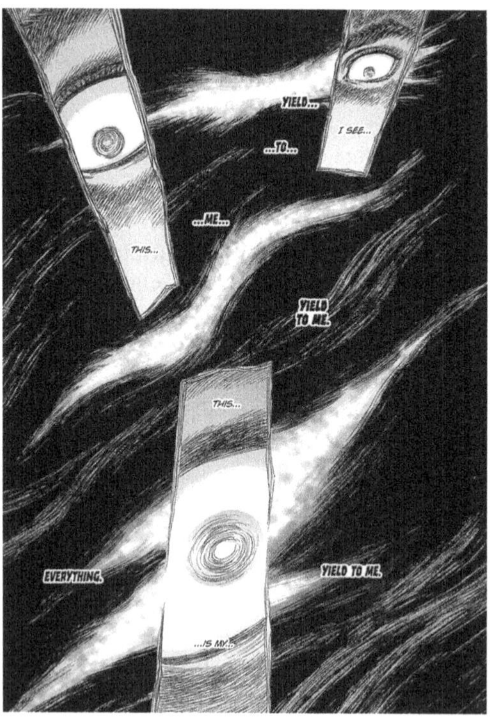

Illustrations 1, 2. Miura, Kentarō. Berserk 26, translation D. Johnson, Dark Horse Manga 2008, 206–207. © Kentaro Miura, © English-language translation Dark Horse Comics

Upon the reception of the armor, Guts meets Schierke—a little witch whose power he will henceforth need in order not to lose his humanity. Schierke, thanks to the ability to travel astrally, is able to reach Guts when he is wearing his armor and to reason with him. The combination which takes place through the armor and in a way against it—Schierke prevents the curse from operating—is an interesting example of a connection between transhumanism and transgenderism. The amalgamation of a human and an object, turning Guts into a kind of medieval cyborg, is based on negative emotions only. The little girl, by moving emotions different from those which feed the armor, that is feelings of care, concern, love and responsibility, restores the warrior to rational thinking. The picture (Illustration 3) of the little witch on the back of the mad warrior (where Schierke, holding Guts tightly by the neck, removes darkness from the berserk's eyes and enables us to see a human eye instead of the wolfish lightning) has a special symbolic meaning to me: on the one hand, it helps to reduce the negative effects of technology; on the other hand, it supplements certain values. Through technological mediation, a woman and a man, a child and an adult, can really work *together*, not only *next to* each other. They achieve unity both on the physical and mental level. It is a paradox of the story that his feeling for Casca makes Guts become a monster while protecting their humanity. As was mentioned before, the operation of the armor is based on the contrast between nature and culture, attributing to them accordingly emotion and reason. The armor, in order to be able to really work (and gradually take control of the user), has to prey on something. Something which in European culture over the ages was connected with the animal part of human nature, namely emotions. Emotionality in Western culture has been associated with such concepts as: femininity, nature, irrationality, subjectivity, chaos as opposed to what is male, cultural, rational, objective, orderly. At the first glance, the little witch who understands the nature of the world and the elements fits into this division. Because of the mediation of technologies, it is for a while abolished. Guts's way is the way of the sword, violence and physical strength. Schierke's way comes from the knowledge of the nature of the world, which is a huge living organism, and over which you gain rule by requests, apologies and thanks. Little Witch removes darkness from berserk's eyes restoring harmony. I am tempted to say that this illustration refers to the graphic representation of yin yang, where darkness and brightness overlap and complement each other. The bright figure of Schierke and the dark figure of Guts drawn with fluid lines intertwine with each other.

Illustration 3. Miura, Kentarō. Berserk 31, translation D. Johnson, Dark Horse Manga 2009, 93. © Kentaro Miura, © English-language translation Dark Horse Comics

In this case, transgenderism seems to consist in connecting stereotypical perception of male and female, not in transcending differences between them. Schierke's actions are based on *reminding* Guts about his promise to protect what he loves. It is she who upholds rationality by pointing out to the positive emotions. However, we should remember that the art of pop culture, especially comics, likes to use exaggeration and stereotypes in order to be understood. On reflection, I would like to argue that the true unity of male and female is possible in the context of the whole story, where, in my opinion, appears one of the most interesting and unusual feminine heroines: Casca. Before she lost her senses, she was a commander in The Band of the Hawk and a skillful swordswoman. The status of a woman in *Berserk* remains problematic. Women are very often depicted as objects of sexual attraction and victims of violence, but the two main male protagonists are also victims of sexual aggression (Guts was raped as a child, whereas Griffith was forced by his difficult economic situation to sell his body to a sleazy earl). This manga also offers one of the few instances of the author depicting a woman having a menstrual period and the female characters are here as differentiated as the male ones. Moreover, subsequent chapters of *Berserk* fulfill the criteria of so called Alison Bechdel's test concerning gender inequality (the Bechdel test).

The next character, Griffith, is a hermaphrodite figure since the beginning of the story. His appearance is feminized (he has long, white hair, alluring lips, etc.), and his personality is so mysterious and variable that it can be determined as neither masculine nor feminine. He is neither overly manly, though he is a charismatic leader and an invincible swordsman, nor effeminate, despite his delicate stature and childish tricks. A year before the Eclipse, he is imprisoned and the tortures he undergoes not only destroy the beauty of his body, but also the ability to independently control it. After the Eclipse, he is reborn as a more beautiful human (even longer hair, even more lustrous eyes, etc.), but he also gets a new form, which, like the berserk armor, points out to another transition: beyond the species.

Guts calls for a connection with a wolf, while talking about Griffith demands recalling another animal—a falcon or a hawk.[1] The leader's helmet is styled as a hawk beak, and when he resurrects as the new demon king, he gains real wings and claws. Griffith and Guts, embracing animal power, do not become animals, but they are no longer human. We witness the interplay between the imagined and the real behavior and looks of these animals. In this way, Miura creates a specific techno-bestiary.

The example of Griffith makes it possible to approach one more problem, present in philosophy for a long time and interestingly depicted in the works of popular culture. The question about the relation between body and consciousness appears in a new light when a consciousness is transferred to a different body. Is Griffith, moving from his tortured frame into a mighty body, still the same person or is he somebody else only bearing the same name? If we treat body in the Cartesian way, only as a carrier of the mind or consciousness, we can feel safe. However, if we assume that body and mind remain interconnected and influence one another, the question

[1] 鷹 (たか) http://jisho.org/search/%E9%B7%B9 (7.05.2015).

whether the change of a "mind carrier" does not involve the change of the mind is a vital one. Griffith's example is all the more thought-provoking because he is a character that has always treated his physicality as a tool: he has used his beauty to gain mercy and favors of women and men either by putting on the charm or, ultimately, by selling himself. His body has been for him a means to achieve a higher goal and he appears to treat it as such also after his rebirth. According to Hayles, a posthuman is somebody who sees his/her body as a prosthesis and there is nothing strange in replacing one prosthesis with another. It is even advisable when the latter turns out to be more effective:

> Indeed, one could argue that the erasure of embodiment is a feature common to both the liberal humanist subject and the cybernetic posthuman. Identified with the rational mind, the liberal subject possessed a body but was not usually represented as being a body. Only because the body is not identified with the self is it possible to claim for the liberal subject its notorious universality, a claim that depends on erasing markers of bodily difference, including sex, race, and ethnicity. (Hayles, 1999, 4–5).

But it is not the attachment to the body as something more than a tool that stops Griffith from excluding others, different as far as species, race or gender is concerned, from the kingdom that he builds after the revival. He unites people of all classes and social standings along with monsters in his New Kingdom of the Happy. And thanks to his newly gained powers, he is also able to connect them with their dead: the spirits of the ones who died can for a moment materialize in his presence to meet their living relations. What is important, despite his appearance of a white man he is not *the* white man who is criticized by feminist, postcolonial and other movements. It is worth mentioning that the author probably plays with the meaning of colors. The story unfolds in a place stylized as medieval Europe, where white color is the symbol of purity, innocence and goodness. In Japanese culture, white is the color of death, and if we meet a character with white hair, we may suspect that he or she is not purely good (e.g. Muraki Kazutaka from *Yami no matsuei*, Jouzou Suzaya from *Tokyo Ghoul*).

Griffith abolishes dualisms from the world. Before the Eclipse he is a beautiful, skillful swordsman and commander, simultaneously charismatic, childish and dangerous. After the Eclipse, he becomes posthuman *par excellence*. He could be perceived as merging contradictory qualities: man and woman, life and death, beauty and ugliness, health and illness, man and beast, center and margin, good and evil.

The aforementioned components create a new inhabitant of the world, combining human, animal and mechanical. All those transitions constitute not a human being, but a subject who answered to the call of history and transformed itself so as to be able to embody the needs of this particular world.[2] According to Hayles, the decisive factor in being posthuman are not elements of the physical world (not everybody has to have a non-organic element in their body), but the open-endedness

[2] The thread that I only signalize in this paper (because of the small amount of information available), however existent and important it is for the story, is the presence of God in this world.

of one's subjectivity, consisting of heterogeneous components that are constantly built and shaped anew.

Both characters have crossed the human level. Guts still struggles, so he is closer to being transhuman. He shows the struggle of a human with technology, with old norms and values, with oneself, he improves himself and becomes a beast in order to preserve his humanity; he craves to be his past self, the self that was taken away from him and that he himself abandoned in order to have his revenge. In all this, he is not a typical hero (nor an antihero). Although his behavior changes with the lapse of time, the reader first encounters him as a warrior who does not hesitate to kill random witnesses of his fight with demons. Griffith, in turn, is like a god, a posthuman. He agrees to modify his subjectivity and his physicality in exchange for the realization of his dream of his own kingdom. Notwithstanding the fact that the reader is never directly informed about his motives, it is worthwhile to speculate about them. Power, wealth, love—these are, in brief, the most common desires of people and all of them can surely be obtained along with one's own kingdom. However, if we take a closer look at Griffith, all this is available to him even before the Eclipse: he leads his own army, he advances in the world, he becomes richer and richer, he gains a position close to the throne by virtue of a princess's favors.

Griffith has already reached the state where his morality can no longer be considered in the old categories. Although during the Eclipse he becomes a new lord of evil, his attitude and actions are difficult to classify, and they are unpredictable. On the one hand, he mercilessly sacrifices the life of his companions, but on the other hand, he does it for the sake of ending a war that has lasted for over a 100 years and building a kingdom of happiness and prosperity. His decision changes the hierarchy of values and is redolent of a reversed decision of Ivan Karamazov from *The Karamazov Brothers* by Fyodor Dostoyevsky. While Ivan refuses to sacrifice one innocent, unknowing girl, Griffith does exactly the opposite: he changes Casca into an innocent, unknowing, mute, insane girl.

The difference between what Guts and Griffith have become results from their initial position: that is, from different moral decisions. According to Braidotti (2013b, 13), our relation to the posthuman is conditioned by the critical assessment of what is human. The emergence of posthuman makes us reflect on the status of human once again. This is not a new idea: in history every *monstrum*, every freak born among people and animals (from a child with six fingers in one hand to a two-headed goatling) was treated as a disruption of norm, which thus needed revision. The human body is a rhetorical matter, a carrier of meanings, and in European culture, a deformed body was an indication of moral transgression of its owner or of his/her mother. A posthuman or a transhuman using an unusual armor or defeating death is something "inhuman," a monster.[3] Nonetheless, despite monsters such as

[3] The world of Berserk is inhabited by monsters known from mythology: unicorns, elves, sirens. Nevertheless, they are separate species, rarely interacting with humans and not possible to be seen by everyone. Their presence plays a role in the novel as a whole, but as they are not important for my considerations, I decided to omit their existence in the body of the text.

Guts as a berserk, there are yet different monsters in this world: the ones that pursue and fight him and the ones that join Griffith's army after the Eclipse. They are called Apostles.

To understand the nature of the beasts that appear in the story, it is necessary to take a look at the procedure of sacrificing (like the one done by Griffith). It turns out that monsters appearing in this world, not so much populate it with new entities, but unveil its inhumanly normative side.

In the course of the story, the reader learns that all the monsters called Apostles were once mortals. They are men, women and children, who lost something or could not stand their situation. A girl who lived in a house full of violence is turned into a blood elf. An earl gives himself to evil, when he despairs after the betrayal of his wife. To become monsters and free themselves from being human (by building their own world and getting rid of the fear of other people), they have to sacrifice something precious. Changing into a monster is preceded by a monstrous moral decision. It could be said that they embody the evil which they choose. The desire to be inhuman is dictated by the desire to be different than human: being something that is not afraid of people, who are seen as moral monsters. Their ethics can be described in one sentence: "Fate transcends human intellect; if that is the nature of truth it is the inevitable result if the children resist fate with evil" (Miura, chap. 09). And this is the way of all Apostles. Due to murder, they gain strength to be able to do even more evil, which they see as justified and legitimate, since they themselves have experienced all kinds of unjustified evil.

Popular culture works primarily on human emotions, allowing to understand the actions of others, and, as such, is very important in ethical considerations. As Stephen King wrote, horror stories (and *Berserk* may be called one of them) express in a symbolic way what we fear more directly. In this case, it is the fear of others, in some way more powerful than ourselves. Monsters still live wearing masks of men and are only a little bit more cruel than humans, because they assume that humans treat atrocities like an amusement. The sub-chapters of *Berserk* called *Retribution: Lost Children* are a story of a girl who wanted to be an elf and who becomes an Apostle by sacrificing her cruel, pathological parents. She steals other children from her village and turns them into bee-like elfish demons. They live under her leadership far from their previous home and wicked parents. Once in a while they play pretending adult people, which means they play war, and they do it very faithfully, really slaughtering, torturing and raping one another.

Horrors, according to King (2010, 198, 2), revolve around the idea of inside evil and free will. This manga can be treated as an allegorical tale of the fight against Fate. Guts, Griffith, the Apostles oppose determinism, and vicious circumstances in which they live. Most of all, they fight economic, class, gender and cultural violence. They resort to the use of tools that are not indifferent for their human condition, their moral standing and their relations with others. Braidotti writes: "Becoming a posthuman is a process of redefining one's own feeling of embedment in the world and one's connection to it" (2013b, 355). It is particularly visible in the context of the so called evil. After the human nature changes, the old structures are no longer valid. The strongest individuals work on their own and this may be the reason they have not yet formed a new culture and thus a new language to describe in a new way what is still classified as "evil."

Berserk, treated as a horror, is, as every other work belonging to this genre, a barometer of fears of a generation. As we can see nowadays in the discussion between transhumanism and neo-luddites, people are still afraid of technology, because they think that it can change human nature so that it becomes evil, while, in fact, it transcends the moral qualification of good and evil. Values and even human him/herself demand redefining. It can be questioned whether without physical and/or mental injury Guts, Griffith, the Apostles, would decide to modify their condition. They choose what is generally called "evil," only after the strong impulse destroying their lives.

So how can we understand humanity in *Berserk*? Most of all, humans are those who are easily killed. Death, in turn, is what a human would desire most to abolish from his/her surroundings. From the belief that "I shall not wholly die," through defining oneself as a thinking reed, to being-toward-death and many more, man has been constituted by his attitude towards death. Last but not least, the question whether human is able to deal with not so much technology but their own demons, remains an open one.

References

Braidotti, R. (2013a). *Punk women and riot Grrls*. YouTube. Retrieved April 4, 2015, from https://www.youtube.com/watch?v=i5J1z-E8u60.

Braidotti, R. (2013b). *The posthuman*. Cambridge: Polity Press.

Clarke, A. C. (1973). *Clarke's laws*. Wikiquote. Retreived July 7, 2015, from https://en.wikiquote.org/wiki/Arthur_C._Clarke.

Fiske, J. (1989). *Understanding popular culture*. London: Routledge.

Hayles, Katherine N. 1999. How we became posthuman. Virtual bodies in cybernetics, literature and informatics. Chicago: University of Chicago Press.

King, S. (2010). *Danse Macabre*. New York: Pocket Books, Kindle.

Miura, K.. (2002). *Berserk—Interview with Kentaro Miura*. YouTube. Retrieved May 2, 2015, from https://www.youtube.com/watch?v=YGLtPxJdfAU.

Miura, K. (2008). *Berserk* 26 (D. Johnson, Trans.). Milwaukie, OR: Dark Horse Manga.

More, M. (2009). *A letter to mother nature*. Retrieved April 5, 2015, from http://strategicphilosophy.blogspot.com/2009/05/its-about-ten-years-since-i-wrote.html.

The Work of Art as a Living System: A Deweyan Approach

Stefano Polenta

1 Introduction

This paper is intended to develop Dewey's aesthetics in view of the epistemology of complexity, proposing a convergence between an artwork and a living system. After presenting an argument on this issue in the first paragraph, this text will go on to investigate the two fundamental concepts on which the theoretical perspective proposal is based, namely those of emergence and organicism. We will highlight how these issues are at the heart of Dewey's aesthetics and the basis on which develop the latter towards the epistemology of complexity.

A secondary objective of this paper is to suggest that organicism can be seen as one of the principal sources of the epistemology of complexity. In conjunction with the examination of the theme of emergence, specific attention will be on the context in which this contribution is placed: posthumanism.

2 The Work of Art as Living System

In *Art as Experience* (1934/1980)—the text Dewey dedicates to art—there are many facets focused on the *dynamic consistency* of a work of art: how it is "made," with the emphasis always on the ability of the work of art to be an integrated whole, a unity in diversity, like any *complete experience*. This entirety is not achieved through a mere juxtaposition of parts nor by virtue of the projection of the artist's feelings but is as a result of a gradual mutual adaptation between the artist's subjectivity and the medium/environment. For Dewey, to understand art, we must start from the

S. Polenta (✉)
University of Macerata, Macerata, Italy
e-mail: stefano.polenta@unimc.it

C. A. Hornbuckle et al. (eds.), *Phenomenology of the Object and Human Positioning*, Analecta Husserliana 122, https://doi.org/10.1007/978-3-030-66437-4_13

experience of any living creature. Here, experience refers to striving to overcome a lack of adaptation in conquering a new adaptation. The reconquest of adaptation is an experience of harmony in that it signifies that the living creature has overcome fragmentation and death, managing to reconstruct an ordered participation within the environment. Facilitating a restoration of harmony, this experience in itself has the characteristics of an aesthetic experience. "Art is […] prefigured in the very processes of living. A bird builds its nest and a beaver its dam when internal organic pressures cooperate with external materials so that the former are fulfilled and the latter are transformed in a satisfying culmination" (24). Such complete experiences enable a sense of precious fulfilment.

The process of the living creature's adaptation to the environment is not one-sided, in that by modifying itself, the being changes the environment in turn. This mutual adaptation between living creatures and the environment is ultimately the same as that existing between artist and medium. The artist is seen by Dewey as a living creature *viscerally* involved in reality, almost like a wild animal that struggles, striving to continually "overcome problems" (138) and recapture adaptation. To this end, "equilibrium comes about not mechanically and inertly but out of, and because of, tension" (14). This is not a mere restoration of the previous context, given that life is *expanding* and the new adaptations are more inclusive than the previous conditions.

This conception of art as the result of man's progressive adaptation to reality recalls the conception of the artwork as a "growth of a living embryo" (137) that—as Pareyson (1966, 20) underlines—is ancient, implicitly dating back to Aristotle. Artwork transforms from seed to fruit as a *continuum*, simultaneously its own regulation and result (Ibid). As man is gifted with creativity and imagination—as per Goethe's thinking—through art, man is able to continue the work of nature in striving for a higher level.

The concept of system is the theoretical tool for understanding how this dense web of interactions between subject and environment: exchanges, transactions, learnings and increased quality can settle and "cumulate" in a structure that is the work of art itself. The concept of *organization* is crucial in this regard. The systemic perspective allows us to see reality not as being comprised of elements but, indeed, of systems of exchanges amongst elements. It is a turning point in thought. The concept of system recalls that of *organization*: a system is organized. This is very clear for Dewey (1929, 254): "Organization is a fact." Yet organization is not a static snapshot of a working system. Systems are processes constantly changing whilst maintaining an organization. In an organism, for example, "the interactions of the various constituent parts of a plant take place in such ways as to tend to continue to characteristically organize activity" (Ibid). We need to determine what is the "force" that maintains the organization over time. In this sense, the epistemology of complexity speaks of *self-organization*, given that organization is not only established once but continually rebuilt. However, this force cannot be—as Dewey acknowledges—an "original organizing force" (Ibid), which would introduce vitalistic hypothesis, whereas it is necessary to follow a "definite factual inquiry" (255). It is evident, for example, that there is a difference "between the animate and the inanimate plant iron molecule", but it "is not that the former has something in additions to physico-chemical energy" (253–254). There is undoubtedly an alternate means of restoring equilibrium in living bodies, including a "complex integrated course or history" (254). But what is it that allows the emergence of the organization? In 1929, Dewey posited that "we may not be able to answer these questions satisfactorily" (255). Dewey later returned to such concepts, fully developing them in *Knowing and the Known* (1949/1981) through the transition from an interactional kind of thought (for which the initial elements and the interactions amongst these are given) to a transactional one (commencing from the field of which the elements are part). There is in Dewey a "field conception" of reality. A system is a "subfield" in the more large field of reality. It is also equipped with a dynamic field that coordinates the parts of which it is composed. A system is a "whole that is more than the sum of its parts." Therefore, it is *one*, a whole comprised of parts. The parts cease to behave as single elements and become "parts of their whole" (Wertheimer, 1924). It is as if "every part 'knows' the stations and activities of every other part and 'responds' to any excursions and disturbances of the collective equilibrium as if it also 'knew' just precisely how best to maintain the integrity of the whole system in concert with the other constituents" (Weiss, 1969).

Dewey stresses this character of wholeness and uniqueness of the experience and also of the work of art, without omitting its articulation in parts. In fact, "that no whole is significant to us except as it is constituted by parts that are themselves significant apart from the whole to which they belong—that, in short, no significant community can exist save as it is composed of individuals who are significant" (Dewey, 1934/1980, 204). It is in the dynamic tension between the parts and the whole that a living system acquires meaning. It must respect the dual task of becoming increasingly differentiated and integrated. If an artist is "rigid" and fails to articulate the overall quality pervading the work of art in its parts—facilitating deep interaction with the medium—it remains undifferentiated, poorly developed, and thus not very meaningful and ultimately boring. This raises the question of how, like a living organism, an artwork can develop its impact with reality while remaining cohesive. In fact, "as an organism increases in complexity, the rhythms of struggle and consummation in its relation to its environment are varied and prolonged, and they come to include within themselves an endless variety of sub-rhythms. The designs of living are widened and enriched. Fulfilment is more massive and more subtly shaded" (23).

To integrate with the medium, a work of art must establish a path of growth leading to being that which it must become. Change and becoming are a key feature of complex systems, conceivable only as *processes*. For Dewey, a living organism is constantly expanding. It is not only the *photograph* of its organization; it is a continuous flow between *self-organized* dynamic (in which the organism is active and able to maintain its integrity) and *hetero-organized* dynamic (in which it must be subject to pressure and environmental constraints). The experience is, therefore, *self-hetero-organized* because it is distinguished by Dewey as an *action-and-passion* of the subject.

The relationship of mutual adaptation between the artist and the medium is also self-hetero-organized. In adapting to the medium, the artist forces it to adhere to his own purposes. The tension characterizing this dynamic, often being acute, can lead to intense states of disequilibrium, on the border between order and chaos, in which the system shows unusual evolution capabilities, crossing phase transitions, bifurcations, sensitivity to even remote contexts and small stresses. The relative predictability of behaviors a living system possessed when nearing a state of equilibrium gives way to an intrinsic aleatory in its evolution. In such situations, the enigmatic connection between free will and determination (Weiss, 1969), between chance and necessity comes into play.

In this regard, mention can be made to the third common characteristic of a living system and a work of art: the presence of a layering of multiple levels of quality, made possible by the *emergence* of new properties. When a living organism reconquers ordered participation in the environment, it does not just restore the prior balance but develops a new and more inclusive equilibrium. Dewey's is far from a mechanistic perspective in which the restoration is a mere return to the previous state. "These biological commonplaces are something more than that" (Dewey, 1934/1980, 14). Rather, there has been a continuous process of growth, marked by expansions of meaning and "intensive changes." The qualitative development introduces a directionality in the growing organism, a before and after. Even "time ceases to be either the endless and uniform flow or the succession of instantaneous points which some philosophers have asserted it to be," becoming "the organized and organizing medium of the rhythmic ebb and flow of expectant impulse, forward and retracted movement, resistance and suspense, with fulfilment and consummation" (23). What's more, these qualitative developments are included in the structure of the system, which thus acquires articulation, even if the parts remain connected to the whole. The artwork acquires expressiveness thanks to "a progressive massing of values, a cumulative effect" (137). Every living organism conserves "the import of what has gone before" (Ibid) and uses such to prepare for the next stages of growth, to "create suspense and anticipation of resolution". A work of art ultimately has a *story*, like that of a living creature adapting to the environment while maintaining its own identity, profoundly transforming in the interaction with the medium.

The concept of *living system*—as per biologist Weiss (1969, 1973)—is well suited to give the idea, even intuitively, of a work of art that is differentiated yet united, constantly in flow, comprised of several layers. Like an organism interacting with the environment, overcoming disparities, resulting in increments of quality integrated into the whole, the work of art is also conceivable as a living system, with its own history of qualitative evolutions, its increasing differentiation impacting with reality, its persistence contained within the identity. The work of art is not a mere surface but structured as a living system.

At this point, we must reflect upon the concepts of *emergentism* and *organicism*, with reference to aesthetic experience.

First, *emergentism*. Stating that Dewey adopts an emergentist perspective means identifying how he acknowledges that human experience derives from a bodily-biological level, excluding any form of *reductionism* between these dual planes. So, if it is true that "art is a continuation, by means of intelligent selection and arrangement, of natural tendencies of natural events" (Dewey, 1929, 389), it is also "the earth in one of its manifest operations" (Dewey, 1934/1980, 3), albeit not only material. In his writings, Dewey does not use the term "emergence," typical of the epistemology of complexity, but rather "continuity." There is a correlation between these two concepts. The Deweyan *principle of continuity* excludes clear breaks in nature and overthrows the dualism between the material and spiritual, the body and the mind, and so on. Similarly, the epistemology of complexity suggests that the differences amongst the various levels of reality (physical, biological, psychological, social, and so on) are attributable to the emergence of new properties. "The distinction between physical, psycho-physical, and mental is thus one of levels of increasing complexity and intimacy of interaction among natural events" (Dewey, 1929, 261). Thus, Dewey maintains that the positions of mechanism and materialistic metaphysics are wrong.

In view of the concepts espoused in the series *Analecta Husserliana*, it is interesting to note that the notion of continuity/emergence includes a creative and constructive perspective: emergence is effectively the creation of something new, not being the addition of extrinsic elements but the intrinsic development of new configurations and possibilities. Dewey does not need to suppose a separate Kingdom of Ends if the evolution of the experience indicates an *inherent* directionality towards refinement. Consequently, he refuses to sharply distinguish *ends* and *means*. Art is certainly imbued with purpose, which is not assigned externally but grows *from within* the subject matter, resulting in the progressive *emergence* of a new balances and meanings. To have an experience with an intrinsic purpose in a precarious world full of hardships—Dewey affirms—is a great gift in itself. It is a *final, consummatory* experience that allows to *intensify* the immediate living.

Dewey's *organicism* is evident especially in his belief that, as mentioned, the experience is a *whole*. When it is truly such, the experience is "one", having "unity." The experience evolves like an organism in which the parts grow, differentiate and interact with the environment whilst the organism remains as a whole. The essence of art is precisely to restore the wholeness of the experience, celebrating such with a sense of harmony. Experience, whole, organicity and aesthetic significance are almost synonymous in Dewey.

It should be remembered that in his early studies, Dewey expands upon organicism in connection with the Hegelian philosophy. In providing a view of reality as an extensive process in which the *true* emerges through the articulation of its many figures, the Hegelian perspective has surely had an aesthetic impact on Dewey.

Dewey laicizes the Hegelian organicism. No more the dialectic thesis-antithesis-synthesis, being replaced by an adaptation where there has been none between living creatures and the environment. Still, the concept of wholeness persists, even if applied to the *experience*. The idea of improving quality contained in the Hegelian dialectic—where synthesis means the overcoming of conflicts, whilst the latter continue to persist, taken out of isolation and positioned within a higher unity—persists in the concept that the experience is continuously expanding, achieving a more inclusive equilibrium where, at any time, there is a *cumulative* accomplished experience and, therefore, a progressive stratification of meanings. This maturing of the meanings is essential for the *expressiveness* of the artwork.

Hence, the real adaptation of a creature is not merely a repetition of the old but "the establishment of a relationship of integration" (Dewey, 1938, 42; 49), which requires the transition to new states of the environment and of the organism. Here, once again, comes the emergentism mentioned prior. Additionally, an organism that interacts with the environment grows in complexity, its organization including resolved problems, harmonized tensions and mediated contrasts "entering cumulatively and constitutively into the outcome" (Dewey, 1929, 368). Thus, there is a progressive growth of an *organization*, the articulating of a *system*, with an artwork springing from this complex development of a structure. Tensions evolve into a new dynamic whole. Consider a tree with a broad canopy. We can grasp its vibrant beauty, having successfully crossed the vicissitudes of time and weather. Rising up before us as an expression of continuity, strength and grace, it is a celebration of life.

Therefore, operating within the work of art as a living system is a harmonized set of tensions stratified on various levels and integrated into a whole.

3 Emergence

In exploring the concept of *emergence* (and that one of *emergent properties*) according to the epistemology of complexity and connected to Deweyan *continuity*, we discover the aesthetic implications and the connection with the topic at hand. If "we move" along the principle of continuity, on the one hand, we can see the artwork as deeply *embodied* and, on the other hand, how it allows to establish a new whole, enriching life.

The concept of emergence is fundamental in the epistemology of complexity as it refers to the ability of complex systems to move from a lower level of complexity to a higher one. Varela (2001) speaks in this context of "ontological levels." Dewey's principle of continuity has much in common with the idea of emergence. Considering the delicate and elusive character of the concept of emergence, come the opinions of two famous scientists: Anderson (Nobel Prize winner for physics in 1977, focused on complex phenomena) and von Bertalanffy (a German theoretical biologist, known for his research on systems and for his critical position towards an equivalence of systems with mere cybernetic mechanisms).

A very clear presentation of the concept of emergence is present in Anderson's famed paper, *More in Different* (1972), which is often referred to as a hallmark on complexity. Its first page contains a concise and effective introduction to the concept of emergence. Anderson shows his skepticism regarding *reductionism*, being the idea that there are some fundamental laws able to explain all phenomena. In his opinion, this hypothesis is accepted without question by most scientists. In his view, however:

> the behavior of large and complex aggregates of elementary particles [...] is not to be understood in terms of a simple extrapolation of the properties of a few particles. Instead, at each level of complexity, entirely new properties appear and the understanding of the new behaviors requires research which I think is as fundamental in its nature as any other (Ibid).

Consequently, "psychology is not applied biology, nor biology applied chemistry" because—and this is the important point concerning the emergence—"at each stage, entirely new laws, concepts and generalizations are necessary, requiring inspiration and creativity to just as great a degree as in the previous one" (Ibid).

Von Bertalanffy had expressed similar positions few years earlier:

> By and large, the notion of emergence appears to be correct: each level of the universe—atom, molecule, cell, organism, society, symbolic universes (with any number of interpolated levels)—has its characteristic properties and laws which cannot be simply derived from or reduced to those of the respectively lower level; in every plane there is also a graduation from lower to higher and we can understand it, not by reduction pure and simple, but by adequate expansion of our conceptual schemes. (1967, 33)

From this consideration, Bertalanffy suggests that "the distinction of animal and symbolic behavior does not mean that they are separated by an empty gap" (Ibid).

Both Anderson's and Bertalanffy's submissions are entirely consistent with Dewey's principle of continuity. As Alexander (1987) correctly points out in his significant text of 1987 regarding Deweyan aesthetics, the principle of continuity "is the key of Dewey's metaphysics" (xvii) and "it is crucial for an understanding of his aesthetic as well" (94). "Most critics, however, have ignored this facet of Dewey's philosophy and therefore they fail to understand some of the most fundamental features of what Dewey says" (Ibid).[1] For Alexander, "Dewey never fully developed his theory of continuity. But he constantly appeals to it at crucial moments in most of his writings" (Ibid). In fact, "continuity refers to increasing levels of organic functioning which exclude either the possibility of being reduced to one identical type or of being utterly disconnected into self-enclosed, autonomous categories" (Ibid). In this regard, Alexander (Ibid) remembers the next crucial step in *Logic, the Theory of Inquiry* (Dewey, 1938):

> The term "naturalistic" has many meanings. As it is here employed it means, on one side, that there is no breach of continuity between operations of inquiry and biological operations and physical operations. "Continuity", on the other side, means that rational operations grow out of organic activities, without being identical with that from which they emerge. [...] The primary postulate of a naturalistic theory of logic is continuity of the lower (less

[1] More recently, attempts to re-read Dewey's thoughts in the direction of the epistemology of complexity are being made (for example: Godfrey-Smith, 1996; Semetsky, 2008). El-Hania and Pihlströmb (2002) further the concept of emergence also with reference to pragmatism and, specifically, to Dewey's thought without drawing the final consequences from Deweyan emergent naturalism. In their view, "perhaps a notion of emergence based on non-reductive physicalism [...] does some interesting, pragmatically valuable work in certain specific fields, e.g., in the philosophy of biology—possibly in accounting for the relation between biological and physico-chemical properties. But when we move on to other, ontological regions, particularly the mental and cultural realms, we do not seem to have a sufficiently clear idea of how the program of non-reductive physicalism could be the carried through (with or without emergence)" (24–25).

> complex) and the higher (more complex) activities and forms. The idea of continuity is not self-explanatory. But its meaning excludes complete rupture on one side and mere repetition of identities on the other; it precludes reduction of the "higher" to the "lower" just as it precludes complete breaks and gaps. (18–19; 23)

Continuity/emergence has a strategic significance in Dewey's aesthetics and acts, so to speak, both *retrospectively* and *prospectively*.

Retrospectively, art continues to be connected with nature, to be *embodied*—to use a term in vogue today—in that, while emerging from nature, it is connected to the latter as a transformation. As Dewey writes in *Art as Experience*, "works of art [...] *are* the earth in one of its manifest operations" (1934/1980, 3). A work of art that loses its derivation from nature is doomed to sterility and mere formalism. In fact, "to grasp the sources of aesthetic experience it is, therefore, necessary to have recourse to animal life below the human scale" (18). In a Nietzschean manner, Dewey declares a "fidelity to the nature to which we belong" (1929, 420). This has no reductionist meaning, but rather obliges man to an elevation and a refinement of one's spirit; it "demands that we cherish our desires and ideals till we have converted them into intelligence, revised them in terms of the ways and means which nature makes possible" (Ibid).

Prospectively, then, the work of art includes bringing forth, integrating disparate elements into a whole. It is mainly thanks to *imagination* that this advancement may occur. Imagination is a central concept in Dewey's thoughts, also in his aesthetics. Dewey distinguishes *imagination* and *fantasy*. In the latter, the "mind stays aloof for the most part and toys with material rather than bodily grasping it" (Dewey, 1934/1980, 268) whilst the imagination is like a creative invention: it includes the ability to broaden our vision whilst remaining connected to reality. Since the higher level is not understandable from the lower level, both Anderson and Bertalanffy suggest in the passages cited above that to make the transition, "inspiration and creativity" and an "adequate expansion of our conceptual schemes" are necessary. For Dewey, it is the imagination that permits this creative leap, allowing us to see contexts not visible in the present. Without imagination, we cannot do or know anything! For example, to understand the other, a dose of imagination is required (Dewey, 1916, 6–7), whilst "not even a useful object is produced except by the intervention of imagination" (Dewey, 1934/1980, 273) and "the conscious adjustment of the new and the old *is* imagination" also (272). Imagination permits us "to move" forward along the principle of continuity, allowing us to sense the evolving of experience in direction of *emergent properties* not derivable from the characteristics of the parts—because they belong to a whole that only imagination can foresee. What's more, philosophy must cease "to be a device for dealing with the problems of philosophers" (Dewey, 1917, 65). It achieves its purpose if it has "faith in the power of intelligence to imagine a future which is the projection of what is desirable in the present" (69).

Ends, *ideals* and *harmonic configurations* are the outcomes of imagination that allow us to integrate vast spheres of experience. They are not something arbitrarily "added" by man, because "the addition is again the doing of nature and a further complication of its own domain" (Dewey, 1929, 422).

In the aesthetic field, imagination "is a *way* of seeing and feeling things as they compose an integral whole" (267), allowing us to overcome fragmentations. The integration of the various parts of a work of art in a whole can be achieved only imaginatively. The intellect fails fatally in this regard. "Coleridge said that every work of art must have about it something not *understood* to obtain its full effect" (Dewey, 1934/1980, 194). Imagination also supplies our ability to grasp the feeling of infinity in an aesthetic way. In fact, every genuine work of art is a *whole* that gives us the sense of

> belonging to the larger, all-inclusive, whole which is the universe in which we live. This fact, I think, is the explanation of that feeling of exquisite intelligibility and clarity we have in the presence of an object that is experienced with esthetic intensity. It explains also the religious feeling that accompanies intense aesthetic experience. We are, as it were, introduced to a world beyond this world which is nevertheless the deeper reality of the world in which we live in our ordinary experiences (195).

The integration of the experience into new wholes which imagination permits introduces man to broader and meaningful significance—to the point of dissolving the mysterious link between *beauty* and *truth*, such that Dewey repeats the concept expressed by Keats, "What imagination seizes as beauty must be truth" (34).

The rapture connected to aesthetic experience could be associated to our physical and mental wellbeing. This is what Jennifer Stellar's and Dacher Keltner's studies (involving hundreds of people) suggest. Indeed:

> Researchers have linked positive emotions—especially the awe we feel when touched by the beauty of nature, art and spirituality—with lower levels of pro-inflammatory cytokines, which are proteins that signal the immune system to work harder [...] That awe, wonder and beauty promote healthier levels of cytokines suggests that the things we do to experience these emotions—a walk in nature, losing oneself in music, beholding art—has a direct influence upon health and life expectancy. (Anwar, 2015)

Concerning the theme of emergence/continuity, focus should specifically turn to exploring the topic of the conference on post-humanism. The question is – since Dewey's *Principle of Continuity* puts man in nature once more, promoting a more "ecological" perspective – is this an argument in favor of posthumanism? In fact, despite the diversity of the themes running throughout, the latter offers a critique of humanism, in particular of anthropocentrism. Posthumanism asks why man is given a privileged status if there is a *continuity* that ties him to the animal? If man is *natural* like the animal, in what sense ask we ourselves about his *essence*? Is man not comprised of *otherness*, as with all that exists? Firstly, we must remember a central aspect of Deweyan continuationism: its antireductionism. The Deweyan principle of continuity seeks to decommission concepts, ideas and positions that derive their existence from a supposed ideal sphere of values separated from nature but it is also a way to conceive the values as emergencies from the bosom of nature. As we noted above, the principle of continuity acts both retrospectively and prospectively. It reconnects man with nature, but then it sees nature as full of virtuosity and emergences: creativity is itself a part of nature. In Chap. 10 of *Experience and Nature*, Dewey deals with *existence and value*, criticizing the modern philosophy distinction between "objective" and "subjective". In this light, the desire, the belief, the choice would belong to the subjective dimension, conceived as something isolated from nature, as a foreign body. The conception of the world as "mechanical or as rational in structure" (Dewey, 1929, 424) is the basis of the assumption that the subjective is not part of the objective. In this case, there is a reification of value, considered as a mere epiphenomenon of a mechanistically-conceived world. Such a position is not shared by Dewey. For him, "it is as irritating to have experience of beauty and moral goodness reduced to groundless whims as to have that of truth" (426), while "these [experiences] are as realistic, as 'objectively' natural, as are the constituents of the object of cognitional experience" (424). A final and sophisticated attempt made by Dewey to account for this *unity in diversity* of nature and its emergencies is to "get rid" of the concept of reality giving value to that of *immediate experience*[2]. The physical and biological mechanisms are "natural", as are man, his dreams and his hopes. Rather than reducing these aspects to a supposed underlying material reality, Dewey adopts a "thorough-going naïve realism" (Dewey, 1917, 55) that legitimates all that is human. "Dreams, hallucinations etc. […] are not something outside to the regular course of events; They are *in* it and *of* it. They are not cognitive distortions of real things. They are *more* real things" (38). Furthering this is the illuminating passage from *The Need for a Recovery of Philosophy* (1917):

> It is often said that pragmatism, unless it is content to be a contribution to mere methodology, must develop a theory of Reality. But the chief characteristic trait of the pragmatic notion of reality is precisely that no theory of Reality in general, *überhaupt*, it is possible or needed. It occupies the position of an emancipated empiricism or thoroughgoing naive realism. It finds that 'reality' is a *denotative* term, a word used to designate indifferently everything that happens. Lies, dreams, insanities, deception, myths, theories, are all of them exactly the events which they specifically are (55).

Referring to the arguments of post-humanism, the abolition of the "ontological" diversity between man and animal does not imply their identity and the elimination of values that are specifically human.

Continuationism is not *identity*, nor *reductionism*.

4 Organicism

The concept of emergence illustrated throughout the previous paragraph pertains to that of *system*, in that emergence is not a characteristic of the individual elements but of the set of elements, of wholes, of systems. For the epistemology of complexity, a system is said to be *complex* when the relationship between the components creates a "collective behavior" that cannot be reduced to the sum of the behaviors of

[2] Even if "a clear understanding of the concept of experience is undoubtedly a more difficult theoretical task for those who turn to the study of Deweyan philosophy" (Calcaterra, 2011, 118, transl. my own), adopting the organismic approach, we can say that the experience, as a whole, *is* its meaning, in an almost phenomenological sense, without further additions. In this regard, please refer to the Sect. 8 in this paper.

the individual parts. The *slogan* of the epistemology of complexity is, therefore, that "the whole is *greater* than the sum of its parts." This *greater* that arising when passing from the dynamic of the parts to the dynamic of whole is an *emergence* of the system assumed as a whole. The parts of a system, whilst maintaining their autonomy, affect the overall organization of the system, becoming *part of their whole* (Wertheimer, 1924). The complex system is self-organized as the result of a dynamic that is intrinsic to the system itself and which cannot be exhaustively inferred by general laws, as suggested by Anderson in the aforementioned article. With reference to living systems, this feature can be defined as *agency* (Kauffman, 2008). The complex system, being self-organized, is not merely passive with respect to the environment. On one hand, it suffers the environmental constraints and pressures but, on the other, as mentioned, it is active and able to "process" the environmental stress basing on its organization. It is *self-hetero-organized*. In this way, the complex approach hypothesizes a universe of *agents* interacting with each other, acting and subjected simultaneously, generating new collective properties and including itself into new wholes they become parts of, in a continuous stratification of auto-hetero-organized levels.

What's more, such systems are not static but constantly changing: they are *processes*, inconceivable outside of their becoming. Such becoming is not linear but is characterized by *leaps*, *phase transitions*, *choices* and *bifurcations*. These occur especially when the system is *far from being in equilibrium* (Prigogine & Stengers, 1979/1984) or *between order and chaos*.

For the epistemology of complexity, therefore, the parts comprising the universe should not be conceived as simple and isolated substances but rather as systems. Even organicism commences, in the analysis of a phenomenon, not with the properties of the individual elements, but the phenomenon as a whole. Organicism can be considered one of the most significant sources of complexity. It concerns the best philosophical, artistic and biological debate since the late-eighteenth century. Consider, for example, Blumenbach's concept of *formative drive* (*Bildungstrieb*) (1781/1971), which sought to give an account of self-learning processes that characterize living organisms, overcoming the alternative between *mechanisms* and *vitalism* and anticipating a complex concept such as *self-organization*. Initially, Dewey approached organicism from a powerful Hegelian perspective, resulting in the idea of a reality that, in its becoming, is infinitely complex and infinitely unitary.[3]

[3] In *From Absolutism to Experimentalism*, Dewey (1930) recognizes that "Hegel has left a permanent deposit in my thinking" (21). We have to recall that, even before knowing the Hegelian philosophy, Dewey had come into contact with the idea of organicism during the physiology course held by T. Huxley. Dewey recalls that meeting as follows: "I have an impression that there was derived from that study a sense of interdependence and interrelated unity that gave form to intellectual stirrings that had been previously inchoate and created a kind of type or model of a view of things to which material in any field ought to conform. Subconsciously, at least, I was led to desire a world and a life that would have the same properties as had the human organism in the picture of it derived from study of Huxley's treatment" (13).

Dewey's work can be seen as an attempt to revisit this concept framing it in a more "scientific" perspective.[4]

Organicism is the resolution to the conflict between *vitalism* and *mechanism* and, more recently, between *reductionism* and *holism*. Such antithetical pairings are "two enemy brothers" (Stengers, 1985, 51). In fact, even holism is a reductionism, albeit inverted since it assumes that the phenomena should evolve towards final structures. With that, it introduces an *automatism*, a *teleology*, a *tendency* of the phenomena towards preconceived solutions,[5] whilst dealing with a phenomenon that is so organic/complex means, in its most virtuous manner, revealing the phenomenon's

[4] For Calcaterra (2011), "if you are willing to accept a certain degree of exaggeration and provocation in the historical work, you could describe the whole evolution of Deweyan thought as an attempt to make the theoretical content of this fundamental youth intuition more and more pervasive and logically compelling" (14, transl. my own). In *Knowing and the Known*, this attempt succeeded. In this book, Dewey comes to develop the concept of *transaction*, with which he ultimately takes the distances from any "ontological reference"—such as "'elements' or other presumptively detachable or independent 'entities', 'essences', or 'realities'" (1949/1991, 101–102), to adopt an organismic logic that examines the dynamics of the whole situation. What's more, in that book Dewey's thought is projected in the direction of those thinkers—from J.C. Maxwell to K. Goldstein, from J.V. Uexkull to the Gestalt psychologists—who were working to develop an organismic (or field) perspective within their respective areas of competence. For example, complaining that, "The new foundation that has been given physics on a transactional basis [...] has not yet been made complete" (107), Dewey comments—citing Maxwell—that "the properties of the field alone appear to be essential for the description of phenomena" and that "[t]he electromagnetic field is, in Maxwell's theory, something real" (108). Likewise, in physiological and biological sciences "against the vitalisms [...] Views of the type called 'organismic', 'organismal' etc., except where they contain reminiscences of the old self-actional forms, stand for the transactional approach intra-dermally" (116–117). This organismic-transactional approach has the "desire to allot the leading adjective rather to the full living procedure of the organism than to minor specialized processes within it" (117).

[5] Dewey knows full well the risk of reductionism inherent in concepts such as *emergence* or statements such as "the spirit and life emerge" (Dewey & Bentley, 1949/1981, 155). It is not simply, as El-Hania and Pihlströmche evidences, that "Dewey did not reject the idea [of emergence] altogether but only what he saw as its magical overtones. He simply required scientific research on the emergence of life and mind", rather he is entirely aware of the risk that a concept such as emergence "still retains an independence and an unnatural isolation of the old type" (Dewey & Bentley, 1949/1981, 155). In this way, emerge results that are *extrinsic* and not *inherent* to the dynamics of the phenomena, incorporating dualisms or containing the remnants of old vitalistic or mechanistic concepts. The new approach is to dispense of any "must be" of the phenomenon, looking at it as a pure relational-transactional becoming. It is no coincidence that, in this regard, Dewey mentions the concept of "field". The direction that Dewey suggests is to make a more careful analysis of the phenomena: "the conception transactional emergency [...] will in the descriptions enriched primary processes of life in their environments and more complex behavioral processes in them". Reductionism implies to stop simplified explanations. To invoke the respect of the phenomenon as a whole, in fact, does not hint at a rejection of the analytic approach, as is sometimes argued, as the true "complex" attitude is to compare oneself with the infinite complexity of reality, renouncing the temptation to introduce extraneous explanations. As Stengers (1985, 50) observes, whilst reductionism says "nothing but ...", the analytical method "could lead to 'this'..., but in other circumstances, 'that', and then to 'that yet'" (transl. my own). Therefore, at every step, upon each discovery, others lie in wait.

sensitivity in the totality of its dynamics and, positioned as such, grasping the possible *evolutionary leaps* and the *emergence of new whole.*[6]

Organicism is a way of looking at the world in a way that is not *linearly* deductible from an interaction of simple parts (reductionism/elementism), nor from final moments (holism/finalism) but as organisms that are always processual in the making, living by virtue of an intrinsic dynamism, capable of joining other wholes (of which they become members or sub-systems, not mere parts mechanically connected to the whole).

This "mysterious" transformation of *parts* into a *whole* characterizes the work of art. From here, arises the need to assess the fallout from the aesthetic organicism.

5 The Work of Art: A Unity in Diversity

In his conception of aesthetics, Hegel compares artistic beauty to an organism. An organism is, in fact, differentiated but unitarian, as its components—even whereby appearing as mutually random and distinct—are part of the organism unity. The purpose of art is to have us seize this "hidden unity to sensitive intuition" (1823/2000, 55, transl. my own). Dewey also endorses "the old formula for the beauty in nature and art: unity in variety" (Dewey, 1934/1980, 161), specifying from an organismic perspective that the latter loses value whereby taken only in an exterior manner, as if there are "many coins in one pocket, and many documents in a safe" (Ibid). Avoiding any associational conception, the work of art must be understood as a system of relationships between the components. This systemic organization is not static but rather dynamic, comprised of harmonized tensions: "the formula [of unity in diversity] has meaning only when its terms are understood to concern a relation of energies […] the unity in variety that characterizes a work of art is dynamic. […] There is unity only when the resistances create a suspense that is resolved through cooperative interaction of opposed energies" (Ibid).

As Hegel affirms, the exteriority of parts in the work of art refers to a *unity*, hence for Dewey it is pervaded from the beginning by a "total and massive quality [that] has its uniqueness" (Dewey, 1934/1980, 192). In fact, an organism is a *unity* that integrates the parts into a whole. For Weiss (1969), within a living system "the patterned structure of the dynamics of the system as a whole 'co-ordinates' the activities of the constituents." It is the *identity* of the organism/system that should be considered in itself, without reducing it to its constituent parts. In Dewey's words, without this "one pervasive quality that remains the same quality in being differentiated […] parts are external to one another and mechanically related" (Dewey, 1934/1980, 192–193).

For Dewey—as we shall see further on in discussing the *expressiveness* of the work of art—this "penetrating quality that runs through all the parts of a work of art and binds them into an individualized whole can only be emotionally 'intuited'"

[6] To this regard, cf. notes no. 4 and 5.

(192), "it can only be felt, that is, immediately experienced" (Ibid). The ability to sense qualities in their individuality has been criticized. For example, as Alexander recalls in his book, Garrett seizes Dewey's error in making qualities ineffable, unique, unrelated and beyond causal relationships, for which we must "jettison one half of Dewey's theory of quality" for its ambiguous, idealistic, contradictory character. Instead, this is a central point for understanding how the work of art can be expressive and *directly* communicate "meanings", by virtue of its *dynamic identity*. An organism *is* its meaning. As Goethe states:

> The highest thing would be to comprehend that everything factual is already theory. The blue of the heavens reveals to us the fundamental law of chromatics. One should only not seek anything behind the phenomena: they themselves are the theory. (quoted in Zajonc, 1987, 223)

Returning to the work of art as an organism/complex system, articulated yet unified, for Dewey the intuition of the *overall quality* precedes the articulation in its parts. It is as if the work was born from a sudden *transition phase* from which a global quality emerges, perfecting itself through the progressive articulation of the overall quality of its parts. In this regard, Dewey turns to the words of Schiller: "What precedes is a peculiar musical mood of mind. Afterwards comes the poetical *idea*" (Dewey, 1934/1980, 191–192).

This peculiar *musical mood of mind* may be compared to a *resonance* establishing relations among the parts, binding them to a whole. It is the latter that should be determined first, rather than the individual elements. Looking for similarities between individual aspects of the work with that already known means behaving in an academic, scholarly or pedantic manner (108). Both the artist and the perceiver should "begin with what may be called a total seizure, an inclusive qualitative whole not yet articulated, not distinguished into members" (191). In addition, "not only does the 'mood' [the peculiar musical temperament of the mind] come first, but it persists as the substratum after distinctions emerge; in fact, they emerge as its distinctions! But not only comes first, but which persists after substrate distinctions that have emerged; in fact, they emerge as *its* distinctions" (192).

6 The Aesthetic Experience as a Whole

Dewey's unit of analysis is, ecologically, the *field* of which the subject and materials are part.[7] Art does not regard certain items ("temple, painting, statue, poem") or only the artist with his subjectivity but rather the *whole* experience that springs from the relationship between these two poles. Therefore, art cannot be "designated by a noun substantive", if not outwardly (214). It does not denote objects but it is a quality of the *process* leading the subject and the object to adapt to one another. There is

[7] Cf. note no. 5.

art, therefore, "*in* the conduct" of certain activities (214). In this respect, when we like the work of a specific artist, we affirm that we like his *style*, his way of working.

As for Dewey, in an organismic sense, the entire context includes the living creature and materials, not as separate poles that interact. This context can be represented in a systemic way by declaring that the living creature is a sub-system within the environmental system.

It is the entire "man-environment" system that is evolving.

Two interesting consequences deriving from this.

Firstly, one consequence of this approach is formulable through the following question: *as per the Deweyan organicism, the aesthetic experience concerns the transaction between the subject and the object, so must you assume that each field variation affects all its constituent parts—hence, not only the subject, but also the objects, materials, things?*

The answer to this question must be to the positive. In fact, as we outlined, the relationship between the parts and the whole of a complex system is such that the "system as a whole 'co-ordinates' the activities of the constituents" (Weiss, 1969). Since the work of art is the result of an experience in which the subject and the object become parts of a new whole, these parts—both the "external" (the medium) and the "internal" ("images, observation, memories and emotions") (Dewey, 1934/1980, 94–95)—are "progressively re-formed" (74); there is a "mutual adaptation of parts to one another in constituting a whole" (134).

A work of art (as with knowledge or science) "confers upon things traits and potentialities which did not previously belong to them" (Dewey, 1929, 381).[8] Dewey claims that this would not be the case in a reductionist conception of physical reality—in which objects are isolated. From a systemic-organismic perspective, this conception is overcome from of a relational perspective, where the parts are always *potential* components of a more inclusive whole. For example, "architecture does not add to stone and wood something which does not belong to them, but it does add to them properties and efficacies which they did not possess in their earlier state. It adds them by means of engaging them in new modes of interaction, having a new order of consequences" (381–382). Prigogine proposes the same example, criticizing the reductionist assumptions of classical science that:

> at some level *the world is simple* and is governed by time-reversible fundamental laws. Today, this appears as an excessive simplification. We may compare it to reducing buildings to piles of bricks. Yet out of the same bricks, we may construct a factory, a palace, or a cathedral. It is on the level of the building as a whole that we apprehend it as a creature of time, as a product of culture, a society, a style (Prigogine & Stengers, 1979/1984, 7).

We need to abandon prejudice in differentiating "subject" and "object", because

> the uniquely distinguishing feature of aesthetic experience is exactly the fact that no such distinction of self and object exists in it, since it is aesthetic in the degree in which organism

[8] "Through art, meanings of objects that are otherwise dumb, inchoate, restricted and resisted are clarified and concentrated, and not by thought working laboriously upon them, nor by escape into a world of mere sense, but by creation of a new experience" (Dewey, 1934/1980, 132–133).

> and environment cooperate to institute an experience in which the two are so fully integrated that each disappears (Dewey, 1934/1980, 249).

In another example, Dewey asks about the sense of reconciliation that tragedy is capable of achieving, reason being that painful content "has entered into a new whole as an integral part of it. In its new relationships, it acquires a new expression. It becomes a qualitative part of a new qualitative design" (113). In this context, wherein the object acquires new characteristics, Dewey reintroduces the controversial concept of "tendency", to be understood not as "prior design" but as "movement in a particular direction, a direction that may be either furthered or counteracted and frustrated, but which is intrinsic" (Dewey, 1929, 373). Similarly, he uses the idea of "energy" to argue that "esthetic effect is due to art's unique transcript of the energy of the things of the world" (Dewey, 1934/1980, 185). These concepts have to be taken in their organismic valence: if we must set out not from the objects taken in isolation, hence interacting and in their dynamical relationship, the objects will always be at the intersection of forces and tensions, manifesting tendencies and energies to evolve in certain directions. Gestalt psychologist Köhler (1938/1939) uses the term *requiredness* to indicate that the intrinsic tendencies of things necessitate development in a certain direction whereby within a certain context.

7 The "Will" of Work of Art

A further consequence of Dewey's organismic point of view is that the "protagonist" of the aesthetic experience is the entire field of experience, not the subject with his own choices and preferences. In fact, the subject is merely a pole of the entire field of experience. Surely, he is active and has affections and thoughts but, when he wants to act aesthetically, he must recognize being part of the experiential field. Aesthetically, he "exists" as a part of the field. It is impossible to avoid the object. The artist "sees and hears in terms of their medium" (Dewey, 1934/1980, 202). Indeed, "The sensitivity to a medium as such is the very heart of artistic creation and of every aesthetic perception" (Ibid).

To act aesthetically means serving the overall situation and promoting its developments. Indeed, since the material components of the situation are unable to think, man—when behaving aesthetically—"is the nature itself that thinks", as Pirandello writes (1990, 402, transl. my own). In fact, man's perceptions, his ideas, his actions, his enjoyment and suffering as a consequence of his own actions are not "added" to nature; "the addition is again the doing of nature and a further complication of its own domain. […] [it] is human, the course which manifests the course of nature" (Dewey, 1929, 422). Human activity allows new forms of emergence. For this reason, Whitehead upholds that "art is the education of nature" (1933, 271).

Hence, artistic creation does not come from the "head" of the artist but from the transaction between subject and object, from the evolution of the overall situation. He does not impose an order to the situation, as if he were external to it, as:

> Order is not imposed from without but is made out of the relations of harmonious interactions that energies bear to one another. Because it is active (not anything static because foreign to what goes on) order itself develops. (Dewey, 1934/1980, 14)

It is not the artist's will to act. Rather, the true artist is able to promote the development of the whole aesthetic situation of which he is part. Pirandello is convinced that "the aesthetic fact [...] begins only when a representation acquires a will within us 'for itself', that is, when it 'wants itself' in and for itself" (Pirandello, 1931/1977, 400, transl. my own). He thus imagines entertaining "speeches with the characters" who come knocking at his door claiming to be represented; he is forced to write, sometimes reluctantly, since their stories are sad. But "born alive, they wished to live" (Pirandello, 1925/1997, 654, transl. my own).

Certainly, the author plays an active role and is able to promote undisclosed developments of the situation thanks to his reflective capacity and imagination. But he cannot "break away" from this and add external elements. In fact, when we feel that "the author, not the subject matter, is the arbiter" or is "calculated intent", then the work disturbs us (Dewey, 1934/1980, 68; 199–200), because we feel that a genuine work of art must come from the intrinsic developments of the content dealt with. Thus, "A work may be much more tragic and yet leave us with an emotion of fulfilment instead of irritation. We are reconciled to the conclusion because we feel it is inherent in the movement of the subject matter portrayed" (68). The artist has to be "at the service" of the artwork. An artist is truly original—Hegel says—when he "represents nothing of his particularity but represents only the Thing" (1823/2000, 113, transl. my own). The genuine artist does not *impose* his intentions to the work of art but promotes its intrinsic development, its "will."

Dewey praises Keats' "Negative Capability", that is the ability to be "capable of being in uncertainties, mysteries, doubts, without any irritable reaching after fact and reason", and therefore not to introduce "rational" or extrinsic elements but to merely trust the imagination and the intrinsic developments (Dewey, 1934/1980, 33). He also recognizes that a central part of the artist is sincerity, that is "the necessity that he shall not fake and compromise" (189). Sincerity, in fact, could be compared to an objective sense, closely attuned to the developments of the situation. It does not rule out an interest—given that nothing can be achieved without an interest (193). Hence, only the sincere artist puts his own interests at the service of the Object not introducing extrinsic elements but promoting the inherent developments of the content, the "will" of the artwork.

8 Expressiveness

One interesting consequence of the organismic/systemic perspective is that the meaning of a system should not be attributed to it externally but it is intrinsic to its self-dynamic nature that binds the parts into a whole that is greater than the sum of its parts. Each system has its own dynamic identity. The system is its meaning, one could say. Additionally, the developments of a system *become meaningful* due to these developments not juxtaposing the system but they being integrated within the system, maintaining its integrity and even changing. The system mediations are integrated into the system and become immediately perceptible along with the immediate quality.

Dewey insists on the possibility of *immediately* experiencing qualities ("quality is quality, direct, immediate and undefinable"—Dewey, 1929, 110), or values ("Values are values, things immediately having certain intrinsic qualities"—396), or the other experiences of life. Dewey also admits, in William James' wake, that mediated qualities can also be immediately perceived, given that:

> only a twisted and aborted logic can hold that because something is mediated, it cannot, therefore, be immediately experienced. We cannot grasp any idea, any organ of mediation, we cannot possess it in its full force, until we have felt and sensed it, as much so as if it were an odor or a color. (Dewey, 1934/1980, 119)

This position taken by Dewey was not often understood.

In addition, the work of art immediately communicates qualities. In this sense "it cannot be asserted too strongly that what is not immediate is not aesthetic" (Ibid) and also: "the opposition of quality as immediate and sensuous to relation as purely mediate and intellectual is false in general theory, psychological and philosophical. In fine art it is absurd, since the force of an art product depends upon complete interpenetration of the two" (121). In another organismic thought, Gestalt asserts the ability to directly perceive expressive qualities. For example, a famous quote by Wertheimer states that "black is lugubrious even before being black." In Gestalt psychology, even the expressiveness of faces or feelings such as sadness and joy are directly perceptible.

9 Conclusions

Despite many topics being excluded from this text (the chief one being how a work of art, as a process, is concretely a temporal structure taking shape among bifurcations, emergencies, reorganizations, etcetera), the intention was to have highlighted the way in which a work of art is a living system. As an organism that interacts with the environment, an artwork develops in its journey of growth in reality, from a cumulation of choices, problems solved, new and more inclusive equilibriums achieved, in its internal organization, creatively balancing the forces to which it is subjected. It has a *story*; it *is* a story. In this sense, a work of art is a living system: it has a structure that is perfected in its growth, interacting with reality. Like a living system, it has its own self-dynamism. This is not exercised in a vacuum but inevitably in contact with reality. Hence, in this interaction with reality, the aim of the artist is to ensure that the overall quality that characterizes the work from the outset may differentiate in the interaction with reality, so long as he fails to fully embody his own creatures, remaining faithful to the overall initial quality but articulating such through all the possibilities of the medium.

From a global perspective, the work of art is not, however, just a product of the author but of the whole "author-medium *field*." The artist, as part of the field, must be of service to the reality to which he belongs in order to restore the harmony of all the elements present therein. This restoration of harmony could be the pinnacle of nature. Indeed, for Dewey (1929, 358), art "is the complete culmination of nature."

References

Alexander, T. M. (1987). *John Dewey's theory of art, experience, and nature: The horizons of feeling*. Albany: State University of New York Press.

Anderson, P. W. (1972). More is different. *Science, 177*(4047), 393–396.

Anwar, Y. (2015, February 2). Add nature, art and religion to life's best anti-inflammatories. *Berkeley News*. Retrieved August 20, 2015, from http://newscenter.berkeley.edu/2015/02/02/anti-inflammatory/.

Blumenbach, J. F. (1781/1971). *Über den Bildungstrieb und das Zeugungsgeschäfte*. Stuttgard: Fischer.

Calcaterra, R. M. (2011). *Idee concrete. Percorsi nella filosofia di John Dewey*. Genova-Milano: Marietti.

Dewey, J. (1916). *Democracy and education. An introduction to the philosophy of education*. The Macmillan Company: New York.

Dewey, J. (1917). The need for a recovery of philosophy. In *Creative intelligence: Essays in the pragmatic attitude* (pp. 3–69). New York: Henry Holt.

Dewey, J. (1929). *Experience and nature* (2nd ed.). London: George Allen & Unwin.

Dewey, John. 1930. "From absolutism to experimentalism." In Contemporary American philosophy: Personal statements, George P. Adams and Wm. Pepperell Montague. New York: Russell and Russell: 13–27.

Dewey, J. (1934/1980). *Art as experience*. New York: Perigee Books.

Dewey, J. (1938). *Logic: The theory of inquiry*. New York: Henry Holt.

Dewey, J., & Bentley, A. F. (1949/1991). Knowing and the known. In J. A. Boydston (Ed.), *The later works, 1925–1952* (Vol. 16). Carbondale, IL: Southern Illinois University Press.

El-Hani, C. N., & Pihlström, S. (2002). Emergence theories and pragmatic realism. *Essays in Philosophy, 3*(2), 3.

Godfrey-Smith, P. (1996). *Complexity and the function of mind in nature*. Cambridge: Cambridge University Press.

Hegel, G. W. F. (1823/2000) *Lezioni di estetica (The Heinrich Gustav Hotho transcript of the 1823 Berlin Lectures)* (P. D'Angelo, Trans.). Roma-Bari: Laterza.

Kauffman, S. (2008). *Reinventing the sacred: A new view of science, reason, and religion*. New York: Basic Books.

Köhler, W. (1938/1939). *The place of value in a world of facts*. London: Kegan Paul, Trench, Trubner.

Pareyson, L. (1966). La contemplazione della forma. In *Conversazioni di estetica* (pp. 20–24). Milano: Mursia.

Pirandello, L. (1910/1990). Leviamoci questo pensiero. In G. Macchia (Ed.), *Novelle per un anno* (I Meridiani) (Vol. 3, pp. 394–405). Milan: Mondadori.

Pirandello, L. (1925/1997). Prefazione a Sei personaggi in cerca d'autore. In A. D'Amico (Ed.), *Maschere nude* (I Meridiani) (Vol. 2). Milano: Mondadori.

Pirandello, L. (1931/1977). Discorso alla Reale Accademia d'Italia. In M. L. V. Musti (Ed.), *Saggi, poesie, scritti varii* (4th ed., pp. 391–406). Milan: Mondadori.

Prigogine, I., & Stengers, I. (1979/1984). *Order out of chaos. Man's new dialogue with nature*. New York: Bantam Books.

Semetsky, I. (2008). Re-reading Dewey through the lens of complexity science, or: On the creative logic of education. In M. Mason (Ed.), *Complexity theory and the philosophy of education* (pp. 79–90). Chichester: Wiley-Blackwell.

Stengers, I. (1985). Perché non può esserci un paradigma della complessità. In G. Bocchi & M. Ceruti (Eds.), *La sfida della complessità* (pp. 37–59). Milano: Feltrinelli.

Varela, F. (2001, January 29). *La coscienza nelle neuroscienze*. Conversation with Sergio Benvenuto. Paris. Retrieved July 10, 2015, from http://www.psychomedia.it/pm/science/psybyo/varela.htm.

von Bertalanffy, L. (1967). *Robots, men and minds. Psychology in the modern world*. New York: George Brallizer.

Weiss, P. A. (1969). The living system: Determinism stratified. In A. Koestler & J. R. Smythies (Eds.), *Beyond reductionism: New perspectives in the life sciences* (pp. 3–55). London: Hutchinson.

Weiss, P. A. (1973). *The science of life: The living system. A system for living*. Mount Kisco, NY: Futura Publishing.

Wertheimer, M. (1924). Untersuchungen zur Lehre von der Gestalt, II. *Psychologishe Forschung, IV*, 301–350.

Whitehead, A. N. (1933/1962). *Adventures of ideas*. New York: The Free Press.

Zajonc, A. G. (1987). Facts as theory: Aspect of Goethe's philosophy of science. In F. Amrine, F. J. Zucker, & H. Wheeler (Eds.), *Goethe and the sciences: A reappraisal*. Dordrecht: Reidel.

Part V
Human/Beast/Object

The Beast vs. Human

John Nelson Balsavich

The idea of being the last man or woman in existence is a significant theme of post-apocalyptic literature. This condition of lastness, or solitude, is ultimately due to the absence of other human beings as a result of the breakdown of human civilization. In other words, post-apocalyptic literature's penchant for the world's complete and utter destruction reinforces the notion of solitude for the last person who must then make sense of his or her solitary situation in order to survive the new dangers that arise due to the absence of humankind. Furthermore, under these solitary conditions, the last man or woman must also contend with the emergence of the animal or beast as the new dominant species. For this paper, I deal with three novels that focus on the theme of being the last person on earth in relationship to the animal or beast that now dominates the landscape. Mary Shelley's *The Last Man* (1882 Reprint 1994), Margaret Atwood's *The Year of the Flood* (2009), and Cormac McCarthy's *The Road* (2006) all explore the transition from the dominance of humans to that of the beast. *The Last Man* places emphasis on the fall of the human race due to a plague and the emergence of the animal as the new dominant species. In *The Year of the Flood*, the character Toby spends the majority of her time on a rooftop spa while also dealing with a new kind of animal that has been brought into existence by human experimentation. These new creatures, the Pigoons, become a primary threat to her existence due to their increased intelligence. Finally, although animals are mostly extinct in *The Road*, large segments of the human race engage in cannibalistic activity, thus losing their humanity and devolving into something bestial. It is these cannibals that the nameless father and son must contend with as they make their way through the devastated world. Overall, the paper examines how the absence of human civilization results in a new paradigm of the beast or nonhuman fulfilling the previous role of dominant species

J. N. Balsavich (✉)
Bridgewater State University, Bridgewater, Massachusetts, USA
e-mail: jbalsavich86@gmail.com

C. A. Hornbuckle et al. (eds.), *Phenomenology of the Object and Human Positioning*, Analecta Husserliana 122, https://doi.org/10.1007/978-3-030-66437-4_14

once occupied by human beings. It is this new paradigm that the last person in each novel must deal with in order to define his or her humanity.

Jacques Derrida's examination of man's relationship with the animal is the primary focus in *The Animal That Therefore I Am* and The *Beast and the Sovereign, Volume II*. Both of these works are crucial to the investigation concerning the theme of the beast found in each of the three novels. Derrida's preoccupation with the animal concerns the notion of how animals were traditionally treated by philosophical thought. In *The Animal That Therefore I Am*, Derrida explains, "Animal is a word that men have given themselves the right to give…They have given themselves the word in order to corral a large number of living beasts within a single concept" (Derrida, 2008, 32). It is not just the fact that humans have given themselves power over the animal, it is also what the animal lacks, or what humans have traditionally denied the animal. Derrida points out that the animal has been denied access to language, death, and finally an ethical treatment. He challenges these notions asserting that, "It is not just a matter of asking whether one has the right to refuse the animal such and such a power…It also means asking whether what calls itself human has the right to rigorously attribute to man, which means therefore to attribute to himself, what he refuses the animal, and whether he can ever possess the pure, rigorous, indivisible concept, as such, of that attribution" (Derrida, 2008, 135). In other words, Derrida questions the traditional idea of how humans have defined themselves through their relationship with the animal. In the three novels, the last person must realize or make sense of his or her humanity in the context of this new dichotomy that favors the primacy of the animal or the beast over that of the human. The essay explores the last person's relationship with the beasts through the prism of language, an attempt to hold on to past social and cultural values that have vanished with the absent human civilization, and finally ethical matters.

In terms of language, Derrida argues that "the animal is deprived of language. Or, more precisely, of response, of a response that could be precisely and rigorously distinguished from a reaction; of the right and power to 'respond' and hence of so many other things that would be proper to man" (Derrida, 2008, 32). In being denied access to language, the animal is denied the right to respond. While it can react to external stimuli, ultimately, the animal is unable to respond and articulate its sufferings because of a perceived lack of language. The ability to accomplish this task, instead, belongs to mankind who see themselves as possessing the gift of language. Human access to language, which is denied to the animal, underscores the human and animal dichotomy as the placement of humans above animals.

Near the end of Lionel's story, the animal becomes the dominant species, while mankind no longer exists. In one scene, after losing his only remaining companions, and thus becoming the last man, Lionel enters the abandoned town of Ravenna. While walking through the deserted town, Lionel remarks: "I saw many living creatures; oxen and horses, and dogs, but there was no man among them…. I stepped softly, not to awaken the sleeping town. I rebuked a dog, that by yelling disturbed the sacred stillness…. The world was not dead, but I was mad…. I was labouring under the force of a spell, which permitted me to hold all sights of earth, except its human inhabitants" (Shelly, 1994, 449). With this remark, Lionel understands that now there are no other humans left except for himself. Instead, animals have claimed the space that mankind has left behind. It is also important to notice that Lionel gives this town a sacred quality and remains silent out of respect for the departed race of mankind. He even rebukes a dog for disturbing the silence. Moreover, the dog does not bark but rather yells, suggesting a human characteristic. The yelling of the dog illustrates the idea that language, which was traditionally denied to the beast, is here given to the animal by Lionel in the absence of other people. There is thus an emerging understanding on Lionel's part that the world is not dead, but rather it is only the human race that is gone. In the absence of the human race, the animal reigns supreme.

The issue of language also plays an important role in *The Year of the Flood* in which human experimentation on the animal results in increased intelligence and other human characteristics being given to the animal, including a sense of language. Derrida explains the role of language and its relationship with the animal: "No one has ever denied the animal this capacity to track itself, to trace itself or retrace of a path of itself. Indeed, the most difficult problem lies in the fact that it has been refused the power to transform those traces into verbal language, to call it itself by means of discursive questions and responses, denied the power to efface its traces" (Derrida, 2008, 50). In other words, the animal cannot express itself through the means of a verbal language; that right belongs exclusively to mankind. However, this idea of language belonging to just man is challenged in Atwood's novel. In particular, there is a moment that takes place at the end of the novel when Toby is awakened by the sounds of birds. As she listens to their sounds, she comes to the realization that, "soon her own language will be gone out of her head and this will be all that's left in there. Oodle-oodle-oo, hoo hoom. The ceaseless repetition, the song with no beginning and no end. No questions, no answers, not in so many words" (Atwood, 2009, 349). This reflection is a moment in which Toby has an understanding that soon her own language, a defining marker of her claims of being human, may soon be gone from the world. This moment of clarity comes as she hears the birds singing in the morning. There is an understanding that the language of man is disappearing while the language of the animal is emerging in its place.

The apocalyptic event in *The Road* has rendered nearly all life, including most of the animals, extinct. While the novel does not place emphasis on the animal the way Shelley and Atwood's novels do, it still draws attention to the idea of the bestial or the inhuman. In this case, the bestial or inhuman are the cannibals that now roam the world. At one point in the novel, the father encounters one such individual who he has to threaten with violence in order to keep from potentially hurting his son. The description of the man is that he: "wore a beard that had been cut square across the bottom with shears and he had a tattoo of a bird on his neck done by someone with an ill-formed notion of their appearance" (McCarthy, 2006, 63). What is perhaps most striking in the father's description of this other person is the image of the bird tattoo on his neck. The image of the bird is described as being incomplete, thus contributing to the idea that the bird has been forgotten because it, along with other species, are now extinct. Human language is, therefore, vanishing from the world because the idea of the bird is now being drawn based upon an ill-formed memory.

The theme of language disappearing from the world also occurs when the son comes across the remains of a dead infant: "What the boy had seen was a charred human infant headless and gutted and blackening on the spit" (McCarthy, 2006, 198). This moment of unimaginable horror renders the boy speechless, and the only thing the father can say in response is, "I'm sorry" (McCarthy, 2006, 198). What the son witnesses is so traumatic that the father wonders if the boy would "ever speak again" (McCarthy, 2006, 199). This encounter with the cannibalized infant highlights not only an assault against humanity but also against language because in this case words are unable to describe the horror of such an unfathomable moment, and furthermore fail to provide any comfort to the son. The transformation of the world has reached the point where the use of language to describe something so inhuman is inadequate. Human language thus becomes ineffectual in describing the current state of the world.

Past social and cultural values are crucial in understanding how each of the three characters copes with being the last man or woman. The appearance of the beast as the new dominant species raises questions for the last man or woman concerning his or her own humanity. It is because of the absence of others and the rise of the beast that ultimately results in the last person attempting to hold onto those values of civilization, even those values no longer hold their original meaning. The question arises, in the absence of all others, what meaning do these social values hold? It is because of the dominance of the beast that there emerges a need on the part of the last man or woman to hold on to those past social and cultural values because doing so allows for an avowal of one's own humanity. It is crucial, therefore, that the last man or woman maintains his or her own sense of self within the new dichotomy of animal over man, or the increasing inhumanity of a world devolved into cannibalism.

In the case of Lionel, there is no hope of ever having his story shared with anyone else. However, despite this knowledge, he still narrates his story but does so in a way that suggests a shared narrative through certain cultural values that once existed. In other words, Lionel narrates under the pretense others exist. As *The Last Man* reaches its conclusion and the survivors realize that they can no longer stay in London, Lionel and his friend Adrian pay their city one final visit where the former observes: "birds and tame animals, now homeless, had built nests, and made their lairs in consecrated spots....Troops of dogs, deserted of their masters, passed us; and now and then a horse, unbridled and unsaddled, trotted toward us...everything was desert; but nothing was in ruin" (Shelly, 1994, 332). The city of London no longer belongs to mankind, as evident by the animals moving in to inhabit the city. Furthermore, there is a mixture of wild animals as well as domestic ones, suggesting that the domestic will soon become wild once the human race that has taken care of them becomes extinct. Additionally, what stands out about this passage is the description of how everything is deserted but nothing is in ruins. This last quote underscores the absence of humanity because the manmade buildings are still standing, but no human being occupies them. Instead, it is the animal that has moved in. The animal has taken over the interior space once belonging to mankind, further revealing the new dominance that the animal has now that human civilization is no more.

The idea of the beast being the dominant species also occurs at the end of the novel in a moment when Lionel comes across his reflection in a mirror. At first, he does not recognize himself which prompts the question: "What wild-looking, unkempt, half-naked savage was that before me? The surprise was momentary. I perceived that it was I myself whom I beheld in a large mirror at the end of the hall" (Shelly, 1994, 455). Lionel has not seen himself in some time, and this causes him to be shocked by his appearance. Upon his initial inspection, he sees a savage creature before realizing that it is own reflection. The idea of seeing himself like a beast is connected to a remark that Lionel makes at the very beginning of the novel. It is during these early moments that Lionel lives in a state of exile away from civilized society. In narrating his life at this time Lionel states: "My life was like that of an animal, and my mind was in danger of degenerating into that which informs brute nature…. I continued my war against civilization, and yet entertained a wish to belong to it" (Shelly, 1994, 18–19). When Lionel is removed from others, he behaves like an animal. It is only after his entrance into society, that the beast like state within Lionel is subdued or repressed. By the end of the novel, with civilization extinct from the plague, the association with savagery and something beast like has returned in Lionel's mind. It is when he is removed from other human beings that he again compares himself to the beast.

At the end of the novel, Lionel narrates that, "My only companion was a dog, a shaggy fellow, half water and half shepherd's dog, whom I found tending sheep in the Campagna" (Shelly, 1994, 467). This dog reinforces his solitary state, as it cannot talk or communicate with him the way another human being can. The dog is also continuing on with tending the sheep, even though its original master is dead. The dog now has the task of taking care of the sheep in the absence of a human being. What was once a job for man, now belongs to just the dog. In other words, the human value of work now belongs to the animal who carries on in man's absence. Lionel now lives with the animals in the sense that he is now part of their world, and this marks a return to nature and the state he was living in at the beginning of the novel. Lionel started off away from civilization, and now, with the world empty of its human population, he comes back to this state. By being alongside the beast, or underneath them, Lionel's status of the last man is accentuated. He is the only remaining member of not only his race but also his own species.

Finally, at the end of the novel, after becoming the last man, Lionel comes across some goats. Seeing these animals causes Lionel to panic: "No, no, I will not live among the wild scenes of nature, the enemy of all that lives. I will seek the towns—Rome, the capital of the world, the crown of man's achievements" (Shelly, 1994, 460). Lionel, realizing the world is absent of other humans, vows not to live among nature but instead seeks to return to the city of man. He chooses Rome because it is the capital of the world and the crown of man's achievement. However, when he does reach Rome, he finds that there are no other men present. Instead, his attention focuses on two statues: "The statues on each side, the works as they are inscribed, of Phidias and Praxiteles, stood in undiminished grandeur, representing Castor and Pollux, who with majestic power tamed the rearing animal at their side" (Shelly, 1994, 461). These statues exhibit the traditional supremacy of man over animal, and further relates to Derrida's argument of man's traditional and self-proclaimed dominance over the animal: "The animal is a word, it is an appellation that men have instituted, a name they have given themselves the right and the authority to give to the living other" (Derrida, 2008, 23). The irony is that mankind is now gone, while it is the beast that has taken on the role of being the new dominant species. The fact that this scene takes place in Rome is further significant because Rome, being the capital of the world, reveals that the world no longer belongs to humans, it instead belongs to the animal.

In Atwood's novel, Toby is still holding on to past values because as far as she is aware is the only person left alive in the whole world. In the beginning of the novel, she is at a swimming pool that has, "mottled blanket of algae. Already there are frogs…The luminous green rabbits, the rats, the rakunks, with their striped tails and raccoon bandit masks. But now she leaves them alone" (Atwood, 2009, 4). The animals make their appearance right from the opening pages of the narrative and while traditional animals are listed, alongside them are the new, genetically created creatures from before the world was destroyed. The animals outnumber Toby who does not disturb them out of respect for the former teachings of a group known as God's Gardeners and their leader, Adam One who Toby once lived with. During Toby's earlier encounter with the group, they ask her, "Surely you wouldn't eat your own relatives." To which Toby replies, "I would…. if I was hungry enough. Please go!" (Atwood, 2009, 40). This conversation between Toby and the group highlights the difference between the two groups when it comes to eating animals. While the Gardener's claim that they would not harm any living creature, Toby argues that she would if she was hungry enough. What ends up happening though is an interesting reversal in which it is Toby who becomes the one not to eat any meat after spending time in the Gardener's commune and being exposed to their teachings and way of life. After the fall of mankind, the reader becomes aware of that fact that the Gardener's teaching concerning the consumption of meat becomes lax to justify their survival, while Toby ends up being the one not to consume any meat. Her decision not to eat meat and thus honor the teachings of the Gardeners, shows the need for Toby to retain a sense of belonging in the absence of others which enables her to affirm her own humanity in a world in which the beast reigns supreme.

The bio-engineered Pigoons, a product of past human experimentation, become the primary threat to Toby's existence because their increased intelligence has placed them on top of the food chain, and as a result of this new placement they end up hunting her for food. When the Pigoons make their first appearance the narratives describes: "Three huge pigs are nosing around the swimming pool—two sows and a boar… She's spotted pigs like this before, in the meadows, but they've never come this close" (Atwood, 2009, 18). Toby's ostensible space of safety, the rooftop spa, is being invaded by creatures that pose a threat to her existence. As the last person, Toby is now at the mercy of the beasts who are hunting her for food; she is, therefore, no longer at the top of the food chain. This threat to her existence causes Toby to shoot at one of the creatures, even though in the past she was instructed not to by Adam One. Toby is ashamed of her actions, and furthermore understands that the other Pigoons will not forget this transgression against them: "Pigs are smart, they'll keep her in mind, they won't forgive her" (Atwood, 2009, 19). The Pigoons are taking on human characteristics, in this case memory. The animal was previously denied memory, but now it has access to this trait. Moreover, it is not just that the Pigoons will not forget what has happened, but that they will not forgive Toby for the act of transgression she has committed against them. The human conceptions of memory and forgiveness now belong to the beasts as they surmount the human race in being the dominating species.

While the Pigoons serve as the primary threat to Toby's life, in *The Road*, it is the father's fellow man that threatens the lives of both him and his son. In *The Beast and the Sovereign, Volume II*, Derrida explains the horrific implications of cannibalism: "For here, what is the worst thing about cannibalism is that these people eat beings of their own species and thus, in a way, self-destruct, by putting to death, to living death, their own species, their own lineage" (Derrida, 2010, 139). Derrida further argues that, "To have lost human dignity by being inhuman is reserved for humans alone, and in no way for the sea, the earth, or the beast. Or the gods…Only humans are said to be inhuman" (Derrida, 2010, 141). To be inhuman is a term usually associated with something bestial or belonging to the animal. Instead, Derrida argues that only man can become inhuman and bestial, the animal itself cannot be a beast. *The Road*, while absent of the animal, becomes the novel in which the beast is most prevalent. The beasts in this case are humans who have forsaken their own humanity by engaging in cannibalistic activity and, through this incomprehensible action, are putting to death their own species.

In one harrowing scene, the father and son encounter the victims of cannibals being kept as livestock. While exploring a house, the father and son come across a secret passage that takes them into a basement. Once inside they discover a horrific sight: "Huddled against the back wall were naked people, male and female, all trying to hide, shielding their faces with their hands" (McCarthy, 2006, 110). This horrific atrocity is further accentuated when the novel directs the reader's attention to a mattress on which "lay a man with his legs gone to the hip and the stumps of them blackened and burnt. The smell was hideous" (McCarthy, 2006, 110). As these victims call out for help, the father knows he cannot do anything for them, and further knowing the danger that he and his son are now in, quickly makes his escape. While the father and son do not encounter any cannibals inside the house, they see other humans being treated as livestock for purposes of cannibalization. They bear witness to the inhumanity of eating human flesh, and such an unspeakable moment, causes the father to realize that this role of victim is a real possibility for both him and his son.

The scene of inhumanity that the father witnesses, along with the understanding that this fate could befall his son, prompts him to command the boy to kill himself if they are unable to get away. He instructs his son, "You put it in your mouth and point it up. Do it quick and hard. Do you understand? Stop crying. Do you understand?" (McCarthy, 2006, 113). The father cannot bear the thought of his son being killed and eaten by cannibals, and as a result of this fear he orders the boy to take his own life. This moment of infanticide, the instruction of the son to kill himself, is an act of self-destruction, as the father killing his son is also a death sentence for the next generation. It is with this moment that the full weight of the potential hopelessness of the world is felt. The world has come so dangerous that there appears to be little to no place for humanity left. It is a world, at least in the father's mind, in which fathers have to kill their sons.

The final issue for this paper deals with the topic of ethics, and here, Derrida's examination of Emmanuel Levinas's ethical philosophy, in particular his idea of the face, is also important to the investigation of the relationship between the beasts and the last person. The face for Levinas calls attention to the ethical responsibility one has towards other people. According to Levinas in *God Death, and Time*: "Someone who expresses himself in his nudity—the face—is one to the point of appealing to me, of placing himself under my responsibility: Henceforth, I have to respond to him. All the gestures of the other were signs addressed to me" (Levinas, 2000, 12). The term the face enables Levinas to describe the relationship between the self and the Other in which the self has an ethical responsibility in recognizing the humanity of the Other. In *Totality and Infinity*, Levinas further explains that "the epiphany of the face qua face opens humanity. The face in its nakedness as a face presents to me the destitution of the poor one and the stranger" (Levinas, 1969, 213). However, the term the Other, for Levinas, refers to other human beings, and when it comes to the animal and whether or not it has access to such a relationship, he was unable to provide an answer. As Derrida explains, "This subject of ethics, the face, remains first of all a fraternal and human face…It is a matter of putting the animal outside of the ethical circuit" (Derrida, 2008, 106). Derrida acknowledges that Levinas does not include the animal in his thoughts, however, he argues that: "If I am responsible for the other, and before the other, and in the place of the other, on behalf of the other, isn't the animal more other still, more radically other?" (Derrida, 2008, 107). With this remark, Derrida questions the idea that the animal is not the Other. In the three novels, the concept of solitude means that the last person no longer lives in a world of Others in the sense that Levinas defined them as, i.e. other human beings. Instead, the Other becomes the animal, and issues of morality and ethics arise as a result of the animal now being the dominant species.

As the plague in *The Last Man* destroys the human population, there is a moment in the narrative in which Lionel explains that foreigners are coming into England as their own countries have become uninhabitable. These outsiders are seen as invading the home space of Lionel and his friends, and he explains that these invaders were given the characteristics of animals and mythical beasts: "Gorgon and Centaur, dragon and iron-hoofed lion, vast sea-monster and gigantic hydra, were but types of the strange and appalling accounts brought to London concerning our invaders" (Shelly, 1994, 298). This army of outsiders is explained in terms as being something bestial and therefore inhuman. Later on, Lionel and his friend Adrian ride out with an army in order to halt the advancement of these outsiders. However, during their battle Adrian calls an end to the carnage, citing claims to a universal brotherhood and recognition of the humanity found within each of them: "You are dear to us, because you wear the frail shape of humanity; each one among you will find a friend and host among these forces" (Shelly, 1994, 302). Adrian reminds the fighters that each of them is a representative of humanity, and that they must recognize the humanity of others, similar to what Levinas promotes in *God, Death, and Time*: "The grandeur of modern antihumanism…. consists in making a clear space for the hostage-subjectivity by sweeping away the notion of the person. Antihumanism is right insofar as humanism is not enough. In fact, only the humanism of the other man is human" (Levinas, 2000, 182). Through his claims of universal brotherhood, Adrian attempts to give this right of humanity towards his fellow man. What is interesting, however, is Adrian's later remarks when he instructs everyone to, "Cast away the hearts of tigers that burn in your breasts…. let each man be brother, guardian, and stray to the other"

(Shelly, 1994, 302). This call to dismiss the animal side demonstrates the belief of man's superiority over the beast, and also shows the attempts made to hold on to humanity and not give into something animalistic.

Toby's relationship with the animals also concerns itself with ethical implications and for this, Derrida's account of Levinas's thoughts on the animal, in particular the notion of the face, comes into consideration. For Levinas, the first commandment is "thou shalt not kill." As Derrida explains: "Levinas promotes 'Thou Shalt not kill' … It is the first commandment to come from the face of the other, being confused, in fact, with the very epiphany of the face" (Derrida, 2008, 110). However, the face for Levinas is strictly a human feature, while the animal is not afforded this right. Again, according to Derrida: "If the animal doesn't die (by killing it) ….it is because the animal remains foreign to everything that defines sanctity, the separation and thus the ethics of the person as face" (Derrida, 2008, 111). In *The Year of the Flood*, before the majority of the human race is rendered extinct, the group God's Gardeners, have, as a basis for their philosophy, a respect for the animal in that they do not eat them. They do not kill the animal, and thus are seen as giving to the animal what Levinas does not—a face and an ethical responsibility not to kill them.

Toby first encounters the Gardeners while working for a fast food chain which the group is protesting because of the restaurant's blatant slaughter of animals. While witnessing these protests, Toby finds herself drawn into an exchange with the group. The Gardeners chants: "Spare your fellow creatures! Do not eat anything with a face! Do not kill your own soul!" (Atwood, 2009, 40). Hence there is a connection between eating an animal and one's own soul evoking a close relationship between man and animal, and producing a sense of brotherhood between species. The Gardeners, therefore, afford the animal the access to the face that Levinas does not, and by doing so they claim that no one should eat them. In other words, the face that gives credence to Levinas's first commandment of "Thou Shalt not Kill" which previously was denied to the animal is now given to them.

The issue of ethics, however, plays out most prominently in McCarthy's novel, as here the issue of ethics arises between the father, who views himself as being human, and the cannibals, whom the father sees as inhuman. In other words, in *The Road*, the bestial is not the animal, as it is in Shelley's and Atwood's novels, but rather other human beings that have given into acts of cannibalism. It is the cannibals that the father and son seek to avoid as they make their way through the post-apocalyptic landscape. An early example of this desire to avoid others occurs early in the story when the novel describes how, "They (the cannibals) came shuffling through the ash casting their hooded heads from side to side. Some of them wearing canister masks. One in a biohazard suit. Stained and filthy. Slouching along with clubs" (McCarthy, 2006, 60). This is the first appearance of other people besides the father and the son, and the appearance of these others causes the father to fear them because of the potential harm they could cause him and his son. In addition, he is frightened by this group because the description of this group of people is one that depicts them as being inhuman and similar to caveman, all of which contributes to the notion that they have regressed away from their prior humanity. Hence, the father has encountered something that is lacking in humanity.

After his encounter with the man with the neck tattoo of a bird, the father reflects on how: "This was the first human being other than the boy that he'd spoken to in more than a year. My brother at last. The reptilian calculation in those cold and shifting eyes. The gray and rotting teeth. Claggy with human flesh" (McCarthy, 2006, 75). The father refers to this man as his brother, while at the same time ascribing to this individual the characteristics of a reptile. In this case, the other man is made to be inhuman by the father who does not recognize the humanity of the other man, and instead sees him as an animal. In his avoidance of others, the father ascribes to them the qualities of an animal, suggesting that it is easier for the father not to communicate with them in any form. The father is not wrong in this situation because the mouth of the man is "claggy with human flesh," meaning that the man has engaged in acts of cannibalism. What the encounter with the other man signifies is that there has been a regression on the part of man back to a state of animalistic nature. It is this state that the father seeks to avoid for himself. However, this also underscores the tragic nature of the father, as he is unwilling to enter into a meaningful and significant relationship with anyone else. According to Levinas in *Totality and Infinity*, "It is my responsibility before a face looking at me as absolutely foreign that constitutes the original fact of fraternity.... The very status of the human implies fraternity and the idea of the human race.... Society must be a fraternal community to be commensurate with the straightforwardness the primary proximity, in which the face presents itself to my welcome" (Levinas, 1969, 214). In other words, humanity is defined by our interaction with others. Tragically, it is the father's refusal to engage in any form of interaction that is ultimately the cause of his solitary condition.

All three novels touch upon the theme of lastness, and ultimately what it means to be the final survivor in a world where the notion of humanity is gone, replaced instead by the bestial. Despite this, the survivors of each story, Lionel, Toby, and the father and son still navigate through their ruined worlds, encountering the bestial in its many forms. Whether it be actual animals that have overrun the cities, supplementing man as the dominant species, or genetically modified beats that have risen to the top of the food chain, or finally, a world in which mankind has given way to cannibalism, these stories all force their respective protagonist to confront the beast and, in the process, attempt to his or her own humanity. It is reading these novels, in conjunction with the works of Derrida and Levinas, that offers insight into one's ethical responsibility to another, and thus what it can mean to be human.

References

Atwood, M. (2009). *The year of the flood*. New York: Anchor Books.

Derrida, J. (2008). In M.-L. Mallet (Ed.), *The animal that therefore I am* (D. Wills, Trans.). New York: Fordham University Press.

Derrida, J. (2010). In M. Lisse, M.-L. Mallet, & G. Michaud (Eds.), *The beast & the sovereign* (Vol. II) (G. Bennington, Trans.). Chicago: University of Chicago Press.

Levinas, E. (1969). *Totality and infinity* (A. Lingis, Trans.). Pittsburgh: Duquesne University Press.

Levinas, E. (2000). In J. Rolland (Ed.), *God, death, and time* (B. Bergo, Trans.). Stanford: Stanford University Press.

McCarthy, C. (2006). *The road*. New York: Vintage Press.

Shelley, M. (1828/1994). *The last man*. Oxford: Oxford University Press. Reprint.

The Situation of Human Being in Nature According to Fedor Dostoyevsky, Thomas Mann, and Robert Musil: A Paradoxical Builder, Self-Enhancing Being and Speaking-Animal

Michel Dion

1 Introduction

We should be very cautious, when exploiting natural resources since we cannot weight up natural phenomena. Although we cannot presuppose that some natural laws are undecipherable, we cannot be ensured that we will know enough about natural processes to prevent disasters (Dostoïevski, 1972, 375). Abused nature impose much more severe punishment than human justice. Human justice could release people (particularly criminals) from despair. However, it partially releases humankind from the atonement of Nature (Dostoïevski, 1973, 883). Dostoyevsky interpreted human rebelling against Nature as an existential fight against death itself (Lamblé, 2001, 113). As a ghastly phenomenon, death is contradicting the beauty of Nature. An anthropocentric attitude toward Nature implies that the whole Nature has been created for the sake of humankind. Is human being the measure of everything (Mann, 2010, 451–455, 763)? Such anthropometrism has been originally asserted by Protagoras (*Theaetetus*, 152a: Plato, 2011, 1904). Plato's philosophy was focusing on God-knowledge rather than the knowledge of Nature and of natural processes.

What is the situation of human being in Nature? From the midst of the nineteenth century to the mid of the twentieth century, three well-known novelists have deepened their philosophical questioning and adopted very different approach of human-Nature relationships. Questioning the place of human and non-human beings in Nature inevitably unveils the relationships between humans and their gods. Fedor Dostoyevsky used two meanings for God's existence. Firstly, God's existence could refer to Divine Life. Secondly, God's existence could be reduced to the idea of the Infinite as it is present in human mind and society. If the Infinite overcomes the

M. Dion (✉)
University of Sherbrooke, Sherbrooke, Quebec, Canada
e-mail: Michel.Dion@Usherbrooke.ca

C. A. Hornbuckle et al. (eds.), *Phenomenology of the Object and Human Positioning*,
Analecta Husserliana 122, https://doi.org/10.1007/978-3-030-66437-4_15

finite, while being present in the finite, then the Infinite should be transhistorically rooted. Overcoming the finite makes impossible for the Infinite to have any historical situation. Otherwise, being-historical would be meaningless. Can we really know the Divine Life? I cannot know what it means to live for a mouse. I cannot seize how a mouse actually feels it own life processes. I will always be unable to grasp to what extent a mouse perceives its own life. In the same way, I cannot understand how God feels its Divine Life, and if God gives any meaning to Divine existence. Some believers claim that in their after-life, dead people will know what the Divine Life is all about. But this belief would imply that knowledge processes are still going on, after one's death. As dead people, do we still have memory, imagination, desires and needs? Nobody knows if dead people are able to know anything in their after-life, and thus if they are able to know the Divine Life. The only thing I know is what it means to live, as human being. And even that remains unclear. I could believe that my own existence is meaningful, while later on I could perceive it as being meaningless. Human existence gives birth to multiple interpretations. According to Dostoyevsky, human being cannot be perfect, and thus infinite. Only transhistorical beings which do not live (like gods) could be perfect and infinite. Referring to the first meaning of God's existence, we could say that God does not live at all, or at least, that we cannot perceive it. If God is a transhistorical being, then God must not be subjected to life processes. Being-historical is required to exist. God cannot be subjected to existential predicament. If there is any Divine Life, then we will never know to what extent it could be compared to life processes in the whole Universe. But we could certainly say that life processes are not expressing perfection at all. Dostoyevsky explained how human being could be the builder who has the power to destroy everything-that-is (the paradoxical character of the builder). Dostoyevsky was not deeply convinced that human being could love Nature, except if he/she looks at it as the Divine Creation. In his *Notes from the Underground* (1864), Dostoyevsky addressed the issue of the paradoxical builder. In a godly-focused perspective, Dostoyevsky asked the following question: Can a paradoxical builder love natural beings?

Thomas Mann unveiled the deep influence of the unconscious as well as the subconscious, especially in *The Magic Mountain* (1924) and *The Doktor Faustus* (1947): both components of human psyche must be taken into account, when exploring the mystery of human being. Human psyche is self-enhancing. In a psychologically-focused perspective, Mann asked the following question: How could a self-enhancing being be champion of life processes?

Robert Musil's literary works (particularly *The Man Without Qualities*, 1930–1933) focused on commonalities between animals and human beings, that is, their similar instincts. Musil adopted a naturally-focused perspective, although human being is defined as the speaking-animal (Aristotle *Politics*, 1253a10: Aristotle, 1943, 54). Musil was promoting a new morals, as it is grounded on instinctive life. Can a speaking-animal overcome its natural instincts?

2 Fedor Dostoyevsky's Godly-Focused Perspective: Can a Paradoxical Builder Love Natural Beings?

According to Dostoyevsky, we cannot design social institutions and regulations, so that we get rid of crimes, once for all. Such idealized society cannot exist, since it would be existentially uprooted. The idealization of any society/social group denies historicity. History is so full of monstrosity and stupidity that it should be erased, as if it would not have never existed. Erasing History implies to perceive life processes as being abominable. Even the existence of a free soul is denied, since it would require a living body. As he/she is historically rooted, human being cannot be perfect, and thus infinite. Only transhistorical beings (like gods) could be perfect and infinite. God does not live at all. If God is a transhistorical being, then God must not be subjected to life processes. We must rather take nature into account: life processes are not expressing perfection. Living soul is basically suspicious and retrograde. Living beings are finite. They cannot be perfectly fair and just. Humankind is historically evolving and gives birth to normal society (Dostoïevski, 1972, 282). Dostoyevsky did not isolate life from human spirit: Dostoyevsky was much more concerned with human fate than with God's existence. Dostoyevsky was promoting a free religion, that is, a religion that is not subjected to any kind of authority (Berdyaev, 1957, 24–26, 37).

Dostoyevsky believed that we must love Divine Creation, and thus, we must love every animal, plant, river or mountain. When loving every component of Divine Creation, we will grasp the Divine mystery in every thing, being, or phenomenon. We will be able to love all living beings and their physical environment. We must love animals, because God has given to them the principle of thought and peaceful joy. Human beings should never perceive themselves as being superior to animals and plants. The way human being is raising himself/herself over all other living beings is dehumanizing human being, since human being is a natural being. Human being as the Lord of Nature is then estranged from his/her real place in Nature. We must love every creature and its physical environment, because any adverse action here and now will have bad consequences on the other side of the world (Dostoïevski, 1973, 432–434). This is an anthropometric view on Nature, and thus an anthropocentric kind of environmental concern.

Dostoyevsky asserted that the Incarnation of Christ was a full redemption of Divine Creation. Human beings as well as animals and plants must be redeemed, since every living creature aspires to be united to God. Every living creature is praising the Glory of God (Hildegarde of Bingen, 1990, 34). Every living creature is participating in the Divine Mystery (Lamblé, 2001, 115–116). However, animals and plants do not commit any sin. So, their status of God's creature remains quite mysterious. Every living creature is basically good and wonderful. Every creature would like to be united to God. However, unlike animals and plants, human beings commit sins. Dostoyevsky did not explain to what extent the redemption process could be quite different for animals and plants and for human beings (Dostoïevski, 1973, 402). Unlike human beings, animals cannot feel mystical anxiety (Dostoïevski, 1977, 510).

Human being is a strange animal (Dostoïevski, 1961, 400). Human being has created the idea of God. What is astonishing is not God's existence, but rather the fact that human thought has created God's necessity. Human being is a fierce and malicious animal. Even though, human being has developed the idea of ultimate wisdom and sacredness. However, the question of God's existence cannot be resolved, once for all. We can believe in God without accepting the notion of Divine Creation. We could believe that existential suffering will eventually cease. At the end of our existential tragedy, an eternal harmony will arise. There will be an event of revelation: a hierophanic self-manifestation of God. Our heart and soul will then be redeemed. God will forgive and justify any human action and decision (Dostoïevski, 1973, 330–331).

Free will is the core concept determining the actual state of our heart and soul. Dostoyevsky used the metaphor of crocodiles to mirror the loss of free will. What is the basic property of crocodiles, if not swallowing people? If crocodiles must swallow people, then their corporeal structure must be hollow. Nature abhors vacuum. Now, what is the basic property of human being? The more our thought is hollow, the less we need to fill it up. Human being withstands to be swallowed. He/she does not want to be transformed in foods, or material goods. Such transformation would deny human dignity and self-transcendence. As long as I am exercizing my free will, I can safeguard my ability of self-transcendence. If I am losing my free will, then I can no longer become who-I-am. My own being then looks like potatoe, or pancake. Human being cannot accept such denigration (Dostoïevski, 2010, 391–395). Nature is not playing music, human being is not Nature's piano. Dostoyevsky believed that human being is not only searching for happiness, but also for something existentially rooted. Although human being owns everything he/she needs and desires, he/she will show ingratitude. He/she needs to have dirty hands: he/she will commit meaningless actions and behave quite dangerously, so that he/she could intermix wisdom and ineptness. Human beings will try to justify their own actions, arguing that such actions are safeguarding their dreams. They will behave in a stupid way, just for proving that they are not constituencies of a universal piano. If science learns us that human being behaves in accordance with natural laws rather than in accordance with his/her free will, then would it unveil that humankind is a universal piano? Such scientific assertion could transform human self-understanding and lead humankind to an absolute despair. If the worst comes to the worst, human being will fall into self-destruction and social chaos. He/she would curse the whole universe. He/she will fight dehumanization processes. Unlike animals, human beings can curse everything that subjects them to its own will. Human beings will oppose to be considered as nuts (Dostoïevski, 2007, 147–148, 156–157).

How could we be ensured that human being must correct his/her own will, in accordance with scientific data and common sense? Why must we raise human will? Could educational learning be useful to human being? Do we believe that human being must never contradict his/her real interests, safeguarded by rationality and mathematics? Such belief is only a presupposition. Dostoyevsky asserted that human being as an animal is a builder, an engineer. Human being is always focusing on a given purpose. He/she is condemned to hit the road. But the road leads somewhere.

What is crucial is not the purpose itself (the direction), but rather the *locus* of the action. Where does the road actually lead us? Dostoyevsky asked the following question (Dostoïevski, 2007, 159–160): If human being likes to build up roads and buildings, why does he/she also like to destroy anything and to create a chaotic world? Human action is not always expressing self-interest and rational deliberation. Individuals could wish something that negates their self-interest. Dostoyevsky asserted that it is sometimes a moral duty to contradict our own self-interest to favour collective interests, or others' interests (altruism). Free will would then be inherently linked to virtues. But how could we reach such conclusion? How could we prove that free will must always be grounded in rational motives and means? The only thing human being needs is an independent will. Human being will pay top dollar for self-reliance, whether the consequences could be. Human reason can only know what it has learned. Since human being has-to-die, then his/her reason will not have enough time to learn countless things. Human action is influenced by conscious and unconscious conditioning factors. Human beings could have very silly desires, since reason could give birth either to wise actions, or to stupid behaviors. Human being is totally free to take wise, or foolish decisions (Dostoïevski, 2007, 67–70).

Humankind is either a mouse, of the 'man of nature and truth'. The mouse is always offended. It is petty and weak. Violence and revenge overwhelms its mind. The mouse cannot identify just and unjust actions. The mouse is skeptical about everything: it is subjected to suspicion and uncertainty (Dostoïevski, 2007, 51–59, 126–127). In his *Notes of the Underground*, Dostoyevsky inverted the symbol of spiritual growth toward the truth, as Eltchaninoff (1998) explained: when we are declining toward the underground, then we become ultimately isolated from humankind (Dostoïevski, 2007, 40–42). We are rejecting any philosophical or scientific truth. We become deeply aware of the contrast between the ideal beauty and the ugliness of our wrong actions. The man of nature and truth rather considers that revenge is a perfectly just action. He/she is morally judging every thing, being, or phenomenon, while giving advices to his/her fellows. The man of nature and truth is convinced that his/her action is basically honest and just, so that revenge could be done, while keeping his/her mind deeply serene and calm (Dostoïevski, 2007, 51–59, 126–127). If God does not exist in human mind and is not part of social life, then everything is allowed. We are releasing ourselves from traditional moral rules and standards. Human being can become a human-god. A human-god does not have to conform himself/herself to any law. Thus, there is no ultimate truth at all (Dostoïevski, 1973, 808–809). There are two basic kinds of believers: those who need miracles, and those who want free faith and love. On one hand, some believers do not search for God. They are rather searching for miracles. They are creating new miracles to replace old ones. Human beings could be subjected to the supernatural (magic). Through their emphasis on miracle, mystery and authority, human beings have completed the Divine Creation. Some believers need masters, so that they could act as slaves, said Dostoyevsky. Masters are the guardians of faith and consciousness. Masters are responding to their servants' need for universal brotherhood and love. Human being is unable to admit his/her weaknesses, cowardice, and vileness. Human being is subjected to anxiety, distress, and unhappiness. If God has come only for the elected people, then free heart and love do not matter at all, said Dostoyevsky. We must accept God's mystery, even against our own will. On the other hand, some believers actually want to be the children of a free faith and of a free love. Very few believers are able to rebel against their God, in the behalf of free will. If they could do so, they could look at God, as the One who is the source of every free will (Dostoïevski, 1973, 355–359).

3 Thomas Mann's Psychologically-Focused Perspective: How Could a Self-Enhancing Being Be a Champion of Life Processes?

Human body actually lives as if it would not bear any soul. It could be disturbing and troublesome to become aware that our body could evolve without being closely linked to our soul. Sometimes, life is tragically orientated. Nature could be so cruel that it hinders someone to reach the harmony of his/her own personality. It is precisely the case when noble soul lives in body that is incapable of supporting life (Mann, 2010, 84–85, 116). Whether it is the cruelty of exquisite pleasure, the cruelty of ingratitude, the cruelty of insensitivity, or the cruelty of mistreatment, cruelty always implies the absence of reflection (Mann, 2009, 517). A body without soul (the rule) is as much barbaric and terrible as a soul without body (the exception). Someone who is critically sick is reduced to a body: that's why his/her situation is so humiliating. Body is the complex network of multiple sub-beings in a given individual: sub-beings are organs, that is, living units aiming at a global life. Having a body means that my self is living. We can easily distinguish between my being and others' being (Mann, 2010, 318–324).

Nietzsche was the Champion of life processes. According to Mann, Nietzsche has deciphered the (Christian-based) process of transforming good instincts into bad instincts (and bad instincts into good instincts). Nietzsche fighted for a radical reversal of values, that is, a transmutation of values. Our own spirit cannot find itself in Nature. That's why it is always rebelling (Mann, 1997, 245). Nature is spirit. Is body the vehicle of soul, and thus a necessary evil? It is God's will that life breeds life. Life processes are both focusing on the past (regressive) and on the future (progressive). Only the past is eternal, since it is dead (Mann, 2005, 96, 2009, 237). Sickness is something that puts harm to the body. Is sickness a depravity of life? According to Mann, human being is existentially sick. Sickness is an existential category, that is, limit-situation that makes an integral part of human existence (Karl Jaspers). Spirit makes possible to distinguish between human beings (as non-natural beings) and natural beings. The transcendental dignity of humankind follows from the spirit. Sickness makes human being a noble species. Sickness and death constitute the real basis of human dignity. Mann (2010, 324–325) asked the following questions: Is there an abyss between living beings (organic) and lifeless things (inorganic)? Is there an insurmontable abyss between material things and immaterial (insubstantial) things? Can material things arise from immaterial realities? Is matter the degeneration of immaterial realities? Is there a conflict between the real and the supranatural? Mann explained the main characteristics of mystical experiences: when a given thing becomes mystical, then either the body (the physical element) turns into spiritual form, or spiritual element is changed into a corporeal/physical form. In both cases, we can no longer distinguish between the physical element and the spiritual element (Mann, 2010, 666). Great historical figures are characterized by the fact that spirituality is not counterbalanced by a negative perception of natural processes and powers (Mann, 2007, 77). Mann agreed with Schleiermacher (1944, 135–138) that the sense of the infinite is something natural (Mann, 2009, 114).

Human beings could perceive their own passions as if such passions would be moral duties, said Mann. For instance, they could put fairness in practice and search for equilibrium. However, fairness and equilibrium are then interpreted as moral duties. We could feel ashamed of our inability to accomplish our moral duties. We should try to reach an equilibrium between altruistic duties (our duties toward other people) and self-centered duties (our duties toward ourselves) (Mann, 2010, 185). Moral duties thus become the replica of human passions. Being passionnate is living for the love of life, and thus being centered on life processes rather than on our own self. Living is having desires, particularly the will (desire) to live (Mann, 2007, 249). But our will (desire) to live is not only determined by our self-consciousness. It is also deeply influenced by subconscious thoughts. We need to escape ourselves into the realm of the unconscious. As Jean Baudrillard (2007, 9–17) said, the Unconscious exists only when we are talking to it: we are living as if the Unconscious would not exist. The obscure realm of human will (Schopenhauer) is identical to Freudian notion of the unconscious (Mann, 1997, 221). Mann preferred the term 'supraconsciousness', since subconscious thoughts provide a knowledge overcoming the limitations of self-consciousness. Supraconsciousness could make basic links between the obscure areas of one's thought and a universal soul (the Omniscient) (Mann, 2010, 748). The feeling of life means self-consciousness (Mann, 2009, 145). At the beginning, self-consciousness was a function of organized matter. Over time, self-consciousness has been impregnated with promise and despair: self-consciousness tries to deepen self-knowledge and self-understanding. This is a useless attempt. Nature cannot be totally known. Life cannot come upon the final word about itself.

Death is the logical negation of life (Mann, 2010, 315–316). According to Mann, Darwin's evolutionary theory leads to the conclusion that the natural intentionality of human being is to improve humankind, from an organismic perspective. Self-enhancement could be perceived as a moral, aesthetic, and cultural duty. Improving humankind is not only an issue of scientific and technological progress. It is mostly the will to get rid of human suffering. Focusing on human dignity and happiness could make human suffering disappear (Mann, 1997, 88). Christian theologians and philosophers have sometimes described the soul as being imprisoned into the body: that's why they felt ashame and confused, when dealing with natural and corporeal processes. Our animal existence is then seen as a slanderous chain, that is, the origin of our pains. Our free will makes possible to release ourselves from such state of debasement

(Mann, 2010, 516). Any aversion towards the body is a way to deny natural powers, and thus a metaphysical revolt (of the spirit) against Nature itself. Spirit then denies any worth to natural processes and to bodies, because it perceives them as reducing human dignity and beauty. The real beauty cannot be purely corporeal. Corporeal beauty is misleading, since it conceals the beauty of consciousness. Human dignity could never be reduced to the dignity of flesh. Above all, it is the dignity of spirit. Spirit must subjugate natural and corporeal processes (Mann, 2010, 516–517). Humanism strongly criticized the fact that the spirit is subjected to corporeal processes. Body is then perceived as a bad and demonic thing, since flesh is an integral part of Nature. Spirit tries to maintain its dignity and to oppose natural powers, although it is a meaningless task. Mann believed that we should rather honour our body. We should glorify the corporeal beauty, the basic freedom of our sensory organs, and the passion for pleasure and happiness (Hildegarde of Bingen, 1990, 123). Body is the main object of humanism (Mann, 2010, 286–288). According to Jean-Paul Sartre, having a body is only possible for beings who are participating in their world: having (and being) body is a basic condition for action (Sartre, 1981, 374–383).

Now, what does life mean? Life transforms matter, while safeguarding the form. Organismic differentiation is the basic condition for life, said Mann. Life is being what-cannot-be. Life is being suspended between disintegration and renewal. Life aims at what-could-arise, that is, new possibilities-to-be. What-is-becoming will be different than what-it-was (in its prior state of things). Life is strictly linked either to existential categories (time, space, causality), or to the moral/rational realm. Religion has nothing in common with reason and morals (Mann, 2010, 316–317). Mann asserted that religion has historically distorted ethics. There is an antinomy between the glorification of life processes (aesthetics) and a pessimist/clear-headed cult of suffering (ethics). Nothing is essentially right or wrong, true or false, beautiful or ugly, said Mann. The universe does not imply a given moral order. Every civilization gave birth to social morality and culture (Mann, 1997, 102–103). Is there any meaning of human existence? This question always remains without any satisfactory answer. Philosophical questioning means that there will never be any final answer, any certainty about the meaning of existence, or about its meaninglessness. That's the essence of philosophical questioning, as the heart of human existence. Truth makes us growing and exploring various possibilities-to-be (Mann, 2009, 149–154). Mann (2007, 181) asked the following questions: Should we search for truth at any price? Should we deepen our understanding of truth? Are there some prohibited truths? Mann was concerned with existential estrangement and anxiety (Mann, 1961, 142, 2004, 69). Human being is existentially estranged from his/her own being as well as from everything-that-is. Everything outside him/her has fallen away.

Human being is the *homo religiosus*. Non-human beings cannot have religious/spiritual experiences. According to Mann, religious humanism is determined by the feeling of our own mystery. Moreover, we become aware that we are not only biological (living) beings, but also spiritually-focused beings. The absolute can be reached through the notions of truth, freedom, and justice. Over time, human beings can grow and improve themselves (Mann, 2009, 331). Does self-improvement be the ideal road to happiness? According to Mann, unhappiness follows from antinomies: law and power, thought and action, freedom and fate, or bourgeois morals and heroic duties. Mann (1997, 65) believed that in each case, the antinomy cannot give birth to harmony. Happiness comes from devotion and self-sacrifice, said Mann. Natural freedom is rooted in self-sacrifice. Mann (1990, 90–96) was convinced that humility implies not to be tempted by pridefulness. Maliciousness is something that could infinitely grow in a human heart (Mann, 1996, 244). That's why forgiveness is always required from everybody. Is understanding closely related to forgiveness? If we are deeply understanding various conditioning factors and individual motives, does it mean that we are necessarily ready to forgive others' fault? (Mann, 1990, 47) did not seem to be convinced that there is such necessity.

4 Robert Musil's Naturally-Focused Perspective: Can a Speaking-Animal Overcome Its Natural Instincts?

Human beings and animals share the same basic instincts: hunger, joy, angry, love, and stubbornness (Musil, 1956b, 566). What is the real origin of human progress? Could it be a complex network of animals instincts? Human beings as well as animals look at love as the beginning of an assumption of ownership. Taking up something as being my property is launching a process of spiritual reflection (Musil, 1956b, 925). Among a given species, animals' language is constituted by shared affects, said Musil. Human being is the speaking-animal (Musil, 1956b, 545–546). Musil perceived that we are pro-jecting over animals a feeling of being comfortable, when being absorbed by the whole reality. If animals actually feel comfortable in reality, then they should have crystal-clear spirit. We are then projecting in animals something that looks like human representations of worldly life, although their ability is qualitatively different from that of human beings. Musil was convinced that nobody actually owns truth-itself. We could improve our representations of truth (our truth claims). However, such representations always remain transitory and fragmentary. We are thus pro-jecting over animals the intentionality of developing transitory and fragmentary representations of the worldly life (Musil, 1956b, 757). What does such pro-jection mean? The pro-jection of human beliefs on animals implies three levels of representations. Firstly, it means that animals' life is analogically analyzed in comparison with human life. In doing so, we are aware that any analogy reflects a gap that could never be filled up. Any metaphor expresses the belief that different things, beings, and phenomena do have commonalities, in spite of their differences. Any analogy unveils the belief that various things, beings, and phenomena will always have distinctive traits, in spite of their similarities. Dogs' life could be described in using human life as analogy. However, only human beings have the feeling of everydayness, the sense of togetherness, and the acknowledgment of the radical otherness. Secondly, the pro-jection of human beliefs over animals is used to express our imagination/creativity, whether it is in literature, painting, sculpture, or architecture. Imagination could make the distantiation between human beings and animals disappear. But insofar as the loss of distantiation is imaginary, the distantiation still determine humans' and nonhumans' life. Thirdly, any pro-jection of human beliefs over animals becomes the source of various emotions and feelings that could be used to favour an anthropocentric as well as a non-anthropocentric view on animals.

Life would then refer to contradictory notions: richness and simplification; harmony, contemplative inwardness; power and aggressiveness (Bouveresse, 2001, 199–200). Human beings are subjected to hunger, sexual drives, and fears. An instinct is a personal and hereditary tendency to choose given actions, and thus to exclude other alternatives. Being both personal and hereditary means that the individual still keep his/her freedom. Instincts do not give birth to immutable patterns of action. Instinctive actions are accomplished without intent, or reflection. However, improving our self-understanding will help us to reduce the so-called immutable character of instincts. Instincts are realized by any individual member of a given species, always in the same manner (identical behavioral pattern). Various species could share similar instinctive actions (for instance, between human being and gorillas). Instincts push us toward action. In the stage of their awakening, instincts are subjected to an indeterminate growth. But every instinct aims at a determined end as well as an indeterminate end (Musil, 1956b, 695). Health is not rooted in our animal instincts. The nobility of spirit, customs, and morals is rather the origin of physical (bodily) resistance. It is not always true for individuals. However, the power of a given people follows from its best intentions and feelings (Musil, 1956b, 452).

Perceiving instincts as the basis of new morals implies that we must reject traditional morality. Traditional morality is centered on what-should-be (ideals), while the new morals focuses on the acceptance of reality as-it-is (instincts). Musil rejected any morals of compromise (Dugast, 1992, 78). Like Nietzsche, Musil did not believe in a clear-cut distinction between good and evil. Like Nietzsche, Musil was promoting a new kind of morality that is totally released from any immutable rule/norm. Morality must always be renewed and should never give birth to eternal truths (Musil, 1956a, 317–318). Evil and goodness are then parallel phenomena: morals is focusing on experiences we could repeat again and again (uniformity, objective regularity), while ethics puts the emphasis on love, inwardness, and humility (subjectivity, self-affirmation, creativity) (Bouveresse, 2004, 153–155). Good and evil are relative and depend on various conditioning factors. Musil believed that good (virtues) and evil (vices) are quite similar, since they come from the same rules and laws, exceptions and limitations. Any distinction between good and evil becomes meaningless. Good and evil cannot be isolated one from the other, for two reasons. Firstly, there is no fact, but only interpretations (Nietzsche, 1967, 301). Human being searches for truth, but is unable to reach it, since he/she cannot seize the contents of truth. Human being is the interpreter of things, beings, and phenomena. So, any truth is a truth claim and should never considered as Truth-itself. Metaphysical and moral truths are only truth claims. Nothing more. We can never be sure that what we perceive as the frontiers between good and evil actually reflect the nature of Truth-itself. If Truth-itself exists, then it can never be seized. That's why there are only interpretations, from a metaphysical as well as a moral perspective. Secondly, the frontiers between good and evil will vary from an ethical theory to another, whether the ethical theory is philosophically, or religiously/spiritually grounded. Of course, most of decision-makers have not read Aristotle, Kant, Hume, or Mill. However, they take ethical decisions in using principles which are quite similar to those following from philosophical egoism, utilitarianism, Kantianism, or any other ethical theory. The frontiers between good and evil will not be the same if the principle guiding ethical decisions is self-interest (philosophical egoism), common good (utilitarianism), or moral imperatives (Kantianism).

From a philosophical viewpoint, we can only choose between faith (certainty about what-is-possible) and doubt (uncertainty about what-is-possible) (Musil, 1956b, 836). Musil was describing a new morals overcoming the limitations of the traditional (religiously-centered) morality, that is, its basic trend to collectivize individual ethical powers and to normalize individual behaviors. Musil fighted such collectivistic way to strengthen social order. Musil endorsed the Nietzschean way of immoralism (Dahan-Gaida, 1994, 125–126). However, unlike Musil, Nietzsche (1967, 250) has explained how to be real immoralists in the daily life. Nietzsche defined immoralism in two different contexts. Firstly, Christian-based morality as immorality. In that case, life-denying virtues are defining an immoral morality. Morality is the instinct to deny life. Secondly, immorality could be the path of self-releasing from the power of Christian-based morality. Immorality is then considered as the only way to make Overman arising in human existence. Nietzsche (1967, 189) said that we must destroy morality in order to liberate life itself. The Overman is an on-going process of overcoming implying to strengthen the Immoralist, that is, the transmutation of values.

Could be apply duties to animals' life? There is only one duty which could analogically be applied to animals. Kant defined the duties of human being to himself/herself as an animal being, that is, the duty to preserve oneself. Kant concluded that as an animal, human being cannot be morally justified to kill himself/herself. As an animal, human being must be prudent, when using food and drink (Kant, 1991, 218–224). If such duties could be pro-jected on human being as an animal, then does it mean that the analogy between animals' and humans' life actually mirrors the gap between human duties and animal duties? If animals really have duties, how could we distinguish them from their instinctive life? Animal duties are animal instincts. Human duties imply free will, and thus the capacity to overcome the real of instinctive life.

5 Conclusion

Fedor Dostoyevsky was confronted to the ideal of the pure builder, that is, the *homo faber* who could use natural resources in a very productive/creative way. Only human being could live mystical experiences. Human being is the *homo religiosus*, that is, the being who has created the idea of God. Human being is also the builder who has the power to destroy everything-that-is. Human being can curse everything. Human being can love natural beings, but in a paradoxical way. Human being can love animals and plants, while exploiting them for his/her own self-interest. Human being is the paradoxical lover of Nature. Dostoyevsky made the ideal of a pure builder totally disappear.

Thomas Mann was dealing with the Nietzschean ideal of a Champion of life processes. Human will to live is deeply influenced by the unconscious as well as the subconscious. Insofar as he/she is led by unconscious/subconscious processes, human being is a self-enhancing being. Could self-enhancing being be a champion of life processes? Not at all. A self-enhancing being is focusing on its self-interest. Being a champion of life processes would imply to radically criticize the situation of human being in Nature, and to get rid of anthropocentric/anthropometric attitudes. Mann unveiled the vicious trend of a self-enhancing being. Any ideal notion of Champion of life processes is disconnected from human existential predicament.

Robert Musil analyzed the ideal self-perception of a speaking-animal. Musil was dreaming about personal ethics that has nothing to do with rationalized and static morality (Dahan-Gaida, 1994, 125–126). Musil was not convinced that evil could be inherently linked to human nature (Vatan, 2000, 68–70). The existential task is not to identify any natural goodness, or maliciousness. Rather, we must understand why human being could be as much good as malicious. Animals and human beings share same instincts. However, unlike animals, human beings actually speak. Human being is a being who is in-language. He/she is the speaking-animal. Musil was promoting a new morals, grounded on instinctive life. Can a speaking-animal exploit its natural instincts to deeply understand every natural being? Not at all. The ideal self-perception of speaking-animal could give birth to a new morals. However, unlike animals, human being is the speaking-animal, so that he/she cannot avoid the multiple interpretations of natural instincts. The gap between human existential predicament and the situation of non-human beings in Nature cannot be overcome.

Dostoyevsky, Mann and Musil agreed that human being is prone to consider himself/herself as an infinite being. Firstly, human being destroys Nature and justifies himself/herself through technocratic and anthropometric/anthropocentric worldviews (Nietzsche, 1966, 215–221). Secondly, human beings forget the situation of nonhuman beings and put the emphasis on their privileged position in the Cosmos. Thirdly, human being looks at language as the supreme quality of living beings and makes human language the ultimate parameter for any other form of language. The way those three aspects of human being and existence are idolized strengthens a mindset focusing on the infinite character of humankind, when compared with the situation of nonhuman beings.

References

Aristotle. (1943). *Politics*. New York: The Modern Library.
Baudrillard, J. (2007). *Fragments. Cool memories III*. London: Verso.
Berdyaev, N. (1957). *Dostoevsky*. New York: Living Age Books.
Bouveresse, J. (2001). *La voix de l'âme et les chemins de l'esprit*. Paris: Seuil.
Bouveresse, J. (2004). *Robert Musil*. Paris/Tel Aviv: Éditions de l'Éclat.
Dahan-Gaida, L. (1994). *Musil. Savoir et fiction*. Saint-Denis: PUV.
Dostoïevski, F. (1961). *Les possédés*. Paris: Le Livre de Poche.
Dostoïevski, F. (1972). *Crime et châtiment*. Paris: Le livre de poche.
Dostoïevski, F. (1973). *Les frères Karamazov*. Paris: Gallimard.
Dostoïevski, F. (1977). *L'Idiot*. Paris: Garnier.
Dostoïevski, F. (2007). *Notes d'un souterrain*. Paris: GF-Flammarion.
Dostoïevski, F. (2010). *Le songe d'un homme ridicule*. Paris: Gallimard.
Dugast, J. (1992). *Robert Musil. L'Homme sans qualités*. Paris: PUF.
Eltchaninoff, M. (1998). *Dostoïevski. Roman et philosophie*. Paris: PUF.
Hildegarde of Bingen. (1990). *Scivias*. New York: Paulist Press.
Kant, I. (1991). *Metaphysics of morals*. Cambridge: Cambridge University Press.
Lamblé, P. (2001). *Métaphysique de l'Histoire de Dostoïevski*. Paris: L'Harmattan.
Mann, T. (1961). *The holy sinner*. Harmondsworth: Penguin Books.
Mann, T. (1990). *La mort à Venise*. Paris: Le Livre de Poche.
Mann, T. (1996). *Altesse Royale*. Paris: Grasset.
Mann, T. (1997). *Les maîtres*. Paris: Grasset.
Mann, T. (2004). *Six early stories*. Kobenhavn/Los Angeles: Green Integer.
Mann, T. (2005). *Sang réservé, suivi de Désordre*. Paris: Le Livre de Poche.
Mann, T. (2007). *Lotte à Weimar*. Paris: Gallimard.
Mann, T. (2009). *Le Docteur Faustus*. Paris: Le Livre de Poche.
Mann, T. (2010). *La Montagne magique*. Paris: Le Livre de Poche.
Musil, R. (1956a). *L'homme sans qualités* (Vol. 1). Paris: Seuil.
Musil, R. (1956b). *L'homme sans qualités* (Vol. 2). Paris: Seuil.
Nietzsche, F. (1966). *Thus spoke Zarathoustra*. New York: Penguin.
Nietzsche, F. (1967). *The will to power*. New York: Vintage Books.
Plato. (2011). Theatetus. In L. Brisson (Ed.), *Œuvres complètes de Platon* (pp. 1891–1975). Paris: Flammarion.
Sartre, J.-P. (1981). *L'être et le néant. Essai d'ontologie phénoménologique*. Paris: Gallimard.
Schleiermacher, F. (1944). *Discours sur la religion*. Aubier: Montaigne.
Vatan, F. (2000). *Robert Musil et la question anthropologique*. Paris: PUF.

Objects and "Objects" in the Historical Narration of the Humanities and Jean Baudrillard's Semiotical (Structuralist) Contribution

Piotr Mróz

The history of mutual relations between human beings (the subjective side) and the unique results of the latter's intentional, conscious endeavours—functional objects have been rendered in diverse ways. Irrespective of the type of narration—a mythological, poetic, aesthetic, sociological or philosophical one—it has invariably been grounded in the celebrated distinction between nature and culture, between man and his/her extensions of the senses and organs between something natural, primitive, primordial and something construed hence artificial, forged or produced. Thus in the realm of culture and civilization (if one is still allowed to resort to this once marked opposition) things/objects as ontologically distinct from the natural, the organic, the biological have been endowed with magic power, the access to which has been strictly reserved for those who were to deal with the sphere of the sacred (*sacrum*) or for those who were anointed (chosen), pronounced the spiritual and/or political leaders of the given communities.

Such objects as totems, amulets, talismans, scepters, wands, crowns, speaker's staves, coronation swords, maces to mention only few, played the role which cannot he underestimated. They were the living symbols of the unity, law and order, abiding in primitive and civilized communities throughout the ages. Apart from the material symbols of power, things/objects were also instruments, tools, implements whose history as the monumental work of L. Mumford (Prown, 2000, 12–20) makes us aware of was a fascinating narration. Not only does it show the oft dramatic manner in which we have overcome so many obstacles and hurdles on our way to conquer more and more "territories of nature," to humanize them—the so-called ideal of civilization, to satisfy the five basic needs Malinowski enumerates, but first and foremost to exercise if one may view the human development in such an idealistic, non-pragmatic terms, our inventiveness, creativity or technological acumen (Ingold, 2000, 50–71).

P. Mróz (✉)
Jagiellonian University, Krakow, Poland
e-mail: piotr.mroz@uj.edu.pl

C. A. Hornbuckle et al. (eds.), *Phenomenology of the Object and Human Positioning*, Analecta Husserliana 122, https://doi.org/10.1007/978-3-030-66437-4_16

As composite entities (of artificial provenance), each and every thing/object was made up of two ontological components: its form (a kind of a pattern, a "stencil" or a "mould") an abstract, formalistic element and its matter—or still better—its material. If one resorts to such diverse concepts as the Aristotelian metaphysics, the Heideggerian fundamental ontology or to the Wittgensteinian axiom of the Form (in the logical sense), one can say that the first component makes things be specific carriers of meaning and signification; they will be regarded as symbols or signs—the latter will be viewed in the contexts of semantics of denotation and connotation.

These are, as will be shown by many of the structuralists, the differentiating mechanism in the linguistic system. Put simply, things as members of a system have the language of their own. This motif will tinge most of the structuralist and post-structuralist analyses of *objets comme signes*, including Baudrillard's writings. Owing to this onto-logical aspect of *things-as-such*—operations as primitive taxonomies leading up to more sophisticated classifications (*genus proximum* and *differentia specifica* type) of dealing not only with the natural, but also with the artificial, the "artefact" reality, a full, multi-dimensional, many faceted meaning was established. As may be rightly surmised, it has not the status of something permanent, unquestioned.

On the contrary, as the history of those mutual relations reveal—the meanings/significations of things are ever open-ended, prone to changes and transformations, depending upon the cultural context. In terms of human development, it first came along with primitive tools and weapons or totems. The latter were mainly natural "objects" of organic (animal world) or non-organic (fauna, flora) provenance.

The totemic power, however, lay in its symbolism (the arbitrary choice, convention honored by a given totemic community) (Grimes, 1990). All those paws, teeth, claws, skulls displayed around the totem poles stood not only for unity, wealth and well-being of the given community (the totem of) but provided each member with the sociological, (tribal) and cultural identity of the group he/she belonged to. This animalistic, animistic symbolism (Cassirer, Frazer, and Levi-Strauss) is important from two points of view. First and foremost, totems introduced—as Levi-Strauss underlines—a preliminary form of knowledge (discriminatory taxonomies); secondly, totems were based on the idea of difference (binary oppositions). The totem of the snake's slough did not want to be the same as the totem of the lion's claw and thus preventing an eventual confusion in terms of desacralization.

Moreover, natural objects lost in the process of—as it were—*totemisation*—their naturalness, their directedness. This will be a significant tendency of *today*. They started to speak an "unnatural" language, as the objects chosen to play the symbolic role because of having been endowed with unnatural powers. One should also mention the phenomenon of magic correspondence of every element with all other elements of nature and culture closely tied by a magic knot (*magia sympatica*). This idea—like many other primitive concepts—will paradoxically enough return under diverse guises in recent days of twenty-first century (Maffesoli).

But things, *objects for*, as classic works of Frazer, Mauss, Malinowski, Durkheim, Levis-Strauss, Greimas, Bachtin, Girard or Kopytoff indicated, apart from their magic and totemic functions played yet other ones. Such habits, practices like *kula*

(Malinowski), *potlash* hardly understandable and explicable from the modern angle apparently seemed to totally disregard all economic value of objects. The latter, more often than not, were specially chosen precious and unique items. Both *kula* and *potlash* resorting to specially designated objects turned the latter into symbols of the given community's affluence, good will towards the others ("aliens"): friends or quite contrary to ward off (put off) all potential (eventual) enemies. It stands to reason that such practices were not representative of the only values in the primitive communities (tribes). The inchoate consciousness of functional, transactional, economic values was entering the scene and began to form a more "modern" (technologically oriented) awareness (Glassie, 1991).

In the Marxian, Weberian or Veblenian understanding of the axiological finance significance or meaning of things, it was a sheer lunacy to make apparently unjustified gifts, to throw away material goods which could have been handy, useful and might have played a concrete role in accruing and preserving the wealth of the given community. To our mind, those allegedly irrational, blatantly unpragmatic habits challenging the economic structure were unquestionable signs of dealing with different types of reality—the everyday versus the abstract, the natural versus the cultural. Primitive *bricoleurs* (Levi-Strauss) resorted to all possible means in order to make all ends meet, if one can paraphrase the idiom—to feel secure under all possible circumstances. To sum up: totems and sacred items communicated with the Unknown, whilst lavish and extravagant gifts appeased and/or put off those engaged in the strife fuelled by envy and competition (Rene Girard) (De Certeau & Giard, 1998, 133–148).

As to *magia sympatica*, it introduced the idea of unity of all elements. What is important in all those primitive approaches and interpretations of things-objects (of such a great variety as may already have been seen) is a kind of duality in the perception of their nature. Let us propose three pairs of terms: direct and indirect, literal and symbolic (semantic), objective and subjective in order to distinguish between the opposite (if not mutually exclusive) ways of "using" things by human beings. Those ways will inevitably be related to a certain stage of culture, history or the Foucaultian *episteme* (Featherstone, 1997).

Although the European antiquity (Greek and Roman) influenced by the overpowering Logos, an ubiquitous narration—which according to Derrida was to exert an enormous influence till the time of deconstruction—superseded the mythological world views, the division between *Sacrum* (*Mythos*) and *Profanum* not only had been retained but in certain areas (domains) even strengthened. With a constant and steady progress, the most advanced nations—taking advantage of science and technology (Heideggerian *techne*)—achieved a high level of development.

Things-objects "participated" in this long march in the direction of the promised land of prosperity, well-being and more leisure. It seemed that those extensions of human bodies realizing our potential began to share human fate, destination and good or bad luck. At one stage of our history, some of them enjoyed a high rank on the social list while at the other they found themselves at the very bottom—they were discarded, forgotten, found useless, supplanted by more modern and handy ones. Their oft-dramatic histories (or rather scripts of vicissitudes) were depicted, presented in masterworks of literature, drama and poetry (Auerbach). They were deemed useful but also beautiful, charming and mysterious. From unforgettable, immemorial pages of Hesiod, Homer, Sophocles, Ovid, the morality and mystery plays, Dante and Shakespeare, the nineteenth century realistic novels till twentieth century art and poetry, things-objects (tools along with artefacts) attracted our attention, were sought after, coveted and constituted the *raison d'être* for our actions and activities.

It was as early as in the Antiquity when the cardinal distinction between tools treated as ordinary ("everyday" objects), which were associated with manual hence denigrated menial labor and artefacts, had been firmly established. Such noble *objects-things-for* like pieces of armour, weapons of various sorts, as well as ornaments were strictly reserved for the highest social spheres of the ancient Greeks and Romans and later the distinguished members of the medieval society. One may add that it was mostly the Greek sense of beauty (later on called "a classic" model) understood as the harmonious relation of transcendental qualities that formed the essence of things. The Judeo-Christian approach and understanding of things—due to ingenious adoption of Platonic and ancient philosophy (*Logos*—now the spirit and the Absolute knowledge) juxtaposed with the revealed word of God—produced concepts and ideas distinctly and univocally prefiguring most of modern interpretations of things-objects.

Although handy, useful, credited with their potential in everyday toil, "things" were the epitome of the material aspect of reality (the vain worldliness). So, the worldly possessions (among them *things—objects for*) could turn out to be inimical to truly human goals (the eternal Bliss and Redemption). Moreover, things themselves—fragile, transitory and prone to all vicissitudes epitomized the human condition. Things were in sharp contrast with the ideal of life in the blessed state of poverty preached and promulgated by the official and unofficial representatives of the Christian Church. Irrespective of fundamental ideological differences, the communist doctrines harked back to this messianic message of the poor in mind and body but potential candidates for the paradise (Popper) (Joy, Hui, Kim, & Laroche, 1990, 146).

What is universally referred to as Modernity initiated yet another attitude in so far as the understanding of things-objects is concerned. As Weber aptly observed, things (the material side of existence) were no longer renounced, denigrated or discarded; quite the contrary. They were nonetheless (as the end-products of human labor) the display not only of human acumen, inventiveness, technical skills—all of which helped to conquer Nature, discover new lands, confirm our high position in the Universe (Pico della Mirandola), but a very important factor in itself—they responded to our ever growing, now acceptable material needs. It seemed that our existence, our human condition was delimited by the all-pervasive (active) presence of material objects the variety of which was staggering in the human environment. Not only tools, utensils were valued as useful, functional, simply handy—but first and foremost things (along with special class of craftworks and works of art) acted as carriers of transactional. Economic values. The ideals of modesty and poverty were decidedly discarded by the economically affluent but conceited new class—the middle class, bourgeoisie, whose style of life made flaunting, sporting their wealth and influence part and parcel of their existence.

The invaluable testimony of art: British, Dutch and the Northern School of realistic (naturalistic), painting with. Its unsurpassable meticulous depiction of everyday activities—"business as usual" gives us a unique insight into the mystery and mystique of workshops, home interiors, pieces of furniture and various utensils displayed in offices, churches, alms-homes, infirmaries or abattoirs. The European drama from *Everyman*, Shakespeare, Calderon, Moliere, Webster, having created iconic figures of men deeply involved in the process of acquiring material goods—things tinged with economic, not always functional value then not so important in everyday toil or industrial context, provides us with dramatic and fascinating, even today, comprehension of how mutual relations of *things—objects for* and human beings were being formed, of how the very fact of possessing them influenced human consciousness, exerted a kind of stigma on those who did have and those who did not.

Last but not least, the novel of the seventeenth, eighteenth, nineteenth century—a new genre—is of special interest to us in so far as revealing the true nature of those interrelations in question is concerned. Thanks to special, rich and hitherto unknown potential descriptions, length and duration of narration and fictional dramatization, novels could intimate a kind of knowledge, other genres, let alone other human-cultural activities were not capable of performing. From French and Italian *conts*: Lafayette, Scudery, English: Swift, Defoe, Richardson, Smollet, then the greatest—Balzac, Dostoyevsky, Dickens—the realistic tradition continued in twentieth century, let us into—so to speak—all those psychological states begotten by the desire to use, possess and first of all store, stash away, *things-objects for* these works of art displayed masterfully, vividly depicted psycho-ontological states of degenerated, reified souls and minds in the most adequate way in the characteristic intensity of feelings, emotions which made an aggressive possessor (i.e. Scrooge) feel superior in relations with all other human beings, let alone the feeling of being more attached to (rather overpowered by) the material side of human reality than to fellow travelers in life (existence). The stories of Balzac's, Pushkin's, Dickens', Turgenev's or Dostoyevsky's and Tolstoy's heroes, the iconic *The Death of Ivan Ilitsh* if read from this angle dramatically show how things (coveted and desired, sought after by hook or by crook) made us temporarily happy if we could possess them even for a while and inversely unhappy, depressed, dejected and worthless if we could not afford them at all.

But there is yet another essential issue here. These great masters of literary narrations prefigured in the visionary, unique way the subsequent philosophic analyses by Hegel. Such great figures in philosophy as Hegel, Kierkegaard, Marx, Schopenhauer and Stirner, each in his own way deciphered the nature of possessing and being possessed. This dialectic is a high point in Modernity, of which the Industrial Revolution constitutes a landmark. One can be justified to some extent in regarding the advent of it, the unprecedented and unparalleled in the human history, dynamic growth of mass production, as a completely new phase in the relation of human beings and *things-objects-for*. For the first time the supply of all kinds of products became so massive, so staggering. But it was not the sole consequence of dirty, smoky factories churning out millions of "items" as British Romantics so aptly observed (that was a foreboding of what the Frankfurt School would sharply and scathingly criticize) (Maffesoli, 1997). What was at stake was the birth of a new consciousness, a modern awareness of the very fact of presence of things-objects in the human environment.

Things began to take on a significance former generations could not have dreamt of conceiving and imagining. The unquestionable turning point in understanding the true nature of the relations between things and men came with perspicuous analyses of Hegel and Marx. Having adapted while transcending the Hegelian panlogic, firmly believing in the total identity of idea (thought) and reality, both aiming at the same goal: the Absolute Freedom of the Consciousness, Marx accepted the Hegelian dialectical method—to wit—a constant interplay of contradictory forces. The latter were, however, not of an intellectual nature, but of a social and economic one. Hence, it was Man who in an unrivalled move pointed out the enormous importance of human labor in the process of humanization of the whole mankind. In other words, each product (a thing of man's provenance) as the end result of our intellectual and manual activity showed successive stage in conquering the brute, blind forces of Nature.

But what was thought of as a favorable process for the whole of mankind in general soon turned out to have been one of the most atrocious practices of social inequality, the unmistakable sign of deepening economic exploitation. Those who possessed the means of production (a special privileged kind of *tools-objects-for*) had the upper hand over those who were not in such a favorable position. It seemed obvious and taken for granted that the social superstructure of such forces as ecclesiastical, legal, educational, military and political ones did everything to strengthen this awareness that the owners of the tools, the latter pertaining to more and more sophisticated and complex structures, could not only organize the production but dictate most of the conditions related to it thus forming and informing the social reality to their own advantage, it goes without saying.

The universal motto was: more and more products, the shorter the time of production, the lower the cost of labor. Due to these inhumane practices each product gained a magic trait: the added value. The latter was the sole possession of the class of owners who accumulated the capital and re-invested it in lucrative enterprises. The active, productive side—so to speak—of those who carried out the actual process of production namely the working class (more adequately—the proletariat) was simply forced to sell (at the lowest price) the only possession they had: their labor. This wild, unhampered accumulation of capital robbed the workers of everything. Human beings have a natural right for: health, education, social and economic stability and dignity. No wonder the revolutionary "dialectical" clash was imminent, and the roles (so vividly depicted by Hegel) of the Master and the Slave was to be reversed.

The growing awareness of the horrible, denigrating *anomia* and deprivation was to lead to the creation of a new class *in itself* and *for itself*. What is of interest here is the Marxian blueprint for a new society consisting in the revolutionized relations of men and things. In order to better render the essence of this utopian—as it is—project we must pay due credit to Marx and his analysis of the two phenomena: having and being. (The concepts in question were to turn out to be one of the most seminal ones in modern and contemporary philosophy). Unsatiated desire to possess—underlines Marx—acts like a double edged sword. Once a subject gets his/her own way—that is—finally comes into the possession of the desired object, he/she, paradoxically enough, is being possessed by it, becoming its slave as it were. His/her thoughts, feelings, emotions, projects, in a word the whole being is so strongly attached to the possessed object that it would be impossible to take his/her mind off it. It is the very object that rules, dictates its conditions, leaves no free space for any movements. Moreover, the possessing and the possessed (note the ambiguity of the term in most of the European languages) subject is unable to conceive of, to perceive reality otherwise than in the categories of things. In other words, he/she reifies the whole world. Like the mythical king Midas, we turn everything into the substance, a being—in itself—to be possessed, to be had solely by us.

This inhumane mechanism described and analyzed by many thinkers of diverse intellectual persuasion from the nineteenth century; Kierkegaard, Stirner, Schopenhauer and later on Berdyaev, Marcel (the latter devotes special attention to the phenomena of *Être et Avoir*), Sartre and Heidegger seem to have gathered a capital significance in this respect as well (Lash & Urry, 1994).

Reification in itself leads to the total atrophy of human relations. We treat others as if they were things: unfree, non-transcending, deprived of will and volition. As Sartre will aptly say, they are what they are and nothing else.

The second phenomenon is alienation: the feeling of strangeness in the human world, the acute awareness of not belonging to the world of our own "products" of our realized projects and ideas. Thus, things revealed the other side of the coin. Being handy, useful (the Heideggerian *Zeuge*) they escape from our control and become a menace to our authentic, true existence. It has been generally assumed that highly developed societies with a technological mind-set are prone to be "enchanted," fascinated by the alluring mode of having, discarding the sphere of Being-identified with the notion of human spontaneity, creativity and transcendence.

The latter although marking off our condition as uncertain and fragile existences is an indelible trait of each authentic being-in-the-world. Apart from philosophical investigations into the nature of relations in question it was Psychoanalysis (in its many versions and orientations) that threw a harsh new light on those issues. The desire (*Lust*) is mainly associated with the unconsciousness overpowered by libido—sexual power "ruthlessly" trying to satisfy its aim (pleasure) and release the energy along with the inner tension of the living organism. Possession is like a sexual act—it gives similar pleasure although it is a kind of substitute, it is not the real thing.

The Freudian proposals were soon to be modified and widened by such analysts like Fromm, Horney (the mode of Having, according to Fromm, leads humanity to the state of modem slavery and dependence upon factors which slip out of our conscious control), Bachelard (the psychoanalysis of objects) and quite recently by Deleuze and Guattari (*désir* is the overpowering mechanism of our lives, *désir* however—add those two authors in the famous *Anti-Oedipe*—does not need a concrete object to be pursued). The aforementioned mechanisms of our psyche, modes of being (authentic and unauthentic) reveal an oft-dramatic aspect of our relations with things-objects.

The latter in the analyses and descriptions of contemporary thinkers, theoreticians and artists are grouped here into three (it is only a theoretical proposal) categories. It cannot be argued that what has been understood as technological (as opposed to primitive) societies perceive *things-objects-for* (or complexes of them) as a datum taken for granted, something so "natural," so obvious in our environment that we hardly notice them or pay any attention to in the process of performing their job. (Heidegger was absolutely right when he remarked that the essence, the *Wesen* of things-instruments would be best seen at the very moment of their failure, malfunctioning).

But, and this is a growing tendency today—the so-called functionality of utensils, instruments and various implements apparently so important in our every-day life has been dramatically diminished/lessened. These objects apart from their being handy and useful (the Heideggerian *handiness*) begin to take up another function: that of representation of: "a social marker of" or an actor on the social scene. Ordinary, tinged with triviality and banality of the everyday toil, instruments become more and more stylish (Sudjic) aesthetically more refined and sophisticated all in order to play the divisive role in human societies; briefly, to express the social status of their owners. As will soon be explained in relation to Baudrillard's concepts, *objects-for* will tend to transcend their very instrumentality and acquire new significance (more precisely new connotations). Stylish and fashionable objects more often than not the luxurious substitutes, versions of ordinary utensils, will become symbols of "Better Life," "Dream Life," unfortunately accessible only for those who can afford getting them (Baudrillard, 1981).

So, functionality will cede place to semiotism: objects will not only perform their functions, but additionally they will do another job: that of speaking in their own unique lingo: If you somehow manage to get me, I promise to represent You in society. Thus the new area of social life must be included in all analyses of the relations between objects for and human beings—that of communication of "psychological" interaction, psychological impact of things upon our existence. Maybach cars, Armani suits, Faberge knick-knacks, Castorini gadgets—the most coveted for (Kopytoff), the wildly desired items invade aggressively our private space, our inner domain of unconscious dreams, stifled desires, and unrealized projects. Such things allure us, fascinate and whet our consumeristic appetites but first of all they make unrealizable (in most cases) promises: with us your life will be transformed, you are certain to climb on the social ladder and be counted as one of those whom you admire and envy.

As interpersonal psychoanalysis has observed our deep personality is being channelled into a dangerous narrowed rut: the overpowering desire of possessing, having more and more, instead of being in a more and more authentic way. In the case of unfulfilled projects, dreams dramatically come untrue, we feel the increasing, growing, and the accumulating frustration leading to the most disintegrating mechanism of destruction: aggressiveness towards ourselves and outer reality. Things as objects of our desire may act in this manner as well.

This liaison of things and men, dangerous, inimical to our life and affecting most of the aspects of the social reality has been masterfully, ingeniously shown in the novels of Capek, Bulgakov, Steinbeck, Dos Passos, Hesse, Andrzejewski, Beckett,

Perec (the iconic *Les Choses*). Whereas the Modern Theatre—especially the theatre of the Absurd represented by Beckett, Pinter (especially *The Dumb Waiter*), Ionesco, Mrożek display yet another *modus existentiae* of things (Baudrillard, *Semiotext(e)*). They—like us, human beings—are characterized by ontological fragility, they are prone to destruction, annihilation. Being contingent after all like most of the entities in the world their standing in the society they are supposed to serve is nothing but precarious. After their service (mission) has been accomplished, they may lose all the significance, meaning and usefulness they used to cherish, enjoy and share all common fate of all "has beens." In a word, they may be discarded and thrown away. (The junk scrapyard is a prevalent motif in the contemporary art).

This process of reduction shows the murky, dark side of things: our accomplices, our status keepers of simply the objects of our desires (wanting). Once proud, unique they lose all those qualities (traits), and become poor shadows of themselves. Moreover their semantic aspect has also been questioned. Things—let us point to some passages in Sartre's *La Nausee*—become items in no man's land, experience the spookiness of some of the Surrealist Art, look at Belmer's dolls, Duchamp's utensils, Switerz (*Merz*) projects, some graphic samples of Minimal Art, Manzoni "objects," Pop Art gadgets and you will understand one disquieting fact: things cannot be taken at their face value, taken as being in themselves endowed with the constant essence making them be what they are, and staying in such a permanency for ever. Quite the contrary, luxurious items (objects) taking pride in their uniqueness, studied, subtle forms, precious materials they are made up from, regarded as the last cry of world fashion, may all of a sudden lose their privileged position. Like their proud, self-conceited owners and users they are the subject of the "general" law of social, cultural entropy. The latter embraces indiscriminately people-things and things-people so closely entangled in a mutual *danse macabre*. Look closely at Bacon's amorphic, crumbling, decomposing forms of human non-human, enter into the interactive installations of Boltansky's scrapyards, "junkies" where the items are completely deprived of any claims to self-identity, permanence and even the travesties of social or cultural status.

The message is quite clear. In our late modern, postmodern and post-industrial society both the proletariat of the things (hastily churned out mass-products for likewise clients), the end result of production process attuned to the laws of supply and demand juxtaposed with the aristocracy of "hand-made" artefacts liberated from the laws of economic growth face the imminent death-recycling or in the case of "aristocratic" items' *thesaurization*, stashing away, for the better or the worse days. In this way things and men confirm as it were the vicissitudes of the same condition.

In the perspective of so many diverse interpretations, approaches to this vast domain (narration) of things and human beings there is, however, one leitmotif recurring in most of the theories. It is an indubitable fact that things are as carriers of senses and meanings, connotations that could be imparted to their owners and users. In other words, *things-objects-for* can "tell" their various stories. It seems that the issue of the semantic and semiological aspect is a crucial one. So the contribution made by the French structuralism in general and Jean Baudrillard in particular cannot be overrated in so far as the attempt at enlightening this problem is concerned.

Let us now focus our attention on one of the most fascinating, riveting conception of things-objects (*les choses*) worked out by the French social thinker, philosopher of culture Jean Baudrillard. Although the Frenchman rightly regarded as one of the most significant and important intellectuals of the twentieth and twenty-first his profuse and seminal work under no pretext can be essentialistically unified and treated in terms of a coherent system (irrespective of some recurring leitmotifs of concepts and ideas visible and present in many of his texts). The author of the iconic *Simulacres* is almost unanimously thought to have been the distinct and independent voice of the French Poststructuralism, Postmodernism and to some extent his own version of philosophy of consumerism. The latter constitutes the source of inspirations for Baudrillard's constant, unremitting searching for the hidden (under the thick layer of the everyday texture of modern existence) leading to revealing some phenomena of which we seem to have had no clue whatsoever.

It is probably his critical, uncondescending attitude along with his somewhat reader-unfriendly style that produced the inimitable Baudrillardian way of thinking and writing. As far as the discourse itself is concerned one must be aware that the Frenchman challenges the obvious, the allegedly self-evident and the taken for granted by the majority of society.

Hence, his texts bristle with paradoxes, riddles and the unprecedented innovativeness of views and analyses. Of all the traits or qualities of form and matter it is probably his enormous critical acumen that makes his *oeuvre* so unrivaled. Recalling the concept of the so-called genetic context of explanation, once so widely accepted and used in the humanities, is in a way justified in his case. Born in a very modest family, Baudrillard had no what the Frenchmen deem so highly: *la tradition de la formation*. He obtained (as the first member of his family) an academic degree after having majored in German philology. He translated Weiss's and Brecht's plays and published a monograph on a practically forgotten figure of German revolutionary mystic Wilhelm Muhlman (off the mainstream figure—a move characteristic of Baudrillard's whole philosophical and sociological output). But it was the cooperation with Henry Lefebre—a Marxist sociologist that proved so important for his intellectual formation. Like his close, strict contemporaries Barthes, Blanchot, Derrida, Serres, Deleuze, Guattari, Sollers, Baudrillard is very suspicious of and toward the traditional intellectual currents which formed and informed most of the outlooks, the attitudes of the post-war generation of the whole Western Europe. As is well known these were mainly phenomenology and existentialism, the latter enormously popular in the first two decades after the Second World War. But the young generation of French intellectuals—philosophers and sociologists along with writers and theoreticians of the so-called *nuoveau roman* was fed up with the idealistic remnants still felt in the above-mentioned formations. The new generation in its arduous search for the certain, the unquestionable, turned their attention to the complex phenomenon of language analyzed by the founding father of the whole movement—the almost obscure for his generation Swiss linguist Ferdinand de Saussure. His revolutionary ideas in the field of linguistics (semiology) were critically continued by such prominent linguists (grouped in the so-called Prague School like Jacobson, Mukarowsky and continued by many philosophers in the world).

It was the language—underlined the structuralists—that could and indeed did serve for a couple of decades as a universal, comprehensive, hermeneutical model. The latter's explicative and interpretative, unbiased and neutral, clear and intelligible nature, freed from all ideological presuppositions and metaphysical pre-notions could be applied to almost all phenomena. In other words, language understood as a highly organized system of various, disparate elements and rules (norms) governing this very organization—to wit—the structure on various discernible levels and planes could be adapted to all entities of astonishingly diverse nature.

From a philosophical perspective, structuralism posing the fundamental questions concerning the Reality as such in its various aspects was seen as a real game changer. Not only did it promise to resolve age long antinomies and *aporias* to get rid of devastating philosophical prejudices and myths blocking the way to the world as it is, to its nature (treated as the objective phenomena) but also it corrected, rectified one of the most inimical views (ideas) concerning the subjective pole of Reality.

What is meant here is the celebrated notion of creative, spontaneous, lucid and transparent consciousness. In the Cartesian myth *ego cogito* was the unquestionable "master" of the world. The *je pense* dictated and imposed its conditions upon the world thus making the sphere of the interior life: thoughts, feelings and emotions the only domain worthy of investigating. In brief, the structuralist movement, basing its research methods on the neutral, objective and impartial tenets, was in the position to overcome most of the idealistic (metaphysical) remnants. Of the projects enumerated above, one seems to have gathered a great prominence. It concerned establishing the universal structure (form) capable of embracing almost all areas (domains) of both Nature and Culture—to resort to the distinction of one of the most important representatives of the Movement—Claude Levi-Strauss.

Language then was to be applied to all aspects of Reality which then could be adequately described and analyzed, still better, univocally translated into the terms, into the categories of the linguistic (*la langerie*). Thus language was to become a universal point of reference.

What is—according to structuralism, his point of reference—language as such? As may be rightly surmised there cannot be one decisive, clear answer to this pivotal issue. The model understanding of the language presented in *Le Cours* was reconstructed on the basis of de Saussure's lectures by a group of his former students, and then modified by Jacobson. It counts this complex entity as a system (containing many subsystems) in which it is the structure—of strict, disciplinarian nature which is of the utmost significance. The structure or the form is related to all elements, "building blocks," primary elements, organizing themselves into more and more complex units. In other words, all elements making up language must abide by the rules and norms otherwise the system could not properly function. The rules constitute that which the Swiss refers to as *la langue*—a set of abstract, universal rules and norms obeyed unconsciously on the part of speakers of the given language. The norms are, however, present (reflected) in each linguistic, individual act, each performance or realization of *la langue*. Let us remind ourselves of the basic "ontological" makeup of each language. Its first layer is *la langue* while the second (the realization, concretization of the first one) is called *la parole*. Two points are of essential importance here. First and foremost, the rules organizing the functioning of low, ground level of the language (e.g. sounds, phonemes) leave not much room for freedom in so far as the combination of those elements is concerned.

It stands to reason that we are not offered any options when we face the strict distinction (the pre-given division) between e.g. the voiced and the voiceless consonants, the position (sintagma) of certain elements in the system. Hence the users of the language (*la langue* along with actualization *la parole*) are subject to as it were all those "depersonalized" (alienated) norms. This brought some of the fierce critics (if not detractors) of structuralism to a declaration that this concept of reality and culture, men and their conscious, spontaneous, free activities deprived us of the human qualities and traits. We have been turned into automatons, ego-less and robot-like. We do not speak, but we are spoken by the language as the apt idiom coined by Lacan had it. Along with this understanding of the language came an articulated postulate to leave out language history: its origin, the subsequent phases of development or even its structure. Language was to be synchronically scrutinized here and now in relation to its basic functions, Jacobson enumerates five of them (Revel, 2008, 124–125).

Let us point to the intrinsic quality of language as system. It has been underlined that even at the fundamental, ground level (let alone the more complex ones: morphemes, meaningful units) language as the communication system of exchange—the system of coding (sending the message) and decoding (receiving and understanding it) reveals its nature as an ever-moving mechanism of constant differentiations, the system of internal distinctions. If we consider the so-called Jacobsoniain distinctive qualities (like in the "voiced" *versus* the "voiceless") introducing substantial changes into qualitatively different units e.g. *pat* versus *bat*, or if we change the single element, a single consonant as in *duck* versus *muck* we see at once that language is nothing but a rigid system of interplaying differences.

The most essential, vital for the functioning of this process of communication elements (often compared to series of exchanges between various, diverse "parties" not reduced to speakers only) are of course signs. (They epitomize in many ways all traits, qualities of language as the system). They are of dual, binary ontological and semantical composition. One part (level) of a sign is that which signifies (*le signifiant*), while the other is that which is signified (*le signifié*). It is taken for granted that there is no natural correspondence between a sign and a given part of reality, it grasps and embraces as it were.

Their arbitrariness, like the differentiation itself, is a cornerstone of the structuralist theories along with their diverse applications. Thus "A"—meaning, signifying something concrete (one has to distinguish between the orders of denomination—designation—and connotation) in an arbitrary, conventional manner finds itself in the presence of other signs meaning something else in such a way that there immediately ensues a semantic "conflict," a strife or disparity inside the system. In a word, each act of communication must be geared up with the series of binary oppositions reflecting the very nature of language as a means of those five functions of Jacobson. Structuralism introduces, operates and applies a couple of very important pairs of oppositions (each functioning on a given level of language). Of these the syntagma and paradigm (the fixed order and the possibility of modifications within the scope of the system) a metonime (a part "standing for" a given whole) and a metaphor (such a usage of word signs which means something different from their original meaning), diachronic *versus* synchronic are invaluable tools—hermeneutic instruments for all the possible applications of the structuralist method. One of the strongest tenets of structuralism is the belief that the idea of a linguistic system (its semantic as well as ontological structure) can be applicable to every domain of what constitutes the second element of the binary opposition between Nature (the natural) *versus* Culture (of the human provenance). In this approach to human reality (ranging from social behavior, e.g. the language of gestures to human artefacts—language of things—to wit, instruments) it was the linguistic-like quality that could explain most of the problems that could point to eventual, possible solutions (Revel, 2008). To put it differently: the cultural, in its rich variety of diverse phenomena "behaves"—maintained structuralists—like language, the system of differences and meanings (denotations and connotation) while the real, the concrete substantial items of human reality begin to lose their directness, their sensory experiential impact. Briefly: they transform themselves (more precisely: are transformed or are treated likewise) into highly complex systems and subsystems of signs (the signifying and the signified).

It stands to reason that the above was the obvious consequence of the resignation on the part of Structuralism from the age-long conviction that signs do correspond with things, with reality as such. Quite the contrary—the dual (binary) nature of each sign (its significant—the matter—acoustic or visual layer, and its *signifié*—the invisible idea or notion) confirmed the fact of final divorce of language from reality. As there was no metaphysical connection (substantial relation) between the world and the language, apart from the semiotic and semiological mediation, a great number of myths and mystification could now be definitively and decisively removed from many discourses. By bringing the social, the cultural phenomena, entities closer to *la langiére* one could hope to see a glimmer of light in the vast dark space, enveloped by the darkness and the thick fog of ignorance, which shrouded the relation between things and words. Although early Baudrillard time and again claims to have followed the main tenets of structuralism, to wit, its linguistic methods of analysis along with a specific and unique "metaphysics," rather anti-metaphysics, (e.g. most of the domain of the human world can be translated into terms of language as a system, the universal structure underlying the social, the cultural thus delineating its "substance," finally the revision of the "strong" Ego, consciousness, now limited in its prerogatives and claims) the Frenchman introduces such new elements as Marxists social and economic ideas concerning the status of a class society, the Freudian theories of desires and frustrations, the psychoanalytical meaning of possession and appropriation.

All these motifs in combination with the French version of structuralism did eventually lead to his gradually more and more iconoclastic, hence controversial position in the discourse of social sciences and the philosophy of culture. Having accepted the fundamental concept of structuralism (reality is translatable into a sign system), so ingeniously worked out, interpreted and reinterpreted by other French structuralists and social philosophers, Baudrillard focuses his critical attention on the phenomenon of everyday life (more precisely on such its elements that make it distinguishable and identifiable in juxtaposition with other aspects of social life), this vast, yet mystified area of research. The latter will embrace every man, human beings engaged in social (cultural) activities such as a production, exchange and consumption of various goods and the goods themselves in this case—things, objects supposedly performing diverse tasks. Although the language spoken by things—declares Baudrillard—cannot be ideally analogous with a linguistic system of say French or English, it bears out all characteristics of *la langue* (Baudrillard, 1981). To state that all of its elements—things-objects will display system-like behavior, will comply to the set of rules (the structure or the code) which will, in turn, influence the way they are used by those who are in this kind of communication process, communication exchange equals with saying that Baudrillard—on both the theoretical and practical level of investigation into the texture of social reality (megasystem), describes and analyses the system in terms of *la langue* and its *parole*.

Following closely in the footsteps of de Saussure, but also taking into consideration the works of his teacher Lefebvre, Baudrillard is fully aware that each structural organization, apart from displaying this sign-like nature has another very peculiar trait. If each system can rightly be treated as a set of elements and the rules governing the ways we combine them into more complex unities then under no circumstances the given system can be identified with the sum total of its elements and rules. It is a propensity closely related to the nature of the very elements making up the whole system. Namely—signs. The latter constitute the world of their own—so to speak—having no natural correspondence with transcendent reality. In other terms, there appears a kind of elusive characteristic of transcendence: the sum is being something more—semiotically, semantically, logically—than a mere collection of the elements. Still better, such "mechanisms" like metonimes, metaphors, connotations, and significations overcome mere *signess* of the signs.

In what way then things-objects will behave like signs? It stands to reason that we must treat them in terms of arbitrariness regarding their senses and meaning. Moreover, a clear cut distinction should be introduced between the order of denotation (a given sign pointing to an "object") and the order of connotation (many senses and meanings are attached, as it were, to a single denotation). As will be seen in due course of the present paper Baudrillard will present quite a significant number of everyday objects, from home appliances, pieces of furniture, utensils and tools, which apart from fulfilling, as it were, their basic form, essence—being functional, hence being denoted as concrete objects: lighters, cookers, tin openers, beds, carpets; may also enter into the domain of complex connotations, like signs to which structuralism reduces things-objects (Baudrillard, 1985).

The latter being members of a given system may assume an unpredictable number of extraordinary shades of meaning, establishing various connotations which transcend ordinary, standard denotation formally delineated by the definition of a given object. One of the reasons is that signs (*respective* objects) are always juxtaposed with other signs-objects appearing in a certain perspective and in certain contexts.

That is the cultural framework—a kind of *a priori* which will inform one's way of looking at them, of interpreting or misinterpreting their sense and meaning. As Baudrillard has it himself "each of our practical objects is related to one or more structural elements, but at the same time are all in perpetual flight from technical structure towards their secondary meanings, from the technological system towards a cultural system" (Baudrillard, 1996).

This motif will play a crucial role in Baudrillard's reflections on things-objects. In social, cultural and human interplay they (as later analyses carried out throughout six decades were to prove) tend to more semantic, more sign-like and less and less object-like. In other words, desubstantialized things will start, or rather will be launched into a process of departure from their "true" nature-as-beings-in-themselves and will take on a complex cultural (language-like) subsistence complying with the rules of the linguistic codes: oppositions, figures of speech, oxymorons, metonimes and metaphors, all accelerating, as it were, the process of transforming the substances into language forms. It is the human (cultural) reality that regulates the eventual, possible translations, bringing substances—in this case, things-objects—into language as the communication system. Things begin to speak to us at the same time revealing their *signess* (as will have been remembered, constituted by the two planes, levels of *the signifying* and *the signified*) while entering into the intercourse of coding and decoding. This structuralist *langerie* seems to affect two parties: us—so to speak—the things-objects we use, we resort to in our day-to-day toil, namely us: the users who participate in the process of communication. It is a kind of a game whose rules pervade all areas of social life.

According to structuralism (Baudrillard seems to fully accept this tenet) human beings are subject to all these norms and rules—to the form which we must abide by. Thus, human consciousness loses its privileged status of a free, transcendent and spontaneous being. It is quite otherwise: instead of creating its world, consciousness (once the proud Ego, acting like a ruler of the whole reality) is being created itself. This view will lead to the drastic announcement of no less and no more than the death of the subject. Likewise things, men share the same destination and fate.

Like his predecessors and teachers, Baudrillard—the structuralist—focuses his attention on everyday reality, bringing the structural hermeneutics to the bustling phenomenon of social and cultural existence. Keeping in mind the basic tenets of structuralism, the Frenchman meticulously distinguishes between the level of the abstract (the form), hence the depersonalized aspect of the linguistic system, and its individualized realization (its *la parole*). The realm of objects is seen by Baudrillard in terms of technology (the latter might be compared to such ideas and concepts as those worked out by Jacobson—the poeticality). The literateness unifying diverse texts or the mythological (Barthes).

Technology, closely related to the production process is regulated by matrices—"elements"—called by the philosopher the *technems* (an equivalent of morphemes in the linguistic system). We find the philosopher saying: "every transition from a system to another, better-integrated system, every commutation with an already structured system, every synthesis precipitates the emergence of a meaning, an objective pertinence that is independent of the individuals who are destined to put it into operation. We are in effect at the level of a language here, and, by analogy with linguistic phenomena, those simple technical elements—different from real objects—upon whose interplay technological evolution is founded, might well be dubbed technems" (Baudrillard, 1996, 7).

In the opening chapters of the *System*, Baudrillard attunes, as it were, the very subject matter to the form he is employing to best render the analogy of things and signs—linguistic elements. One is fully justified in detecting in this a film technique reminiscent of shooting a commercial. In the foreground of the text the philosopher focuses his camera-eye on a chosen object (a single element belonging to the system). Be it a thing-sign, like for instance a coffee mill. All other coffee mills are kept at bay. This particular item is a kind of *la parole* of technology (that is, *la langue*. The universal megasystem of norms and rules governing the production and possible application) a concrete realization of the abstract, depersonalized layer of language. It should be noted that in accordance with the structural approach, Baudrillard leaves out, discards the diachronic perspective (the history, development, modifications of coffee mills) solely concentrating on the synchronic aspect—the here and now of the object along with its relations to essential attributes distinguishing and differentiating one class of objects from another, as well as one individual from another. This distinctive propensity is referred to by Baudrillard as "objective." The latter is always juxtaposed with the subjective. What is characteristic of our technical culture of today—adds Baudrillard—is a growing tendency of perpetual flight from technical structure, towards secondary meaning. The result of this flight is a direct threat to the objective status of the object itself.

From the perspective of the philosophy of culture one must discern the process of a gradual shift. Things-objects depart, as it were, from the rigid system regulated by *technems* ("technicality" and "functionality") and drift towards the abstract system. The appearance of a new object of its kind is of no significance. The innovative new products will share the same fate. Because of the subjective, oft-irrational inconsistency, the irrationality of needs, things begin to forcibly populate (if one may use such a metaphor) a cultural system. In other words, de-objectivization supersedes such a strong *technem* as functionality (by the way, the latter is open to all modifications and transformations, as even the structure itself may be prone to culturalization). Baudrillard's penetrating analyses point generally in one direction. The technological sphere of human world from the moment of its birth, throughout rapid and uncontrollable growth (especially in the capitalist society) aiming at fulfilling the unrestrained needs on the part of consumers faces the "inextricable complexity" (Baudrillard, 1996, 9). Apart from pure technical objects with which, as subjects, we never have anything to do—underlines Baudrillard—we can say that the two levels of objective denotation and connotation are not separable. It is due to this process of imparting of meaning and sense on things-objects that the two parties are changed, are transformed. If the analogy to *la lingerie* is right, then we can say that the acquired meanings/senses of things make them much more similar to signs. The latter of intrinsically dual, binary nature informs us, communicates with us, while some are even capable of expressing emotion (recall the functions enumerated by Jacobson).

As Baudrillard has it, it is the denotation and connotation whereby the object is commercialized and personalized, whereby it attains utility and enters into a cultural system. In other words, it is the fruitful moment when the object begins to speak to us.

In many examples produced by the philosopher (e.g. furniture, the items decorating the interior of modern homes) we see a series of binary oppositions (e.g. traditional—the old language addressed to the conservative and moneyed classes of capitalist society versus the modern, but less affluent), functional versus non-functional, hand-made, unique artefacts, having atmosphere versus mass-produced atmosphereless. As one can see, things-signs may use a variety of *la parole*. The language spoken by antique pieces of furniture, in traditional bourgeois interiors, must by definition differ widely, substantially from that of modern households. In the first case—Baudrillard makes his case more precise—the response on the part of receivers of the message passed on in the language of tradition is expressive of old symbolic order. These responses are none other than acts of communication with traditional values, legends, myths or history of a given social class. It is no wonder that modern—less glorious, glossy—says the philosopher—pieces of furniture must, like signs in the linguistic system, be in natural, to wit, cultural opposition to the former. Modern items liberate us from old rituals, conservative ceremonies, the entire "past" ideology we, modern people using modern codes, would never be able to understand. As modern items—to resort to the structutralist terminology—display different, and more often than not, less intricate sets of connotations, the understanding of their message requires different hermeneutical means of grasping. But this process, distinctly discernible in the modern society, does not liberate us from the growing tendency of all-powerful *la langerie* rolling in our lives.

According to Baudrillard, objects become less substantial, and more like "flying signifiers," superseding and removing their ontological originals ("hard" beings-in-themselves, endowed with matter and form). It seems to be one of the consequences of transferring things-objects into the sphere of the linguistic system of assigning to them a semantic role. Their *la langue* is technicality, functionality, along with the given, highly restricted rules regulating their position in the system (the structure). It seems only rational, and Baudrillard does not contest this point that we resort to coffee mills, lighters, beds, chairs and other pieces of furniture or home utensils, to perform concrete functions. To cut the long story short—they "do" certain jobs for us. But functioning in a cultural (social) milieu—in everyday register, these items undertake many linguistic functions contrary to their ontological make-up, but analogous to those of signs, figures of speech or even rhetoric. Thus one thing-object-for-living (e.g. a house), is rightly treated on the plane of *la langue* as a carrier, receptacle of such abstract qualities, traits which may be grouped under such terms as "provider for safety," "comfort," "protection against" versus the "lack of," "discomfort," for instance, "the state of homelessness." But due to the process of *la langerie*, i.e. the mechanism of connotations, the house's *la parole*, the actual realization may take on a variety of meanings and senses. In short, all objects, things—classified as "posh," "unique," may stand for (an order of metonimes) personal or class well-being, wealth, a high social status of its owner—the qualities not directly associated with houses—as buildings performing assigned roles.

Let us carry out this analysis in the manner a structuralist thinker may have done. The parole of this house should be the concretization of houseness (the essence) of all objects belonging to this subsystem. The object in question occupies an allotted position among other objects similar to it. At the same time, it finds itself in a "natural," that is cultural (based on social way of mutual exchanges) opposition to other items of its kind. Denotation starts to spread around itself, as the case may be, series of unprecedented, unexpected connotations, which the object does not have or evoke. Moreover, it may be used and perceived in the social perception as a metaphor, the sense of which is not to be literally taken, or a symbol of something which is not usually linked with the concept (*la signifié*) of houses, or it may act as a metonime (one house standing for all the richness in the most prosperous quarter of the town).

Would we be justified then in concluding, due to the process of semantization that the concept of *housness* has taken on some new meaning, having been superseded by some other un-house-like "accidents" (in the old Aristotelian sense)?

The Baudrillard of the latest works will use a colorful metaphorical term: *flying signifiers*. Whereby things become so "unstable," so unlike things or objects. As a matter of fact, their ontological status will point to the culturally transcendent status of signs constituting not the reality, but hyperreality, so characteristic of the age of post-modernism, post-technicality.

References

Baudrillard, J. (1981). *For a critique of the political economy of the sign*. St. Louis, MO: Telos Press.

Baudrillard, J. (1985). *Semiotext(e)*. Cambridge, MA: MIT Press.

Baudrillard, J. (1996). *The system of objects*. London: Verso.

De Certeau, M, & Giard, L. (1998). *The practice of everyday life* (Vol. 2) (T. J. Tomasik, Trans.). Minneapolis, MS: University of Minnesota Press.

Featherstone, M. (1997). *Undoing culture, globalisation, postmodernism, and identity*. London: Sage.

Glassie, H. (1991). Studying material culture today. In G. L. Pocius (Ed.), *Living in a material world: Canadian and American approaches to material culture*. St. Joim's, NL: Institute of Social and Economic Research.

Grimes, R. L. (1990). *Ritual criticism: Case studies in practice. Essays on its theory*. Columbia: Columbia University Press.

Ingold, T. (2000). *Making culture and weaving the world*. London: Routledge.

Joy, A., Hui, M., Kim, C., & Laroche, M. (1990). The cultural past in the present: The meaning of home and objects in the home of a working-class Italian immigrants in Montreal. In J. Costa & G. J. Bamossy (Eds.), *Marketing in a multicultural world: Ethnicity, nationalism, and cultural identity*. Thousand Oaks, CA: Sage.

Lash, S., & Urry, J. (1994). *Economies of signs and space*. London: Sage.

Maffesoli, M. (1997). The return of Dionysus. In P. Sulkunen, H. Radner, & J. Holmwood (Eds.), *Constructing the new consumer society* (pp. 21–37). New York: Palgrave Macmillan.

Prown, J. D. (2000). *The truth of material culture: History or fiction?* Michigan: Michigan State University.

Revel, J. (2008). *Dictionnaire Foucault*. Paris: Elipses.

A Criticism of the Current Subjectivist Totemism in Werner Herzog's *Grizzly Man*

Victor G. Rivas López

> *In the future, no creation of any kind will have an author, a foundational totem animal.*
>
> Vilém Flusser

1 Introduction

Whether it has been brought about by the threat of an imminent ecological disaster of unimaginable consequences or, on the contrary, by the hopeless tendency towards mysticisms of every kind that reappears time and again in history, the fact is that one of the most intriguing sociocultural phenomena of our time is the *sui generis* denial of the natural or substantial superiority of man over animals (whether wild or domestic). This has led to the vindication of the animals as beings whose vital potency always is in harmony with nature, which they can share with man when he is worthy of it, namely, when he renounces willingly the abstractedness of reason and the shallowness of civilization (including the coexistence with his fellow creatures). This idea is what I call "totemism" and implicates at least three things: firstly, that the vital potency of animals has been disfigured by human violence, particularly during the modern times; secondly, that that potency is neither circumscribed to a certain territory nor manifested through elaborate rituals (as in the ancient totemic religions) but through the intimate affinity that the animals arouse in every tenderhearted person; thirdly, that (contrary again to the historic totemism) the animals will finally redeem us from any kind of emotional unbalance or evil penchant that are somehow or other the outcome of our selfishness and narrow-mindedness.

Now, independently of the fairness of the criticism against the fearful ecological unbalance that in a few lustra have brought about irreparable changes in the climate

V. G. Rivas López (✉)
Meritorious University of Puebla, Puebla, Mexico
e-mail: cupio_dissolvi@prodigy.net.mx

C. A. Hornbuckle et al. (eds.), *Phenomenology of the Object and Human Positioning*, Analecta Husserliana 122, https://doi.org/10.1007/978-3-030-66437-4_17

all the world over, it is worth wondering if the right way to solve it lies in the aberrant desire to become a quasi-animal or, rather, *in a deeper comprehension of the philosophical unity of existence that includes animals as well as natural environments within a world with a human sense*. This would doubtlessly be more decisive because, oddly enough, most of the totemism at issue goes hand-in-hand with a subjectivism that, despite its efforts to disguise itself as a radical discourse, ends up agreeing with the unbalance that it is supposed to fight. In fact, subjectivism (as I understand it) takes existence as a permanent drama that demands that the whole cosmos is the background of the individual performance, above all when that drama leads to the would-be mystic overcoming of any negativity in existence. The absurdity of this is precisely brought to light by Herzog in his picture, which follows the work of Timothy Treadwell, an amateur filmmaker and out-and-out ecologist who spent most of the summer seasons of his last years in the Katmai National Park in Alaska in order to shoot the habitat of the grizzly bears. Treadwell considered them Edenic creatures, endowed with supernatural insights, that suffered because of human cruelty and to which he tried to protect and emulate because he considered himself as a being halfway between man and animal, i.e., a *grizzly man*. Nevertheless, what Herzog reveals on editing the videos of Treadwell is not the totemic powers of grizzly bears, the primeval harmony of nature or the mysterious metamorphosis of Treadwell into a bear-like being but the contradictoriness of the subjective approach to existence. All of which is embodied in the man who tried to fuse with nature though he never could understand it. In other words, what Herzog shows through the videos of Treadwell is how a miscomprehension of the existential difference of man and animal leads inexorably to a deep unbalance not so different from the one that Treadwell himself so eagerly criticized.

2 The Sense of Animality

In the middle of the immense Alaska tableland, surrounded by mountains and under a bright summer sun, two male Grizzly bears seem to repose indifferently to each other on a muddy site, next to a river that murmurs pleasantly. All of a sudden, the larger one of the animals approaches the other grunting threateningly, and they start to fight; after a first attack, both of them move slowly and give the impression of being again in peace until the smaller one seizes the other by the snout and they roll over the mud, firmly intertwined; for a moment, one could believe that the larger one will be the winner, since it has succeeded in flooring its enemy and has its paws over its belly, but the latter has driven its fangs in the ear of the other and does not release it. The two beasts remain unbearably taut and motionless and the larger one defecates because of the terrible pain of the bite of its antagonist; then they start to wallow again and several pieces of flesh fly around until the bear that started the attack moves away from its opponent while the latter gives the impression that nothing had happened. Although the whole scene hardly lasts 2 min, its brutality is overwhelming, and the spectator cannot help wondering how it is possible that such beasts are usually considered the very embodiment of tenderness and limitless benevolence. In fact, bears, whether of the grizzly species or of any other, embody more than any other animal the purity of heart that is so hard to find even in children, which is doubly peculiar in the light of the ferociousness that springs when least expected and that belies the liking that people are wont to have for them. Bears are not peaceful and gregarious herbivores, they are, on the contrary, terrifying predators that do not hesitate to kill their congeners. In view of which, one could wonder if the symbolic value that they inspire does not reveal a peculiar trend in man, *id est*, the useless attempt to exorcise a natural power that is by principle completely beyond his control by caricaturing it. Bears, and concretely the grizzly ones, seem to incarnate *an original experience of vital strengths* that is, however, disfigured by the want of securing an anthropomorphic and teleological interpretation of nature and of existence in general—all in order to make up for the continuous weakness that carries us away. There is, however, a possibility of overcoming such a fake interpretation and discover the experience of animality as an existential that we have not been able to comprehend; in other words, the doubtless primacy that bears have as the animal symbols of human sentiments could lead not to an edulcorated vision of nature and, by and large, of humanity but, on the contrary, to a more critical perception of the phenomenological integration of both of them.

If we follow carefully the development of the brutal fight just mentioned, we see that there are three elements in it that the vulgar childish image of bears tends to mystify—*total strength, bestial aggressiveness and incongruous rest*—which make up a unity fundamental for explaining how man tries to make sense of his presence in a world subjected to forces that express themselves in cycles that are mostly unfathomable to him. This capacity means that independently of the way the animal experiences the vital potency, it somehow or other is for man to penetrate it through the image of a being that embodies simultaneously instinct and self-command. The brutality of bears can change as if by magic into the most touching tenderness, and the process is reversible: this tenderness or benevolence can for its part become that limitless power that man enjoys in his experience of nature, however much he had symbolized it by an exigency imposed by the deities (let us think of the importance of a bloody sacrifice in any religion, Christianity included, according to which sin would be unforgivable without the sacrifice of the Lamb of God). The animal plays so a decisive part as the medium to span forces that make us feel anguish because they ride roughshod over our rational determinations, which are expressed in the vision of nature as a process aimed at our personal happiness (Deleuze, 2010, 28–29).

This mediating and symbolic function that the animals carry out among so much opposed existential planes brings to mind the idea of schema, that is to say, of the structure for making perceptible the transmutation of an element into another or its integration into a field other than the one to which it originally belongs (Ferrater Mora, 1994, 1106–1107). Defined this way, the schema always works when there is at hand an insurmountable difference as it happens, for instance, between the everlasting cycles of nature, their unfathomable crisis and the human want of making sense of it. Without the schematic image of a becoming that integrates the most violent oppositions, it would be preposterous the image of an animal that can be quiet in the middle of turmoil and that rushes instead at an opponent when there is no apparent provocation. What compels it to act that way is incomprehensible for us, so that we need to schematise or mediate the process in order to integrate it within our experience of reality, and the schematic element is, in this case, the absolute determination of instinct that fixes the animal to an environment through a series of wants, such as survival or as reproduction, which can be satisfied in the most variegated ways (with preys that can indifferently belong to this or that species or with casual matings where the female is just perceived as a duct for the physiological discharge). The schema is then an image that links a universal determination such as the instinct with the variation of reality such as the diversity of habitats and of animals. That allows us spanning the abyssal dissimilarity between the all-embracing order or nature and the organic multiplicity of the vegetal and animal life. Still more, the schematic function would allow us to pass from the compactness of the animal body that solely moves by immediate wants to the emotional human complexity that projects instead symbolically those wants to new sentimental and creative senses. Thus, thanks to its analytic potency and its imaginative adaptability, the schema is for understanding why it is so popular the caricature of the animal as a human being endowed with a childish attitude or with a benevolence that belies the brutality of the natural environments where every animal, whether powerful or not, is equally subjected to the iron law of survival.

Of course, the caricature is not the only way to schematise the link that man has with animals and with the inexhaustible potency of nature; a lot more important is another form of schema, the *totem*, which works not through the childlike reduction of the animal strength to a playful or sentimental illusion but through the identification of the human being with the pleiad of vital forces that the animal embodies within a certain environment. Beyond the substantial opposition of man and animal, there is a hidden affinity between them because they are equally subjected to nature, however much man had tried to overcome that servitude with his industry, which is, nonetheless, useless when a cycle reaches crisis point and the existence of every living is threatened by hunger and misery (Heidegger, 2000, 180–182). Viewed, then, as a schema, *the totem merges the human being and the animal within the vital diversity of the environment but shapes it through the emotional expressiveness in order to give man the possibility to participate in an ontological fullness that would otherwise be almost unbearable for him.* As already mentioned, although nature acts more times cyclically, it breaks that regularity all of a sudden and then man has to face his weakness and his belonging to the common fate of the livings: scantiness and want. Even more, the totem is not merely an object like any other that someone manufactures or creates; it is rather a being with a *sui generis* consciousness that acts in favour of man provided that the latter respects its power and does not trespass the domain ruled by other natural forces and by their respective servants. This explains that although the totem is not precisely omnipotent (for it is fixed to a territory), it has at any rate a power that man can learn to master by himself, in a process during which his own body will become as strong or as resistant as the totem and, through it, as the animal, which endures the harshness of existence all the time. Unlike the plush doll that provides above all a sentimental schema for the psychological development of a child or for the emotional fusion of a love relationship and also unlike the religious image in the strict sense of the word, which furnishes a metaphysical and moral schema to identity the will of the believer with the will of God, the totem supplies a thaumaturgic, operative or technical schema to merge the own being directly with the force of nature (Rivas López, 2010b, 181–196). It prevails upon the ominous untimeliness of the vital forces and acts through the animal that embodies them, which for its part allows man to release himself from their tyrannical unevenness. Even more, the totem makes possible for man to incarnate the animal schematically and to overcome at the same time the physical (or psychological) weakness of human condition on a natural plane, which is why the totemic images always represent the most powerful or dangerous animals, whose strength becomes magically or mysteriously human without losing its devastating violence, particularly when weather or environment compels the human being to hinge upon the unforeseeable upheaval of the natural cycles (as it happens, for instance, in the desolated northern regions where the bears inhabit and where man is at a disadvantage even in the middle of a group that can be dispersed or annihilated by a storm).

It is comprehensible now why the bears have for us such a symbolical relevance among all the animals independently of the apparent contradiction existing between their, on the one hand, ferocity and, on the other hand, the childish feelings that we project on them. Bears are terrible predators for they have sharp claws that are usually concealed by the fur that covers their body and by their quadruped position but that spring as soon as the animal rushes into the attack, which proves that they are more frightening than any member of the family of the felines (which are instead usually thought of as the worst carnivores). The ferociousness of the species is more often than not passed over because of the deceitful appearance of a harmless creature that can at best be domesticated and entertain us in the parks as it was so frequent not so much time ago. Furthermore (and this is worth emphasising), when bear is standing on its rear paws it resembles a fat but peaceful person, which is, obvious in one of the first scenes of the film that is our thread, when an impressive male appears standing under the top of a tree and scratches himself against the trunk, grunting softly, as if he were experiencing a great pleasure; the movement is decidedly funny, for the animal seems to be dancing like a puppet or a drunk man. However, hilarity vanishes at once when one sees right in the centre of the shot the sharp claws of the bear extended toward the camera, while Treadwell comments that this apparently amusing animal can kill, bite, and decapitate. The image is then a lot more perturbing than that of a tiger or a lion: the apparent harmlessness of the animal makes you believe that he is able to feel and perhaps even, why not, to requite the devotion of his self-appointed protector, whose remarks spoil nevertheless such an idyllic presumption. What the viewer sees here is not a funny relaxation, an expression of an easiness that is equivalent to the acceptation of the human presence; it is instead a force at rest that who knows why lets man go on unscathed instead of crushing him. Why does the bear act as if he were unaware of the man or, even as if he were ready to coexist peacefully with him? The most intriguing feature in this scene is that this behaviour of the animal reappears time and again in the presence of Treadwell, inducing the idea of an ineffable supra-rational community of beings that allows man to coexist with the most dangerous animals as if he were with his fellow creatures or, more than that, with beings that are by far preferable to the humans because they enjoy a direct intuition of nature. They can share it with those that have become worthy of it by spending all the summer seasons in the solitude of a dangerous habitat to film how the animals live. This is what Treadwell proclaims exhilaratingly at a certain moment when celebrating the consistence of the faeces of one of his so-called quadruped friends as if they were an almost sacred relic or the sole keepsake of his beloved. As a matter of fact, the short films of Treadwell show time and again the *sui generis* closeness that he succeeded in establishing with the several bears that inhabited in the park—in almost every scene the bears appear a few steps away from him, as when he speaks pleasantly next to a female that looks restless in the background or when he gets into a river where a male swims and even dares to stroke slightly the head of the animal before the latter turns threatening to him without, however, trying to attack him. As if the totem were not necessary at all as an objective mediator for it is somehow or other always present thanks to the human goodwill and the animal mysterious intuitions (which is precisely what I called in the title of this essay the "subjectivist"). Treadwell and the bears act all the time as if they recognised each other and accepted their abyssal differences in an edenic harmony that unfolds through the whole of nature.

Independently of its function in the film, this affinity or totemic community props the schematic sense that the animals and concretely the bears carry out and corroborates moreover that it has a value in accordance not only with the physical appearance of the species or the individual at issue but also with the imaginative or sentimental possibility that every person projects in certain circumstances. That is to say, the animal embodies certain dynamic configuration of the world within a natural environment, such as the Katmai Natural Park, just like someone embodies a certain emotional experience within a historical medium, such as the modern and specifically American society, where the existence of everyone is supposed to be beyond the natural determinations but also beyond the historical restrictions. This is why the bear stands out once again among all the other wild animals as the one that is able to agree with the ambiguousness that rules the complexity of the human sensitiveness, from the protectiveness that a strong being always embodies (which could at worst be nevertheless the other side of a fearful aggressiveness) to the tenderness of a child, including the capacity of making a fool of oneself. Thus, the bear provides by its deceitful appearance and behaviour the perfect schematization of the human existence and of the contradictory nexus that man has with nature, which, needless to say, is the ground of whatever metaphysical interest that nature or animals have to us (Schopenhauer, 2003, 79–80). For however much science integrates any vital cycle in a regular rationality that comprises points of crisis, it cannot embrace the mysterious affinity that everyone perceives in his experience of animality as a determination that acts through the own desire and that makes imperative to look after the irrational beings more eagerly than what we could care for our fellow creatures. And although this feature could be attributed to an unbalanced personality or to a wrong appraisal of the vital dynamics of existence, it is undeniable that the animals always arouse in man an intuition that does not aim at a theological transcendence but at a perturbing immanent fullness that imposes upon us and upon every living being. In other words, *the animal brings to light the total potency of nature and compels man to recognise it through the fascination that he feels for a way of being that he has nevertheless fought through history*. For although nature, according to Hegel, has nothing to do with human spirit because it is subjected to the endless reiteration of the vital cycles whereas man always is projected through the lens of the historical becoming of his liberty, it always finds, thanks to the mysterious spell that the wild animals exert upon man, the way to remind us of our unsurpassable defencelessness, which is the very essence of the tragic sense of existence (Hegel, 1993, 47–48).

3 The Sense of Human

Bathed by a delightful afternoon sun and placidly lying down on his side in a nook surrounded by undergrowth, Treadwell tells a she-fox that is next to him why he reveres so obsessively the bears:

> I love you, look at you. You are the best little fox. But how did I come into his world, Iris? Did ever get the story? I was troubled, I was troubled, I was troubled. I drank lots, Iris, you would not even know what that is, but I used to drink to the point of that I guess either I was gonna die from it or break free from it. But nothing, nothing could get me to stop drinking, nothing; I went to programs, I tried quitting myself, I did everything that I could not to drink, then I did everything I could to drink. And it was killing me. Until I discovered this land of bears. And I realized that they were in such great danger that they needed someone to look after them. But not a drunk person. Not a person messed up. So I promised the bears that if I would look over them would they please help me be a better person. And they became so inspirational and living with the foxes too, that I did it. I gave up drinking. It was a miracle, an absolute miracle. And the miracle was animals. The miracle was animals. I live here. It is very dangerous. I run wild with the bears, I run so wild, so free, so… like a child with these animals. It is really cool and it is very serious.

While Treadwell emphasises the thaumaturgic powers of bears, the image changes all of a sudden, and one sees a group of them in the middle of an uneven brook; one of the animals approaches calmly the camera as if it were aware that someone is speaking of it, and then Treadwell reappears. The movement of the image is very soft and gives the impression of a placid integration of the two elements, the human and the animal, that merge in a natural environment that nothing could disturb. The speech that Treadwell delivers, his languor and the calm of the group of bears form so a complex unity that represents that totemic fusion that we have so far examined, which confirms that the bears must not be seen as a species among others but, rather, as a particular embodiment of the vital forces that goes hand-in-hand with the protectiveness and tenderness that can, however, change abruptly into the most frightening aggression. As a schematic image of the identification with nature, bears are the best instance of the unfathomable dynamics of life, of that all-embracing potency that has nothing to do with reason or with human intentions no matter how much metaphysics tries to convince us of to the contrary (Nietzsche, 2002, 5–7). Instead, we can only aspire, at the most, to an image of certain affinity between reason and nature so as to satisfy our want of making sense of the multiplicity of the living forms, of the mysterious adequacy that is at first sight perceptible between them and the harshness of the vital cycles as well as of the brutality of the most powerful predators and the best adapted organisms, the human one included (Kant, 2007, 195–196). Of course, if there is no way to get a rational harmonization, a teleological compensation for the recurrent unbalance and the sudden breaks of the cycle, there is at least a sentimental projection to overcome the human tendency toward the conflict, which is what Treadwell expresses clearly when referring to his problems and the opportunity that he discovered to get through them because of the totemic protection of bears, beings that for their part also wanted someone to protect them from the wickedness of people. *The reinsertion in an existence with a sense of its own is then equivalent to the discovery of a supra-historical community that even resembles the ancient image of the great chain of being but on condition that every alien being is excluded* (Eliade, 1969, 158–160). The encouragement to change was not however in this case in the contact with others but in the perception of the helplessness of bears before the abstract human heartlessness. Consequently, the sentimental projection serves as the nexus with a vital balance through the image or totemic schema of a personal metamorphosis, that of the human into the animal and, by and large, into every living being of the wilderness that is victim of man, like the foxes, which are, so to speak, the alter-totem for Treadwell. Moreover, this points at a quasi-Manichaean or rather neo-romantic vision of existence in accordance to which there are two definitely separated realms, that of the natural fullness and that of the human wickedness, and there is no middle point to span their opposition bar the totemic protection of the animals and the readiness for transcending the human presence. This is emphasised in the following sequence of the film, where Treadwell shows that he lived completely alone in the tableland, enjoying solely the company of his thoughts and of his animals, which for their part embodied at least two different sentimental configurations: the protectiveness and sudden aggressiveness of bears and the playfulness and proverbial cunning of foxes, which made up the two flows of a vitality and steadiness that he has not found in society but in nature, which, as I have just mentioned, reminds us of the romantic criticism against the essential perverseness of man (Berlin, 2013, 219–221). And this is perfectly visible through the Edenic image of the bears in the brook, of the individual in a dialogic soliloquy with a she-fox, in the delight wherewith the camera follows the features of the animal that reposes at her ease together with a man that has understood the falseness and absurdity of human beings.

Needless to say, Treadwell belies such an existential harmony when, on the one hand, he insists time and again on the vital chaos that he experienced before the "revelation" that bears are endowed with portentous powers and, on the other, when he acknowledges at the end that he is in danger on doing what he does because both of the ferociousness of the animals and the solitude wherein he is. Independently of the intentions of the speaker, this feature shows that the basic attitude of Treadwell did not vary at bottom, for just like alcoholism was before a barrier between him and the others, his devotion for bears was the perfect justification for spending several months away from any human contact. Absurdly, Treadwell does not seem to have sought for allies in his crusade although he admitted that bears were at risk and although he had at least two girlfriends that assisted him in planning his expeditions and accompanied him during them (one of whom, Amy Hughenard, was killed together with him) at any rate, he did not seem to have taken them into account as the members of the transcendent vital community that he was so fervently after. Regarding this, it is meaningful that he had not devoted even a single shot to the women with whom he spent so much hard months, and the three moments when Amy appears in his short films are rather disappointing because she is just an incidental presence, as when she is ranging some bundles at the side of the light aircraft that carried them to the national park, when she is sitting next to a bear or, which is the most impressive of the three shots, when she appears stooping down in the foreground so as not to cover up a terrifying bear that rests in the background among the undergrowth, a few steps away from her. This image is very impressive because it shows without the slightest doubt that Treadwell had at least one person that were ready to help him and literally to die for him, which he, nevertheless, does not seem to have recognised. Thus, when he speaks in one of his soliloquies of his love experiences while he walks, he says that he has not been fortunate enough with women although he is gentle and funny, and that he would even like to have been a gay man for in that case his relationships would have been a lot less demanding than those that he had had with women. And it is this passage, together with the way he filmed Amy and with the remarks of the people that knew and loved him (included Jewel, a former girlfriend), what reveals that Treadwell did not find the elemental way to communicate with others, that he always was essentially an outsider, someone who shared the company of others without really being with them, and that only identified himself with beings as impervious as bears or as slippery as the foxes and perhaps not even with them, since he gives the impression all the time of being projecting on them his own daydreams and emotional wants, which is the psychological expression of the subjectivist totemism. Oddly enough, although Treadwell talks ceaselessly and brings up in his soliloquies all that is between heaven and earth, he never properly dwells upon anything and confines himself to perorate, criticise or insult. His vision is that of a resentful person that has withdrawn from others not because they have rejected him but because he does not stand them, which is not the same. And that would merely be a question of personal temperament or valuation if there were not behind the discourse of Treadwell, despite the egotistical framework, a constant reference to the exemplary sense of his enterprise for vindicating the so-called harmony with nature and bring to light the human wickedness. That is why, however much he struggles to merge with an all-embracing vital potency, he remains at the end alien to the deepest dynamics of existence that can solely be experienced (and, why not, enjoyed) through the communication with other human beings.

The distance that Treadwell kept regarding the persons that backed him up is the reflection of that that he kept regarding himself and of his failure to attain a personal balance: according to what he says time and again, the others were completely incapable of understanding how marvellous and sensitive he was, which is the reverse of their blind cruelty towards nature. However, his belief that there is a more pure communication with his animals shows his own consciousness: the elemental interchange of natural wants and the sudden attack against everyone that prevents or seems to prevent one from getting what one is after. The problem is not precisely that he uses bears and foxes as representatives of his vision of existence, since animals have from the beginning played for man an axial part in the symbolic or universal configuration of existence; rather the problem is the kind of animal wherewith Treadwell identified himself. Unlike, for instance, the serpent and the eagle, which are for Zarathustra the elements of a critical but deep coexistence with men because both of them are animals that embody superiority and a wisdom able to challenge the omnipotence of God, the bear and the fox are animals that embody respectively a strength too overwhelming and a craftiness too slippery so as to change them into an experience of a really free spirit. Thus, when Treadwell says: "It is so weird though, when it sinks in how alone you are," he is not solely expressing an experience but an attitude that has, as far as I see the issue, a structural value that goes beyond the incidental vital unfolding and points at the philosophical sense that totemism can have as a subjectivist experience of nature. This is all the more evident when one understands that the final aim of totemism is not the liberation of nature from the human brutality but the conversion of man into a being able to exert his strength and his craftiness with the would-be innocence of the beasts that are nevertheless subjected to the ineluctable violence of instinct. The most curious feature of this structural attitude is that although it seems to be against the shallowness and the cruelty of the common subjectivism, it is at bottom a much worse variety thereof: Treadwell slights others because they spoil nature and shows himself as a "free spirit" in the Nietzschean sense of the expression that demands to subvert completely the abstract values that average people vindicate but he does not realise that he acts throughout on the level where the weak ones are located, that of the shallow criticism and the overwhelming violence that the free spirit is supposed to set aside. Is it really so different the defying attitude that Treadwell exhibits during most of his shots from that of the average individual that looks for opportunities to give vent to his anger? Contrary to the Nietzschean "free spirit," who experiences nature as a way to carry out a "transvaluation of all the values" that he has until then taken for granted (Danto, 1980, 114–116), the individual that Treadwell embodies takes nature as a thing in itself endowed with substantial potencies that will lead the "chosen ones" to the new epiphany of being that is at any rate a regression to an abstract consciousness, which belies the ideal of overcoming the metaphysical vision of nature and also makes impossible at the end the would-be liberation of man. It is not a case of recovering a lost Edenic fusion, let alone of changing oneself into a natural being (whatever this means), but of experiencing a new relationship with reality and with ourselves that demands to forget the Manichaean opposition of man and nature.

I insist that subjectivism takes instincts as transcendent forces that ought to be unrestrictedly expressed (Marcuse, 1991, 11–12). This approach, which Treadwell took to its extreme, goes hand-in-hand with the idea of the totemic protection of some wild animals, which strengthens the animist conception of habitats and environments that endows them, moreover, with benevolence towards every living being and particularly towards man (which the latter is nonetheless mostly unaware of). Although this is not a fatalistic determination as for example that of the biblical expulsion of paradise, man can and even must realise and fulfil those instinctive forces in himself, which would mean a return to nature. Now, although this seems *prima facie* for everyone, the fact is that only a few people are able to carry it out because it demands an unyielding resoluteness and, which is a lot harder, the total oblation of one's humanity, a condition *sine qua non* for the process to be accomplished because humanity is, after all, the very historical root of every evil, since it always tries to overcome nature. Therefore, solely those ones that are ready to face the utmost risk deserve to attain the final metamorphosis of man into a bear-like, into a fox-like or, in general, into an animal-like being literally beyond good and evil (Nietzsche, 2002, 109–110). However, the individual that is at least imaginatively after this metamorphosis remains on a selfish plane of consciousness and does not realise how he participates in the configuration of his own existence: just like the wild animals embody and express limitlessly their instincts, the man that tries to transcend the limitations or rather the phenomenological finitude of humanity has got to dispense with the elemental want of coexistence. Alone as nobody has ever been but without being tormented by it, this *sui generis* antihero will be the final aim of an evolution towards the fullness of the Edenic genesis where there was no opposition between man and nature, which is at least questionable because there is no creator to secure it.

Surprisingly, when the unheard-of free spirit that has merged with the totem should suit the action to the word and show men how wrong they are on vindicating all that that separates them from nature, it just vanishes, and in its absence one perceives the most potent and despicable of all human emotions—that odd mixture of fear and mistrust that is the inevitable consequence of a subjectivism that has denied the reality of everyone else and of humanity itself. In one of the most amazing scenes of his short films, we see how Treadwell hides behind some bushes while a group of members of an expedition keeps a bear at bay by stoning it. When the out-and-out apostle of the new gospel of nature should come into action and fight the intruders to the Eden, he seeks for a shelter and prefers to be safe instead of taking up the cudgels for his so-called protecting divinity, which confirms that the paraphernalia of the totemism are a simple strategy of the subjectivism to reassert itself not beyond the conventional values but before them, that is to say, in the mental realm where everything is equally subjected to the arbitrary opinions and beliefs of oneself. Instead of carrying out a transvaluation of his personal way of experiencing the vital dynamism, the individual has limited himself to idealize nature as an all-embracing harmony where the worst expressions of instincts and wants (violence and unconsciousness) are interpreted as the effusions of a mysterious, inexhaustible entity that acts from a blurry transcendence that is at bottom a kind of caricature of the ancient providence, which releases in a way the individual from whatever responsibility because his enterprise does not spring from a conviction but from an emotional outbreak (or, why not, from a blind desperation). This explains that whereas animals are even for the historic totemism elements for man to recognise himself within nature and trace a field for his action, they become the final aim of the subjectivist vision that changes man into an appearance and nature into the sole reality. Metamorphosed so into the ground of existence and without having been attuned to the passional and symbolical framework of human existence, the animals end up being the derisory guides of the free spirit instead of embodying the phenomenological sense that they have as the living manifestations of the irreducible plurality of being whereby existence is never explainable as a mechanism or as a pre-established harmony. Still more, this real value of the animals, by which the free spirit has time and again to reshape experience, is all the more embodied in other human beings who compel the former to modify the emotional framework of the own identity and the relationship between what he wants and what he is able to attain so as not look for redemption in the most aberrant vital manifestations, namely, the blind strength or the craftiness (that is however hidden behind the bushes!).

In the light of all this, and contrary to what some people could think on reflecting on the ardent wish of Treadwell to become a bear, it is more than doubtable that the subjectivist totemism stands for a real option in the search of a new religiosity that could overcome the so-called indeterminate characteristics of, for instance, Christian religiousness. The reason thereof is not hard to find: *man cannot experience himself as a natural being, let alone as an animal because he has a really subjective or reflective framework that prevents him from merging with reality the way animals do*. And on saying this I must emphasise that "subjectivism" does not perforce mean what hinges upon a particular way of feeling or an incommunicable perception. On the contrary, the word alludes to the original finitude of existence that solely man experiences and that is alien to the dynamics of sheer instincts that go on in an undifferentiated temporal circularity where there is no possibility of metamorphosing or overcoming the origin of the cycle (Schopenhauer, 1966, 541). Thus, whereas animals are immersed in the everlasting push of instinct and pass from the satisfaction of a want to another without minding any of them, the least perceptive variation demands a transformation of the subjective ground of experience and of the corresponding consciousness (Merleau-Ponty, 1945, 261). From this standpoint, "subjective" agrees with a vision of existence ruled by a would-be divinity that should consequently have a certain similarity with man and vice versa so as to allow the mutual recognition beyond the blind terror to an overwhelming power or arbitrariness (as it happens with the Christian ideal of providence); for the same reason "subjective" is incompatible with a vision of existence sunk at the bottom of a nature that becomes indifferent to man and that does not allow him to make up for it imaginatively or through the resignation that coexistence always brings about on showing the scant motives that man has to be at his ease in a world where animals are the first to suffer the unevenness of nature. As I have said, animals strengthen the phenomenological community of being thanks to which man recognises himself together with them, but they cannot be the unconscious props of a subjective projection.

4 The Sense of Art

In one of the final shots of his, taken just a few days before his frightful death, Treadwell appears in the foreground before the camera and in the middle of a beautiful landscape encircled by a blue river and high mountains. While he perorates, he approaches the camera or moves it back and lets the spectator perceive his zestfulness. The whole scene looks like that of a perfect integration of man and nature, although what Treadwell says belies that idyllic impression:

> Let me tell you, honestly, camping in grizzly country is dangerous; people who camp in grizzly country must camp in the open to let the bears know where the tent is; my camp is unseen; it is the most dangerous camping, the most dangerous living in the history of the world by any human being. I have lived longer with wild brown grizzly bears without weapons, and that is the key, without weapons, and I have remained safe, but every second of every day that I move through this jungle or even at the tent I am right on the precipice of great bodily harm or even death. And I am so thankful for every minute of every day that I found the bears and this place, the grizzly maze. But let me tell you that there is no, no other place in the world that is more dangerous, more exciting than the grizzly maze. Come camp here, try to do what I do, and you will die here, you will die here. They will get you. I found a way to survive with them. Am I great person? I do not know. We are all great people. Everyone has something wonderful. I am just different and I love these bears enough to do it right. And I am edgy enough and I am tough enough. But mostly I love these bears enough to survive. And I am never giving this up, never giving up the maze. Never. This is it. This is my life. This is my land.

This shot is interesting because it shows clearly the whole psychological engine of the totemic subjectivism, which does not spring from a kind of religious respect for nature or from critical care for an animal species in risk of extinction but from the simple excitation of challenge. Treadwell flaunts outright that his enterprise is the most dangerous one that had ever been carried out and that he has endured what none else has. Needless to say, he does not speak of those feats that, like the historical ones, demand the participation of a lot of people and that are abstractedly attributed to the so-called heroes; instead he speaks of a feat that has been accomplished without the support of anyone else. There is magic indeed in his extraordinary link with bears, but there is above all the decision of compelling them to obey his will. Of course, everyone else could in principle do the same, since subjectivism upholds that everyone has an absolute right to everything, including to run the utmost risk, if he wants, or to make the most brutal instinct give in. At any rate, there is a meaningful difference among people, and that lies in their dauntlessness, for a very few people would be able to remain tranquil before wild animals or in the middle of an overwhelming solitude. And the relevance of this dauntlessness must be emphasised because it is the only criterion to value the uniqueness of someone in particular, which is not then a substantial condition but, rather, the inevitable outcome of an uncompromising effort. That effort must last to the end so as to prove, that that person has metamorphosed into a superior being and that he is beyond fear and, of course, beyond the want of company that most people are after instead.

The most singular feature of this setting is so its dramatism, which at all costs demands that the subjectivist antihero always appears alone in the centre of the shot and that his would-be totems are relegated to the background. Independently of the fact that the material of the picture is made up by all the short films of Treadwell and that they could as a whole project a vision completely different from the one perceptible in Herzog's work, the fact is that the latter shows a series of shots where Treadwell appears as the lead of a feature film and where the sobriety and distance usual in a documentary are conspicuous by their absence. And to make that evident no other art is more appropriate than cinema, for it brings to light without mediation the megalomania usual in subjectivism, the insatiable want that the individual experiences of expressing himself before that silent camera that works as a loudspeaker through which he can shout at the top of his voice that he has achieved what none else has been able to and that all the others are a band of idiots. As a matter of fact, the short films are like the various moments of an endless soliloquy where Treadwell unfolds the emotional framework of subjectivism that he embodies to perfection, from the rage that he exteriorises against the cruel divinity that lets bears starve to death when the level of brooks and rivers are too low for salmons to swim into the park to the almost lyrical reflections on existence and love that we have already analysed, so as not to mention the childish joy that carries him away when the divinity that he has scolded finally sends rain to the park and the bears can feed their fill. And what surprises here is not precisely the inversion of values that allows him to mock whatever superior providence could be whereas he reveres animals that attack at the first opportunity; it is that that agrees with a way of thinking characteristic of a very unintelligent or very unbalanced person that takes his daydreams as things in themselves. For instance, when he exults because rain has finally fallen, he looks as if he were before a wonder, which confirms that nature never was for him the regular framework of life or the limit or the ground of human historical experience but a kind of picture whose would-be harmony is the outcome of one's wishes, which is probably why there are so much scenes where he appears soliloquising. And this tendency to daydream is fully confirmed during the interviews with the parents and friends of Treadwell, who time and again allude to the extraordinary histrionics of his, to his constant efforts to get a personality, a history or even an accent other than the one he had, and, last but not least, to his frequent problems with alcohol and perhaps with drugs. But what is more revealing is the fact his father reveals—that the straw that broke the camel's back was his failure in the attempt to become a member of the cast of a TV programme. The histrionics are not then the sign of an exhilarant emotiveness that sought to integrate all the elements of the existential whole into the image of an everlasting animal or superhuman potency, it is the projection of a human, all too human reaction before the incapability of finding one's way in the world. It would only be Treadwell's predicament if he were not a prototype, which is clearly visible in the aesthetical disposition of the shots already analysed, where he always appears alone and free in the middle of wilderness. And although Treadwell all the time apostrophises that imaginary listener that is behind the camera and reveals so that he is somehow or other aware of the interpersonal dimension of his words and actions, the fact is that that is for him just the indispensable background of an experience that he could not share with anyone else bar bear and foxes. The very aesthetic condition of the work makes obvious how subjectivism (and more concretely the totemic one) starts from the untenable vision of man free from any natural constriction, which despite what Treadwell thinks, is not the symbol of a free spirit but, oddly enough, of a biblical creature that lives alone in the Eden. Yes, the origin of this image does not lie in a critical approach to existence; it lies, on the contrary, in the most metaphysical or rather absurd myth of human freedom, that of the golden age when every living being enjoyed an everlasting bliss and man was beyond good and evil because he did not need to take care of anything.

It would certainly be bewildering that this image went hand-in-hand with subjectivism (in the vulgar sense that belies completely providence whatever) if it were not for the mutual reference to a temporal fullness that does without the phenomenological sense of existence, which implicates the correspondence among every individual consciousness and an interpersonal world of beings irreducible to our wants or wishes (Merleau-Ponty, 1945, 406–408). Existence itself belies how subjectivist totemism takes others (namely, as fleeting images) and animals (i.e., as real potencies of wilderness exerting a mysteriously benevolent influence on man). Treadwell's shots show him to be in an absolute or metaphysical solitude, preaching his unheard-of gospel of the supernatural redemption through bears. This is why I have said that *the film is the best mean to penetrate into the real subjectivist vision of existence* for it shows the vision someone like Treadwell has concerning existence. Of course, this would have been impossible without the talent of Herzog who articulated the heterogeneous material that Treadwell left and endowed it willy-nilly with an identity of its own and a clear sceptical approach to totemism that attains its acme when he comments off at the end of the picture: "what haunts me is that in all the faces of all the bears Treadwell ever filmed I discovered no kinship, no understanding, no mercy. I see only the overwhelming indifference of nature. To me, there is no such thing as a secret world of the bears, and this blank stare speaks only of a half bored interest in food. But for Timothy Treadwell this bear was a friend, a saviour." And the sense of these words that so outright belies that of Treadwell's is the thread of the visual development of the picture, where the odd metamorphosis into a bear-like being that Treadwell pursued comes to the fore not as the expression of a poor megalomaniac but as the symbol of a whole epoch, that of the subjectivist totemism.

How does the formal framework of the picture show this? First and foremost, by the way it tells Treadwell's life story, which contradicts how he idealised his devotion to bears and how, furthermore, invented an identity that was according to the image of a superhero that he mistook for the free spirit that overtakes whatever moral or cultural limitation so as to save no one. On cutting the different shots within a new narrative unity, Herzog brings to light the contradictions of subjectivism without changing Treadwell's personality into a simple caricature; far from that, the film makes see him as a man that in spite of his delusions and dramatism tried to experiment the unpolluted essence of nature. In other words, the picture allows seeing with a relative clarity the opposition existing between subjectivism and the humanity of Treadwell. Particularly revealing is the final sequence of the film: it starts with the shot that we have analysed at the onset of this section, where Treadwell brags about his daring to live in the grizzly maze and continues with some others where you see the bear that very likely killed Treadwell and Amy. The bear looks for salmon in the middle of a pebbly brook and seems to play a little later in the bottom of a mere; this shot is followed by one where Herzog comments the absurd error of Treadwell while we perceive in close up the impervious sight of a bear. The next scene contrasts masterly with this one, for it shows in the morgue the forensic surgeon that carried out the autopsy of Treadwell and Amy, who has a very strange physiognomy and speaks emphatically with a shrilling voice and weird gesticulation to tell in full detail the last moments of the life of the victims. Then in the last shot that Treadwell ever filmed, just a few hours before the attack, he reminds us for the last time while a soft rain bathes him that he loves bears wholeheartedly and that he only wants to remain with them. Finally, we see the image of three young bears playing in the distance among the mists of an autumn evening that spread over the beach. The interplay of the elements of this sequence, punctuated as it is by the sober remarks of Herzog, has a tremendous effect over the viewer because it shows that if Treadwell was doubtlessly wrong concerning nature and concerning its would-be thaumaturgic potency, he was not a lunatic either. Before a cloudy landscape whose limitless beauty reverberates in the apparent joyfulness of the animals, it is very easy to believe in an all-embracing harmony that must be preserved from the so-called human brutishness, and with all the more reason when this idyllic image is contrasted with the odd appearance of human finitude, as, for instance, that of the surgeon, who looks as if he were a farcical character, or as Treadwell's himself, who challenges everyone time and again. The sequence coordinates so the multifariousness of a historical world that subjectivism reduces for its part drastically either to the representation of a Manichaean conflict or to a kaleidoscope through which we contemplate the endless reconfiguration of sentimentalism.

On this first plane of the formal framework of the film, which could be called the vertical one, the totemic subjectivist idealization blurs the existential sense of the historical world that man shares with his fellow creatures and with animals and where nature is just an elemental dimension of consciousness. This plane works together with that that could be called the horizontal one, which pierces right through the temporal unfolding of the personal story and brings so to light the contradictory way of defining the corresponding identity regarding its idealizations. This agrees with the interplay established between Treadwell's work and the narrative of his life: the people that appear in the interviews that Herzog alternates with scenes of the short films (above all Treadwell's parents) show that the limitless devotion to bears was not in this case the expression of a really critical interest, let alone of a coordinated enterprise, but the extreme development of an egotistical vision of existence. That is to say, this interplay shows that *subjectivism can vary of objects and perspectives but that it always starts from an untenable reduction of experience to a mental representation in the worst sense of the word that is not so different from the sheer burst of madness, and that is fully opposite to the phenomenological comprehension of existence*. Of course, there is an abyssal difference between being a rebel without a cause and being a rebel for a cause with a historical transcendence such as environmentalism; nonetheless, what is at stake here is not that but the way the participation in a socio-historical determination can endow existence with a sense of its own and make man more conscious of the impossibility of mistaking reality for an almost mystical transcendence.

Finally, the two planes that I have mentioned intersect on the very singular Treadwell's utilization of cinema as a means to express nature beyond our human intrusion, although the fact is that it was more to express his own obsessions. Absurdly, he who flaunted his passion for an Edenic nature, was incapable of perceiving the overwhelming beauty thereof, and the wonderful landscape of Alaska remained for him just the background of a personal drama with a lead, he himself, and mute listeners, bears and foxes. Regarding this, there is a shot where the blindness toward natural environment is particularly evident: before the camera, a very narrow path surrounded by shrubbery ascends a hillside. All of a sudden, Treadwell comes down the path running and approaches the camera; after a moment he decides that the shot is not good and repeats it as if he were an actor rehearsing a scene of a picture. While he climbs the path, he rubs against the shrubbery and when the path is alone again it appears an image of extraordinary simplicity and beauty, like those that are the core of so many romantic paintings of rural landscapes and that inspires so much metaphysical thoughts (Kant, 2007, 76). Or, to return to a shot that we have already mentioned, when we see how the likely killer of Treadwell seems to play in the mere although he is more likely to be desperate and hungry, we cannot help wondering how grace springs from the iron grasp of instinct. However, Treadwell never saw that, or at least did not seem to have seen it, and he preferred to film his soliloquies, his diatribes or his jeremiads, where he always took somehow or other a subjectivist standpoint. The short films are so the best ground for understanding Treadwell's very questionable conception of nature and existence and to link it not with a superhuman experience or with a critical stance against the current exploitation of environment, but with the most elemental disinterest of the subjectivist regarding the true originality of nature and the personal complexity of the dullest individual. And because of this, it is a lot more amazing that Herzog had been able to change what was mostly a series of shots threaded by anger into a film of a great dramatic depth and of extraordinary coherence thanks, I think, to the consciousness that *cinema is more than any other art the one that clarifies the plastic unity of existence and the emotional dynamics of consciousness that coincide in the precise instant that the camera focuses on the way someone carries out the fleeting sense of existence* (Rivas López, 2010a, 95).

5 Colophon

Finally, I just want to reconsider briefly one more time the part that the subjectivist totemism plays in a possible solution to the current environmental unbalance. Inasmuch as it starts from a conception of animality utterly alien to the historical development of human consciousness and of the corresponding valuation of existence (since it upholds that animals and wilderness possess a thaumaturgic potency that they can use in favour of man, provided that the latter is ready to adjust his behaviour to their mysterious will), totemism cannot lay the foundations of a new unity of nature and culture, and it is rather a hindrance for attaining it. Why? Firstly, because it just subverts the terms of the problem and imposes an abstract animal force upon man; secondly, because it passes over the specific condition of individuality, *i.e.*, the phenomenological embodiment that distinguishes everyone from others, however much he shares with them the totality of a historical determination; thirdly, because it sparks off a sentimentalist link with nature that oscillates between a blind submissiveness and a childish merriment. Thus, totemism in all its variants, above all when it allies with a belligerent misanthropy, just worsens the opposition of man and nature and fosters the solipsistic approaches that go hand-in-hand with utilitarianism, which is what for his part denounces a man that appears at the end of the picture and that belongs to the Inuit natives that have inhabited Alaska from time immemorial, who says that, contrary to what Treadwell did, they always have been able to keep a solid balance with nature and concretely with bears because they always have been conscious of the absolute dissimilitude of them with man.

References

Berlin, I. (2013). *The crooked timber of humanity* (2nd ed.). Princeton: PUP.
Danto, A. (1980). *Nietzsche as philosopher. An original study*. New York: CUP.
Deleuze, G. (2010). *Nietzsche et la Philosophie* (6th ed.). Paris: FUP.
Eliade, M. (1969). *Le Mythe de l'Éternel Retour*. Paris: Gallimard.
Ferrater Mora, J. (1994). *Diccionario de Filosofía* (Vol. 4). Barcelona: Ariel.
Hegel, G. F. W. (1993). *Introductory lectures on aesthetics* (B. Bosanquet, Trans.). London: Penguin.
Heidegger, M. (2000). *Introduction to metaphysics* (G. Fried & R. Polt, Trans.). New Haven: Yale University Press.
Kant, I. (2007). *Critique of judgement* (J. Creed Meredith, Trans.). Oxford: OUP.
Marcuse, H. (1991). *One-dimensional man*. Boston: Beacon.
Merleau-Ponty, M. (1945). *Phénomenologie de la perception*. Paris: Gallimard.
Nietzsche, F. (2002). In R.-P. Horstmann & J. Norman (Eds.), *Beyond good and evil. Prelude to a philosophy of the future* (J. Norman, Trans.). Cambridge: CUP.
Rivas López, V. G. (2010a). *El Cine y el Mal. Una Ontología del Presente*. Puebla: BUAP/El Errante.
Rivas López, V. G. (2010b). On the phenomenological problematicity of the religious image. In P. Trutty-Coohill (Ed.), *Art inspiring transmutations of life*. Dordrecht: Springer.
Schopenhauer, A. (1966). *The world as will and representation* (Vol. 2) (E. F. J. Payne, Trans.). New York: Dover.
Schopenhauer, A. (2003). In S. González Noriega (Ed.), *On will in nature* (M. de Unamuno, Spanish Trans.). Madrid: Alianza.

Part VI
Human/Nature/Cosmos

Towards a Hermeneutic of the Artificial

Marie Antonios Sassine

Our daily world, the one in which we form judgments and orient our actions, has grown populated with scientific constructs, technical products, in the form of artifacts that shape our perceptions and actions by molding them and constraining their possibilities. We live in a world where a good part of our experience is mediated. This is most evident in the way we now communicate with one another, in how we receive information and act on it. Who are we sharing contexts with, who are our interlocutors, who and what is affected by our decisions? The answers to those questions grow dimmer and more distant, which is paradoxical given that we are instantaneously connected to the far reaches of the world. We still act, of course, though the initiating movement may be far removed or carried out in pure isolation. The act may simply be to unplug our connections, to delete a request, or to adjust a cursor on a target. Action, however, despite its passage through an innocuous gesture, has not lost its impact. No matter how subtle or distant the impetus, consequences remain. People are affected, lives are changed, environments are altered. The results of our technically mediated actions, though, occur on such a far removed plane from our immediate environment that we frequently only seize them abstractly, if at all. What phenomenology called the life-world is being radically transformed. This paper will describe some of the fundamental features of this transformation, specifically its artificial and veiling character. It will briefly highlight the previous appearance of the questions at play here in the history of philosophy, namely through Plato and Husserl. This is to set the stage for Jean Ladriére's analysis of how science is destabilizing ethics, with a view to illuminating the crucial contribution that he believes phenomenological hermeneutics can make, in the form of a hermeneutic of the artificial, to help us meet the unprecedented ethical destabilization that technology is operating on our world.

M. A. Sassine (✉)
Dominican University College, Ottawa, Ontario, Canada
e-mail: marie.sassine@dominicanu.ca

C. A. Hornbuckle et al. (eds.), *Phenomenology of the Object and Human Positioning*,
Analecta Husserliana 122, https://doi.org/10.1007/978-3-030-66437-4_18

Phenomenologists, beginning with Husserl through to the contemporary philosopher, Jean Ladriére, whose views will be examined later in this paper, have argued that the technological civilization that is ours has emerged out of a specific philosophical history. Its underpinnings are to be found in philosophy and its child, science. Due to the tremendous epistemic success of science, we have come to see certain beliefs as self-evident. These beliefs have to do with the form of most of the knowledge available to us. There is a story to how we have come to privilege certain forms of rationality, the ones best captured in an algorithm, for example, so that we are able to isolate aspects of the real, and then to reconstruct those aspects so they mimic the originating phenomenon. This acquired capability is the result of a sustained trend in our history that has gradually abstracted and perhaps obscured conceptions of the real. The peculiarity of contemporary constructs is that they remain veiled while altering and structuring our day-to-day experience. These hidden constructs that are shaping new relations of community, of political space, of shared interests and friendships, are not seen or experienced in their genuine meaning. Today, in a fundamental sense, the mathematical formal procedure of science has arrived at transforming time and space as basic categories of understanding. The sensible world of action is now radically reconfigured. We inhabit a proliferating and expanding universe populated with artifacts, "which is to say engendered by art, in the broadest sense of the term, in the sense in which one can speak of the engineer's art, or the doctor's, or even that of the economic decision maker. The bearer of this artificialization of the world is undoubtedly science" (Arnsperger, Larrère, & Ladrière, 1999, 150).[1] These artifacts are a product of the mathematical abstraction and reduction, but they are not seen as such. As Lee Smolin, the distinguished physicist whose research is reintroducing the notion of irreversible time back into theoretical physics, asks, "How many users of Facebook are aware that their social lives are now organized by a potent scientific idea?" (2014, xxix). The question is of course rhetorical. The important point, though, is that the very lack of awareness forbids reflection on how the technology is transforming our being in the world and with others.

Plato, though often seen as the progenitor of idealized scientific practice, was not blind to the intimate bond between the simulated and the real. He saw that simulation required a specific kind of knowledge. He hunted it down in the *Sophist*, particularly in the discussion on simulation and mimicry. In the *Sophist*, Plato is particularly attentive to the *techne mimetike*, disturbed by the affiliation, the slippery terrain between copy and semblance, trying to determine at which point the copy, no matter how faithful in appearance, is simply sophistry. The *techne sophistike* is distinguished from the *techne mimetike*; it is an empty kind of *mimetike*, a fictive or a factitious one. "This 'appearing' or 'seeming' without really 'being,' and the saying of something which yet is not true," declares the Stranger in the *Sophist*, "—all these expressions have always been and still are deeply involved in perplexity"

[1] All Ladrière citations in this paper are my own translations. The works cited have not been translated into English.

(1971, 979, 236c). Plato points out the logical difficulty, the contradiction, of saying that what is false or unreal has genuine existence and impact, but cannot seem to resolve how to discuss sophistry and its reliance on rhetoric, on "words that cheat the ear, exhibiting images of all things in a shadow play of discourse," without according power, and therefore reality, to this shadow play (1971, 977, 234c). The puzzle is not satisfactorily resolved. The relation of real to unreal remains clouded: "We profess to be quite at our ease about the real and to understand the word when it is spoken, though we may not understand the unreal, when perhaps we are equally in the dark about both," the Stranger again pronounces (Plato, 1971, 987, 243c). This avowal rings today with a certain immediacy and poignancy. For our real and unreal seem even more intimately intertwined, barely separated by obscure demarcations. We are, in a way, seeing the embodiment of Plato's shadow play of discourse and hearing the echoes of the ancient and troubled enquiry into the nature of simulacra.

Phenomenology, from its inception with Husserl, called for a rigorous attention to the phenomenon, to things as they present themselves, and warned how easily they can become concealed and replaced. It was alert to how the product of scientific activity, an abstractly derived and reduced object, was beginning to be seen as more real, displaying a greater truth than the fullness of the phenomenon. Husserl spoke of a crisis, Heidegger of the reign of technique, Jan Patočka of a situation of distress. Husserl was probably the first philosopher to argue that the scientific rationality that was becoming the dominant form of thought and knowledge in our world represented a historic danger to humanity. He tried to show in the *Crisis* how all other forms of thought were beginning to model themselves after it. Scientific rationality was deemed to be the one most closely aligned, both in its method and in its contents, with truth. The danger was that the scientific mode and technique isolated abstract mathematical properties within the phenomenon, and then replaced the original, finite phenomenon with a constructed, infinite version that had the effect of masking the life world, the *Lebenswelt.* As Husserl explains in the *Crisis*, "In geometrical and natural-scientific mathematization, in the open infinity of possible experiences, we measure the life-world—the world constantly given to us as actual in our concrete world-life—for a well-fitting *garb of ideas*, that of the so-called objectively scientific truths" (1970, 51). The life-world is measured against the mathematical object, the idea, then it is made to fit within it. What does not fit is discounted. Husserl underlines the way in which science loses the *Lebenswelt* and ultimately covers it. The finitude and the variety of the world are replaced by an infinity of endless potential experiences.

Husserl argues that for Euclidean geometry, there are only a number of specific, finite tasks available. The a-priori that Euclidean geometry knows is limited; its sphere of application is also well delimited within the world. Since Galileo, Husserl tells us, this concrete application to real space has been replaced by an ideal space, to be found nowhere, in which experiences can be repeated ad infinitum. The consequence is the emergence of an idealized praxis: a kind of pure thought dealing with abstract forms. This praxis aims at exactitude, at a better representation of the mathematized nature. What we know is what we can measure, calculate, act upon.

We know an abstraction, removed from any particular phenomenon. It is this calculated and abstract form that is then projected back onto the life-world, but in the process ends up veiling the *Lebenswelt* as the original source of the abstract idea's meaning. This is how technique, in its broadest sense, and technology function.

In many ways, Ladrière's work can be seen as a contemporary variant and extension of Husserl's *Crisis* analysis. His is a reflection on contemporary science and its philosophical sources and assumptions. He has sought to deepen Husserl's insight and to situate it in today's world. Ladrière points to the emergence of a new form of rationality, the rationality of construct, which is now subtly destabilizing ethical judgment. He draws a distinction between the temporalities that the body imposes within the historicity of the existent and the temporalities that the artifacts introduce into the life of the existent. The two can easily come into conflict, and it is the temporality of the artifact that will impose on the human being a rhythm, based on an ideal of rapid and infinite mathematical repetition, that is beyond the capacities of the finite being to meet.

The hermeneutic of the artificial that Ladrière proposes is part of that ongoing phenomenological project that has anticipated and highlighted the perverse effects of certain forms of rationality on the life-world. His concern is with explicating the relationship between the scientific approach to truth, the source of the universe of artifacts, and the life-world, with its interplay of ethics, politics and even economy, and he suggests that science and scientific products are destabilizing ethical judgment. This has to do with the nature of truth that science achieves and the manner in which it is ethical.

In line with Husserl, Ladrière insists that science is primarily faithful to a method. In his view, this faithfulness to a method is where science shares with philosophy an articulated concern for truth. The concern with truth in science is expressed in the application of a formal, transparent methodology. This is in harmony with the *object of science*: it seeks to discover or un-cover, as the view may be, the parts of phenomenon that repeat in intelligible patterns. Ladrière draws a clear distinction, however, between science as an activity and science as a form of knowledge. As an activity, science seeks the truth according to a specific method and in a set fashion, and it is only in its faithfulness to this method that it shows an ethical orientation: "And insofar as the scientific enterprise constitutes itself according to an aim for truth as such, one can recognize in it an ethical dimension" (Ladrière, 1997, 71). The ethical dimension is not, however, in what is sought; it is in the faithfulness to a method.

The truth associated with science, then, has to do with how the task is accomplished and not in the knowledge that is consequently obtained. Like Husserl, Ladrière wants to clarify that there is a difference between what science intends, what it believes it is obtaining, and the method it uses to achieve its goals. Scientific thought is convinced that because of the discipline it follows, it actually achieves the "things-in-themselves," and that its interpretive system gives rise to a discourse that is consistent and concordant with reality. As a form of knowledge, though, it describes, it puts forward formal and abstract propositions on fragments of reality. Since a normative principle cannot be deducted from a descriptive one, scientific knowledge as such cannot inform ethical orientation. What science knows in any case is a constructed object that is abstracted from the sensibility of the life-world. This reduced object is in reality a kind of separation—a kernel is extracted from the phenomenon and isolated from its existential significance.

This view is similar to one expressed by Werner Heisenberg regarding the concepts of ordinary language, because they are rooted in reality, being richer than scientific concepts. Heisenberg puts it this way:

> the concepts of natural language are formed by the immediate connection with reality; they represent reality. The scientific concepts are idealizations; they are derived from experience obtained by refined experimental tools, and are precisely defined through axioms and definitions. Only through these precise definitions is it possible to connect the concepts with a mathematical scheme and to derive mathematically the infinite variety of possible phenomena in this field. But through this process of idealization and precise definition, the immediate connection with reality is lost. (1959, 171)

In this process of idealization and reduction, though, Husserl and Ladrière would argue that the originating phenomenon is covered over, lost. The discoveries of science, derived through this method of idealization, are projected back onto it and overlay the phenomenon, making it more inaccessible, obscuring it. The immediate connection with reality becomes more and more fragmented as scientific products and technique proliferate in our world.

Ladrière speaks of science as an idealized interpretation, an abstract and intelligible construct, which inserts itself into the existential situation: "This system of interpretation is the scientific view of reality, which substitutes for the world's own intelligibility an intelligibility of *constructs* through which it succeeds to account for a part of reality that thus becomes the only pertinent reality" (1997, 53). These superimposed *constructs* easily recall Husserl's "garb of ideas" and how this garb clothes the *Lebenswelt* (1970, 51). Of course, the reason that these constructs become the dominant form of interpretation is because they show a capacity to act in the world, to transform it, or add to it. Their epistemic and practical success demonstrates that at least some aspect of the real is being mastered.

Ladrière, however, argues that this scientific realization is destabilizing ethical judgment. He further develops the phenomenological critique by suggesting that these constructs, in the form of technique, acquire an *autonomy* of functioning and, indeed, place *constraints* on the capacity for action. He says of technique that "It is the product of a constructive activity, but it detaches itself from this activity and sets itself up as an autonomous form of reason, bearing its own dynamic and imposing on action the constraints of its own operating laws" (1997, 9). This very *autonomy* and the way that in which it *constrains* possibilities of action end up clouding ethical judgement. We have witnessed recently how autonomous algorithms become in sophisticated social media. An algorithm designed to select news items will do so in way that is almost impossible to control, and will constrain access to other items not included in its parameters. The constructed element is not apparent; the functioning of the algorithm is hidden. The consequence, however, is that an ethical judgement based on the assumption that the facts presented reflect the fullness of a concrete situation is rendered impossible. Awareness of the artificiality involved can inspire a conscious effort to seek out and assess other views of the same situation. The autonomy of the algorithm, however, and the constraints it imposes cannot be controlled, as we have seen in the many recent discussions and debates raging around global social media enterprises and their mode of operation.

Ladrière notes the emergence of new ethical situations today, whose primary characteristic is their artificiality. They are created by science and wear the form of scientific knowledge. He says, using Kant's terminology, that while reason is unified, there is a theoretical and a practical usage of reason, and that they have an analogic way of functioning (1999, 156–59). But the universe of artifacts is governed by another form of rationality, which differs from either theoretical or practical reason: "Since the primary characteristic of this universe is to result from a process of construction, one could call the rationality that inhabits it a 'rationality of construction,' distinguishing it thus both from the rationality—theoretical as well as practical—which is found in the mind and in cosmic reality" (Ladrière, 1997, 9).

The rationality of construction destabilizes the life world since the cycle of abstraction and reduction does not remain and end with the exercise of the scientific method. It results in products fashioned in its image. These products are both abstract and concrete; they are both concepts and artifacts. They acquire an autonomy of functioning and constrain action, as has been said, and since they represent fragments of reality, their dominance displaces and destabilizes the interconnections that remain beyond the grasp of the mathematical ideal that forms the foundation upon which they rest. There is substitution of the abstract object for the authentic phenomenon. The destabilization of the life-world occurs as it becomes more and more governed by technique and populated with scientific products. It becomes a "universe of artifacts" in which what surrounds us is "in short the projection into materiality of apparatus reflecting the image of the world constructed by science. And this projection has the same abstract character as the image which inspires it" (Ladrière, 1997, 84–85). This projection is at the heart of situations bearing an artificial character, not quite recognizable or understood in their existential meaning.

This new universe of artifacts is inhabited by the form of knowledge that created it. This knowledge is an essentially formal and reductionist one, removed from the existential dynamic. And without existential situation and significance, what Ladriére calls the ethicity of a situation, that is to say the ethical question it presents, is difficult to discern. Ethical judgment, which necessarily begins with recognition of an ethical question within a given existential situation and proceeds to determining the desired ethical goal and the action required to fulfill it, is perturbed in the universe of artifacts. That is the fundamental and most glaring way in which the life-word is destabilized through the idealization and mathematization of nature. The artifacts introduce a level of abstraction into ordinary existence: abstraction that is beyond sensibility, that is counter to the temporality of presence and which disorients praxis. The scientifically reduced object that is transforming the life world, and furnishing the universe of artifacts is not some truer version of what is sensibly given. Ladrière says, "The reduction is not a bringing forth of an essence, or an invariant, that is present in the phenomenon as what is irreducible in it, unaffected by the conditions of its phenomenalization." A plant, for example, is not reducible to genetic components. But science can decide beforehand which genetic components would render such a plant more resilient for the global markets, and produce genetically altered specimen, based on the mathematical probabilities of success. The genetic reengineering replaces the authentic plant and its inborn mechanisms of adaptability: "It is an operation of construction which consists in substituting for the authentic phenomenon an abstractly defined object, whose best form of representation is actually the one provided by the ideal objects of mathematics" (1997, 84). This substitution of the original phenomenon by a mathematically conceived object is, as suggested earlier, close to Husserl's *garb of ideas* veiling the life-world. Since the constructed images are the products of science, what they reflect back is an abstraction: reduced and impoverished images of the world. They represent science's transposition into the universe of artefacts.

Certainly, as mentioned previously, science and its products can be said to be proof, epistemological and pragmatic, that something about the nature of the real has been grasped. But the result, Ladrière tells us, is not only troubling to the life-world in general, it renders ethical judgment more difficult. Why? Because the products of science, those embodiments of abstract intelligibility, do not seem so much to be revealing the real, but rather to be providing a superimposition of a different order of interpreted reality. This is to the point where the real, as a totality, a situation, a horizon for action, has become obscured, as though the framework through which it is presented already closes certain possibilities of perception and interpretation, or even imagination, for us.

A recent example is the use of drones by a state to target individual "terrorist combatants" wherever they may be in the world, and in whatever situation they might find themselves in—a wedding, a funeral, a family dinner, all that becomes pure abstraction. What is the common context, situation, between the individual target and the actor state? Who determines the appropriate context and timing of the action, initiates the action, and is responsible for it? Who are the actors? How do we judge victory or failure? Not only will we never know, it has become unknowable—we would have to choose, almost arbitrarily, any number of elements to be included or excluded from any context we want to portray, at the service of the rhetoric that belongs to whoever has the power to control the communication. There is no common space and time that binds all the actors in a shared situation. We are confronted with a projected image of the world: "And it is in the operation of this universe that new situations appear, perplexing for action in terms of what should or should not be done. In these situations, the ethical conscience is struck with uncertainty because it does not find itself in the customary conditions for the application of norms: the evidence which conditions the application is lacking. If this is the case, it is because these 'artificial' situations are deprived of existential significance" (Ladrière, 1997, 85).

One of the most significant and primary effects of these artifacts, then, is the deconstruction of space, the elimination of the limits of the body. But it can be argued, I think, that as virtual possibilities expand, the world can also be said to shrink in its contours as world. It becomes more and more structured by subjective choices, wired to contexts of choice, unplugged from undesirable encounters, functioning outside the immediate constraints of time and space. There is a diminishment of possibilities of surprise and mystery, both of which are characteristic and significant elements of presence. Indeed, these very elements call upon us to also be present and receptive and to accommodate what is other. A good part of our political and social systems are based on ways of accommodations, on creating diverse communities where the notion of otherness is continually enlarged. The more sophisticated the artifacts that populate our contemporary become, the less compulsion there is to shift habitual perspectives. The acquisition of any desired situation is made possible, at least virtually, without any requisite change in the person. If neighbours and friends prove to be unsatisfactory, for example, one can easily substitute virtual fora that can be visited or abandoned without risk or bother, at one's own convenience. Concrete situations can easily be replaced with abstract ones. Where one is compelled to recognize the hurt expression worn by an acquaintance whose greeting has just been spurned, no such signal is available to trouble the ethical conscience when we stop responding through social media. The effect of our action is concealed because it occurs in a disembodied, a-temporal domain where otherness is an image on a screen.

Indeed, Ladrière's suggested response to the destabilization of ethics is a hermeneutic of the artificial, one that considers specifically *embodiment, temporality and otherness* (1999, 186). These three elements are drawn from the insights of hermeneutic phenomenology. He suggests that those are the very dimensions that should inform the "hermeneutic of the artificial" that is called for today. There is no question of a return to the world prior to Galileo's mathematization of nature. On the contrary, this hermeneutics' point of departure is the scientific reduction in its entirety. The existential elements of body, time, and otherness (being-with) are the lens through which a second reading of the reduced phenomenon is conducted. It is a reading which aims at restituting the artificial construct back into the life-world, the world where bodies, time and others co-exist, in order to grasp the existential meaning of a situation. *Embodiment* is the site of receptivity and action; it renders existence actual, sensible, and responsive. It is through the body that we are open to the world and receive it. It is also through the body that we act. The body is what feels, what is affected by the world, and, in turn, what allows for the insertion of a life into the world. *Temporality* is what gives existence a history, a past and a future; it is what makes sense of ethical choices and turns them into a legacy from the past that carries us to a future destination. The unity of an existence is only available through our experience of time. Time for the human being is what offers the possibility of a story. *Otherness* (being-with) is a fundamental function of existence as co-existence—without *otherness* the ethical significance of situation cannot be deciphered; it as much in how it affects others as well as ourselves. For Ladrière, in order for a situation to acquire existential meaning, it has to be grasped in these three dimensions. The question of cloning provides an example. On the one hand, we can see it as a scientific process providing a certain reliable result. We can also look at it from the point of view of the clone, as an individual created through clonage. Clearly, the very determination "clone" signals an artificial construct. "From the perspective of the existential meaning of the situation," Ladrière tells us, "the status of clone is primary" (1999, 193). The ethical dimension would only emerge on the basis of a hermeneutic that would imaginatively try to seize the existence of he clone in a body that is not individual and particular, but born out of a process of replication. The hermeneutic would also include how the clone might experience history, birth and death, all the dimensions of human temporality and, of course, the fundamental relation to others. The three dimensions drawn from hermeneutic phenomenology can light a path to a reading of the situation that moves it from the abstract scientific real, from its artificiality, to its insertion in existence.

These dimensions of existence, however, also have their autonomy and constraints. That is why they form the essential frame for the hermeneutic of the artificial: "Embodiment, co-existence, temporality, as existential dimensions, have their own structures and those structures place constraining limits on what can be undertaken in a positive way by existence" (Ladrière, 1997, 62). Human beings are physically and mentally only able to process so much information at a certain level of speed. They need rest and nourishment and shelter. While technology can go non-stop, human bodies cannot. Human beings live with a past, not only their own, that affects their present and future. They live time as irreversible and as finite. Scientific constructs are founded on the infinitely repeatable and on freedom from the arrow of time. Most importantly perhaps, certainly for ethical choices, humans are born into a world of otherness. The hermeneutic of the artificial is about seeing and understanding what existence as a human being, within the autonomy and constraints offered to such a being, can embrace in a conscious and responsible way. What is at play, in George Steiner's words, is "hermeneutics as defining the enactment of answerable understanding, of active apprehension" (1989, 7). There is, as Ladrière admits, a requirement for an act of imagination. The hermeneutic is an effort to imagine how the implacable temporalities of mathematics inscribe themselves in the finite horizon of the life-world, in an embodied existence that is always lived in community with others.

And not just human beings, or human bodies, as George Steiner reminds us. The life-world contains, in addition to post-human artifacts, other beings that are also affected by its destabilization. The insertion of mathematically constructed artifacts also troubles the life-world in ways that are basic to our situation within it. In an exceedingly poignant example, showing the brutal capacity of powerful artifacts to disrupt the most humble processes of existence, George Steiner recounts that "In Baghdad, during the Gulf War, the cocks crowed in shrill song the night through. But the brilliant light over the city was not that of the sun" (1996, 384).

References

Arnsperger, C., Larrère, C., & Ladrière, J. (1999). *Trois essais sur l'éthique économique et social*. Paris: INRA.

Heisenberg, W. (1959). *Physics and philosophy: The revolution in modern science*. London: George Allen & Unwin Ltd..

Husserl, E. (1970). *The crisis of European sciences and transcendental phenomenology* (D. Carr, Trans.). Evanston: Northwestern University Press.

Ladrière, J. (1997). *L'éthique dans l'univers de la rationalité*. Québec: Catalyses, Artel-Fides.
Plato. (1971). In E. Hamilton & H. Cairns (Eds.), *Collected dialogues*. Princeton: Princeton University Press.
Smolin, L. (2014). *Time reborn*. New York: Houghton Mifflin Harcourt.
Steiner, G. (1989). *Real presences*. Chicago: University of Chicago Press.
Steiner, G. (1996). *No passion spent*. New Haven: Yale University Press.

Apeiron Civilization: The Irruption of Infinity in Science and the Universe

Ion Soteropoulos

The expanding sphere. (Courtesy of Vassarely)

1 Human Progress: The Infinitization of Our Nearby Finite World

We conceive the progression of our human civilization as an infinitization process that consists of expanding the finite sphere of our surrounding observable world to the farthest point in the infinite universe. The infinitization of our nearby finite

I. Soteropoulos (✉)
Apeiron Centre, Paris, France
e-mail: ion.soteropoulos@apeironcentre.org

C. A. Hornbuckle et al. (eds.), *Phenomenology of the Object and Human Positioning*, Analecta Husserliana 122, https://doi.org/10.1007/978-3-030-66437-4_19

world—which we take to be coextensive with our local biological and terrestrial worlds hierarchically ordered by Euclidean time—had already begun in in the fifteenth century with the work of the great rationalist philosopher Nicolas of Cusa. He was followed by the philosophers Giordano Bruno (sixteenth century), René Descartes (seventeenth century), Henry More (seventeenth century), Baruch Spinoza (seventeenth century), Isaac Newton (seventeenth to eighteenth century), and Pierre-Simon de Laplace (eighteenth to nineteenth century).

Regardless of the different conceptions of infinity held by these rationalist thinkers—that infinity is reduced into a finite world varying indefinitely with time (Nicolas of Cusa, Bruno, Descartes) or that infinity is the maximum or infinite universe that has the ontological properties of the Divine Being and is given at once to our infinite reason (More, Spinoza, Newton, Laplace)—they had the following common aim: to open and extend the hermetically closed frontiers of the narrow sphere of the familar world of our finite particular senses via the expansive and creative power of the *apeiron*.

Now, at the beginning of the twenty-first century, the infinitization process of our finite world forces us to enter its second phase: the decline of the finite analytic paradigm on which our finite civilization (including our positivist science, hierarchical society, and finite neuronal brain) is based and its gradual replacement with the infinite synthetic paradigm conceived by our infinite mind (the Greek nous, νοῦς) and its faculty of infinite synthetic reason.

By extending our finite world to the farthest point, which is the limiting boundary of the infinite universe, human existence aims to reach the level of completeness (τελειότητα) the infinite universe principally has without losing its human individuality. The term we use to designate this cosmic end of completeness via the infinitization process is the *Pythagorean project of human development*. However, insofar as human progression takes place within the Euclidean finite analytic paradigm, it is an indefinite progression deprived of the capacity to reach the limiting boundary of the infinite universe. Despite the successive infinitization of our finite world, we are still inside our indefinitely expanding finite world and at an infinite distance from the limiting boundary of the infinite physical universe.

To break free of the incompleteness of our indefinitely progressing finite world and effectively reach the completeness of the infinite universe, we must transform the Euclidean finite analytic paradigm of our finite civilization into its antithesis, the non-Euclidean infinite synthetic paradigm on which the emerging *apeiron civilization* will be based.

1.1 The Analytic Principles of Our Finite Particular Senses

The finite analytic paradigm began with Aristotle (fourth century BCE)—the father of the analytic principles of thought—who postulated these principles by observing the way our finite senses behave in a state of rest. The analytic principles are (1) the *principle of self-identity* (reflexive identity), which holds that everything a is equal to itself: $a = a$; (2) *the principle of contradiction*, which holds that nothing is both a and a' (not-a): $(aa')'$ or $(a = a')'$ and that if $a' = b$, then by substituting a' for b we obtain $(ab)'$ or $(a = b)'$;[1] (3) *the principle of the excluded third*, which holds that everything is either a or b: $a + b$;[2] and (4) *the principle of inequality and irreversible temporal order*, which holds that given any two spatially different things $a \neq b$, we have either $b < a$ or $a < b$.[3]

According to the Euclidean finite analytic paradigm, which constitutes the positivist and dogmatic side of science, the universe (everything) is assumed to be simple or to be made of simple parts—the individuals—arranged in succession and deprived of any immediate contact among them that would allow them to unite and communicate. They are, therefore, isolated individuals, which we geometrically represent by discontinuous, finite, simple points. The universe of isolated individuals, represented geometrically by the line of simple points, the mathematicians call *arithmetic continuum* in which "the multiplicity alone subsists; the unity has disappeared" (Poincaré, 1893, 26).[4] In the line of points, the number and nature of points (individuals) are determinate and verify analytic principles of organization. These simple principles are erroneously considered by positivist science and its finite analytic paradigm as necessary and universal

[1] The principle of contradiction also states the impossible coexistence of opposites in the same individual subject. This is expressed either as the impossible conjunction (contact) of opposites $(ab)'$ or as the impossible unity and equality of opposites $(a = b)'$. The symbol of negation (′) designates "not," "false", "impossible or absurd"; it may also designate the opposite (or inverse) term—say $a' = 1/a$, which we call b—and the opposite (or inverse) relation R' between two terms.

[2] The symbol of exclusive disjunction (+) designates either/or. The principle of the excluded third is a principle of one-way decision and selection. On the basis of this principle, we arbitrarily choose one alternative at the expense of another. Of the two alternatives, one and only one alternative is true. In other words, either alternative a is true or alternative b is true: $a + b = (a = 1) + (b = 1)$, where = designates the copula is, and 1 stands for "true".

The above choice generates the determinate character of knowledge. The price to pay for this determination is the partiality of the determinate knowledge involving a loss of one of the alternatives and therefore a loss of information.

[3] The symbol of difference ($\neq$) designates the relation of symmetric difference between any two things a and b. The symbol of asymmetric inequality (<) designates the relationships of comparison (smaller/greater), succession (before/after), implication (if/then or cause/effect), inclusion (contained part/containing whole), and hierarchy (primary/secondary). Here the symmetric spatial difference between any two things a and b is reduced by our time-conditioned, finite, analytic senses to a relationship of asymmetric inequality, temporal order, and conflict.

[4] Mathematicians distinguish the arithmetic continuum, which is a multiplicity without unity, from the geometric continuum, which is unity in multiplicity. It is precisely this unity or intimate bond among the parts of the geometric continuum that transforms their arithmetical multiplicity into a geometric integrated whole or universe (Poincaré, 1893, 26). Etymologically speaking, the word universe is the composition of the Latin words *uni*, meaning one, and *versus*, meaning the inverse. The universe is simultaneously the *one* and its inverse, the *infinitely many*. The universe is therefore a geometric continuum. However, when we look at the infinite universe with our finite senses, we reduce it to an indefinitely expanding finite universe, which is an arithmetic continuum—that is, a multiplicity without unity.

principles of nature (*physis*) *qua* whole, that is to say, as *physical principles* that govern the physical universe, whereas in reality they express the way our finite senses function in a state of rest.

1.2 Light Is Finite and Travels Empty Space in a Straight Line at the Constant Finite Speed c

Another primitive assumption of the finite analytic paradigm is that light travels through empty space in a straight line at the speed c, which is taken as a constant, simple, and finite unit: $c = 1$. Because of the finiteness of the speed of light, the communication between any two things a and b is delayed and erroneous; it involves a loss of information and a loss of communication along the way. Thus, the signal received by the "subject a" is not the signal emitted from the "object b". In fact, the received signal, which we can designate by a, does not carry information about the object b as it is now—the real object; instead, it carries information about an object b as it was in the past but does not exist now, in other words it carries information about an object that has been modified during the communication process at a non-instantaneous finite speed. It follows, then, that the received signal a, contains less information about the object b than its emitted signal b in conformity with the second law of thermodynamics regarded as the thermodynamic expression of asymmetric time. We have thus the asymmetric temporal order and inequality $a < b$, where a, designates the received signal (or the information content of the received signal) and b, designates the emitted signal (or the information content of the emitted signal).

1.3 The Finiteness of the Speed of Light Generates Asymmetric Time Between Any Two Things a and b

Because of this loss of information and communication, the "subject a" perceives the real "object b" not as it is "now" but instead as it was in the past—before a—or as it will be in the future—after a—depending on whether the communication was directed backward (case of perception) or forward (case of action). Therefore, we reduce the symmetric spatial difference between points $a \neq b$ into an asymmetric temporal order, where the "object b" appears either before or after the "subject a" and the "subject a" appears either after or before the "object b":

$$b < a \text{ or } a < b, \tag{1}$$

The above formula defines the delayed communication between any two things a, b at the finite speed of light c and in a finite universe conditioned by asymmetric time.

The relation of temporal order between the "subject *a*", and the "object *b*" determines the solipsistic state of epistemological isolation in which what the "subject *a*" experiences immediately is the conscious behaviour of its finite brain (that is, the phenomenon or received signal *a*), whereas it ignores empirically the distant real object *b* (that is, the noumenon or emitted signal *b*) even though it knows *b* intellectually via its infinite synthetic reason (Kant, 1991, 187–188).[5] The time difference $b - a > 0$ measures the error, entropy, loss of information or communication that the process of delayed communication involves.

Another conflicting divide produced by the finiteness of the speed of light is the one between the *infinite universe b*, which we define as the sum total of an infinite number of parts, and the *indefinitely expanding finite universe* a_n, defined as the sum of a finite number of parts. With center a_0 (us) and radius a_0a_n of finite magnitude, we draw a finite circle $a_n = (a_0, a_0a_n)$ whose radius increases in proportion to the indefinite progression of our finite observation in deep space.

This finite circle represents our indefinitely expanding universe a_n, which constitutes the finite observable part of the infinite physical universe *b* revealed by the finiteness of the speed of light. The finite speed of light *c* is the finite observable part of light's real speed captured by our Euclidean finite sense organs—or, rather, by our finite brain at rest, which we define as a collection of nerve cells at ground (room) temperature. In this sense, the expanding finite universe a_n is a relative, contingent, observer-dependent universe, and we call it the *observable*, *sensible*, or *phenomenal universe*.

1.4 The Progressive Expansion of the Finite Universe Under a Unique Expansion Force

At the beginning, the expanding finite universe a_n included our nearby biological and terrestrial worlds. With center $a_0 = 10^{-4}$ m, which is roughly the size of a human egg cell or the lens of a human eye, and the finite radius $a_0a_{11} = 10^{-4} \times 10^{11}$ m $= 10^7$ m, which is roughly the radius of our planet earth, we draw a finite circle $a_{11} = (a_0, a_0a_{11})$ incorporating our familiar biological and terrestrial worlds. Subsequently, the finite circle $a_{11} = (a_0, a_0a_{11})$ is expanded by: (1) 16 powers of 10 to incorporate our solar system of finite radius a_0a_{16}, roughly equal to $10^{-4} \times 10^{16}$ m $= 10^{12}$ m; (2) by

[5]According to Kant (*Critique of Pure Reason* 1991, 187–188) the real object *b*—namely the thing in itself (τό καθ'αὐτό)—is not a phenomenon of our senses, or an object of a sensuous intuition, but a *noumenon* of our infinite nous (νοῦς), whose faculty of infinite synthetic reason *intellectually intuits* the real object. Thus, we claim that what appears to our finite senses is the received signal *a*, which is the phenomenon of the real object, whereas the emitted signal *b*, which corresponds to the real object in itself remains unknown to our finite senses but known—a *noumenon*—to our infinite nous and its faculty of infinite synthetic reason. This in turn generates the positive time difference $b - a > 0$, that is, the time delay between the emitted signal (noumenon) *b* and the received signal (phenomenon) *a* involving a loss of information about the real thing (noumenon) *b*.

24 powers of 10 to incorporate our Milky Way of finite radius a_0a_{24}, roughly equal to $10^{-4} \times 10^{24}$ m = 10^{20} m = 10^4 LY (10,000 light years); (3) by 26 powers of 10 to incorporate our local cluster of galaxies of finite radius a_0a_{26}, roughly equal to $10^{-4} \times 10^{26} = 10^{22}$ m = 10^6 LY (1 million light years); (4) by 28 powers of 10 to incorporate our local supercluster of galaxies of finite radius a_0a_{28}, roughly equal to $10^{-4} \times 10^{28} = 10^{24}$ m = 10^8 LY (100 million light years); and (5) by $n = 30$ powers of 10 to incorporate the totality of superclusters of finite radius a_0a_n, approximately equal to $10^{-4} \times 10^{30} = 10^{26}$ m = 10^{10} LY (10 billion light years), which constitutes our local observable universe, equally called the Hubble universe (see Fig. 1).

At the very great cosmological distance 15×10^9 LY (15 billion light years) from the center a_0 (us), the finite universe a_n is expanding at the light speed c perceived as a constant, simple and finite unit $c = 1$, which, according to the finite analytic paradigm of science, places absolute unsurpassable limits to our observation.[6] Because we are incapable of transcending the finite light barrier, we are unable to solve the cosmological problem of whether our finite universe a_n at the ultra-great cosmological scale $\approx 10^{10}$ LY (roughly 10 billion light years) is going to keep expanding without limit or stops and contracts into the limiting point (boundary) b—what cosmologists call the cosmological singularity of zero volume, infinite curvature, and infinite matter-energy-information density—which either begins or ends the indefinitely expanding universe a_n.[7]

[6] If the universe a_n designates the Hubble universe with the Hubble finite age $t = 1.5 \times 10^{10}$ Y = $1.5 \times 3 \times 10^{17}$ s, and the Hubble finite radius $r = a_0a_n = 1.5 \times 10^{10}$ LY = 1.5×10^{26} m, which is simultaneously the finite distance from us d, then on the basis of Hubble's law

$v = kd$,

in which the Hubble constant k is 3×10^4 m/s/1.5×10^{22} m (Jean Heidmann, 1982), we conclude that the speed of expansion v of the finite Hubble universe at this ultra-great cosmological distance from us is:

$v = 3 \times 10^4/1.5 \times 10^{22} \times 1.5 \times 10^{26} = 3 \times 10^8$ m/s.

The above speed of expansion is equal to the observable speed of light c taken as a constant, simple, and finite unit: $c = 3 \times 10^8$ m/s = 1.

[7] As a matter of fact, the general relativity theory predicts that any finite body—for example, the finite universe a_n, which accelerates at the finite speed of light c—has its matter-energy become infinite, its volume become zero, and its matter-energy density become infinite. Cosmologists call this infinitized object *cosmological singularity*. It is a limiting point b of infinite matter-energy density—a fire point—that violently ends or originates the finite body—for example, the indefinitely expanding finite universe a_n. We have here the *evolutionary singularity theory of the universe*, in which the singularity is identified to the chronological end or chronological beginning of the time-conditioned finite universe a_n, and is tantamount to the violent destruction or generation of its finite body.

The limiting point b can also be considered as a complex absolute whole comprising the indefinitely expanding finite universe a_n and transforming it into an infinite universe b, which is constant, balanced, and timeless. We have here the *steady state singularity theory of the universe*, in which the singularity *qua absolute whole* is identified to the ontological first or final cause and unifying principle of the universe. In this case, the limiting point b is not a violent fire point that destroys the finite body; rather, it is a fire point that continuously energizes or animates the energy degrading and eventually inanimate, material, finite body assigning to it eternity and infinity. The negative conception of evolutionary singularity is developed by positivist (or empiricist) science; per contra, the positive conception of steady state singularity is developed by rationalist science.

Indefinitely Expanding Finite Universe

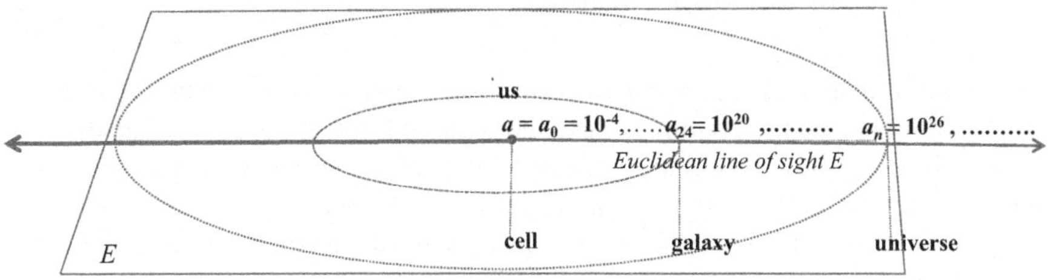

Fig. 1 With center $a = a_0$ (us, as human cell) and finite radius a_0a_n, we draw a finite circle $a_n = (a_0, a_0a_n)$ lying on the horizontal Euclidean plane E and representing the finite universe a_n. Under a unique outward force—the force of expansion (or repulsive force)—the finite circle a_n expands at an accelerating speed proportional to its distance from the center a_0. At the ultra-great cosmological distance 1.5×10^{26} m = 15×10^9 LY (15 billion light years), the finite circle $a_n = (a_0, a_0a_n)$ expands at the finite speed of light c and represents our observable universe a_n of finite radius a_0a_n, equal to 1.5×10^{26} m = 15×10^9 LY. Because it is impossible to perceive via our finite senses speeds greater than the finite speed of light, that is to say, an infinite super luminous speed, the region beyond our observable universe remains unobservable and hence empirically unknown: We do not know whether the expansion of our observable universe is unlimited expanding to an unlimited plane, or limited contracting into a limiting point that halts the expansion

Our inability to decide whether or not our expanding universe a_n has the property of limiting (stopping) its expansion at this ultra-great cosmological distance constitutes the cosmological problem and the source of our unlimited ignorance. Are we doomed for eternity to ignore the ultimate and real structure of our expanding universe, or is there a possibility to transform our unlimited ignorance into infinite knowledge?

2 The Cosmological Problem: Is Our Universe Limited or Unlimited?

The super-system of the galaxies is dispersing as a puff of smoke disperses. Sometimes I wonder whether there may not be a greater scale of existence of things in which it is no more than a puff of smoke.—Arthur Eddington[8]

[8] See Arthur Eddington, 1987. *The Expanding Universe: Astronomy's "Great Debate."* Cambridge: Cambridge University Press.

2.1 *At the Utmost Limits of Distance the Universe Is Indeterminate and Complex*

One of the fundamental problems of cosmology is whether our expanding universe (space-time) is going to continue to expand without limit as a dynamic Euclidean or hyperbolic plane or will stop its expansion and contract into a limiting point of infinite curvature sometime in the future (relative to forward action) and sometime in the past (relative to backward perception). Is our universe (space-time) unlimited or limited, open or closed in regard to space-time, to its remote future or distant past?

According to the finite analytic paradigm, a complete cosmological theory must have in principle the ability to answer these questions in one definitive way because our expanding universe is assumed to be simple and determinate. However, observational data show us that it is impossible to solve the problem with one simple answer. Indeed, measurements of the *current mean density* of matter within the expanding universe at the ultra-great cosmological distance $\approx 10^{26}$ m $\approx 10^{10}$ LY show us that it is roughly equal to the *critical density*, which is about 1 hydrogen atom per cubic meter of space (Abell, 1980; Barrow, 1999). By critical density we understand the density that the universe would have if all the atoms of its stars and galaxies could be distributed with complete uniformity.

According to cosmologists, a universe with a critical density of matter is a universe in true balance between its expansion energy, which stretches the universe into an infinite plane, and the gravitational energy of its contained matter, which contracts the universe into a limiting point that stops the expansion sometime in the past or future.[9] This means that at the very great cosmological distance $\approx 10^{10}$ LY $\approx 10^{26}$ m

[9] The actual mean density of matter in the expanding universe is its universal density if the hydrogen atoms of its total mass could be scattered about with complete uniformity. If the total mass of our expanding finite universe of radius $r = a_0 a_n = 1.5 \times 10^{26}$ m $= 1.5 \times 10^{10}$ LY is roughly 10^{52} kg, then its actual mean density $D = m/r^3 \approx 10^{52}/(1.5 \times 10^{26})^3 \approx 10^{-27}$ kg/m$^3 \approx 10^{-30}$g/cm^3 is roughly 1 hydrogen atom per cubic meter of space; its actual mean energy density is roughly $E = mc^2 \approx 10^{-27} \times (3 \times 10^8)^2 \approx 10^{-10}$ joules per cubic meter of space, which is the total energy locked in 1 hydrogen atom. Now, this actual mean density is the same with the critical density of a critical universe in balance and harmony between its expansion energy and the gravitational energy of its contained matter. Cosmologists designate the ratio of actual mean density to critical density by Ω (omega). If the actual mean density is equal to the critical density, then Ω is equal to 1.

If $\Omega < 1$, then the expanding universe is an under-dense universe in which the expansion energy overcomes the gravitational energy of its matter. This unbalanced and time-conditioned finite universe accelerates its expansion so that eventually it stretches into an unlimited Euclidean plane up to its mater-energy dissolution and death by a great freeze. In a universe dominated by the expansion energy, or the cosmic force of repulsion (repulsive force or repulsive gravity), the universe, life, and communication are impossible. In fact, how is it possible to continue to live and communicate if the universe's matter-energy is infinitely diluted and any two parts of the universe are infinitely separated by the cosmic force of repulsion?

If $1 < \Omega$, then the expanding universe is an over-dense universe in which the gravitational energy of its contained matter prevails and curves the universe. This unbalanced and time-conditioned universe slows down its expansion so that it eventually contracts into a dimensionless limiting point and dies by a big hot crunch. In both cases, the prevalence of one energy or cosmic force

a part of the indefinitely expanding universe a_n infinitely curves into a limiting point—a cosmological singularity b, which is not the violent origin or end of the indefinitely expanding universe but, rather, the limiting boundary enveloping the entire indefinitely expanding universe a_n without destroying its unlimited expansion. This limiting boundary transforms the indefinitely expanding universe a_n into an infinite universe b, which is both an unlimited plane and a limiting point enveloping the unlimited plane. Thus, the cosmological singularity b is not a chronological boundary—a chronological origin or end of the indefinitely expanding a_n—but an ontological boundary—an ontological origin or principle—that converts the indefinitely expanding Hubble universe a_n deprived of unity into an infinite universe b endowed with unity and called real, physical universe (see Fig. 2).

Because it is impossible to decide whether the real physical universe is limited or unlimited, we conclude that the physical universe is indeterminate or undecidable, and therefore neither an unlimited plane nor a limiting point enveloping the unlimited plane. Accordingly, at the cosmological scale $\approx 10^{10}$ LY $\approx 10^{26}$ m, our real physical universe defies the primitive assumption of simplicity and determinacy of the finite analytic paradigm.

Indeed, if according to the finite analytic paradigm the universe (everything) is simple and determinate, then it is governed by the analytic principle of the excluded third:

$$a + a', \tag{2}$$

on the basis of which the universe (everything) is either a or a' (not-a), either limited or unlimited relative to its distant past or future. The analytic principle of the excluded third is then, one that postulates the alternative existence of opposites. Observation, however, indicates that at the ultra-great cosmological distance $\approx 10^{10}$ LY $\approx 10^{26}$ m where the expanding Hubble universe a_n becomes an infinite universe b—a maximum—it is false that the universe is governed by the analytic principle of the excluded third: We write therefore:

$$(a + a')', \tag{3}$$

where (′) is the sign of falsity but also the sign of negation or complement.

over the other generates imbalance, asymmetry, and the eventual death of the universe. To conclude: The coexistence of these opposite tensional inequalities and unbalanced universes generates our balanced infinite universe, which is simultaneously a critical universe whose omega is 1 and therefore is a universe free of asymmetric time and imbalance by comprising the totality of unbalanced and time-conditioned universes: $(\Omega = 1) = (\Omega < 1)(1 < \Omega)$.

The unbalanced universes are therefore time-conditioned parts of a timeless, balanced universe, which on the whole is indeterminate, neutral or impartial by virtue of being complex and comprehensive comprising the totality of parts. Only in a permanent and balanced universe that is simultaneously limited and unlimited, closed and open, neither limited nor unlimited, neither closed nor open is continuous life and communication possible.

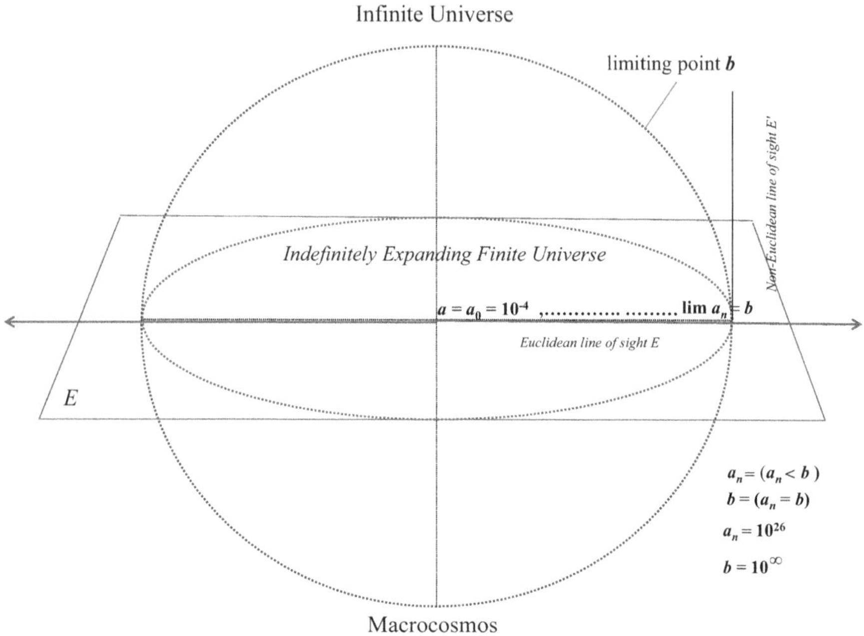

Fig. 2 At the ultra-great cosmological distance $\approx 10^{26}$ m, a part of the indefinitely expanding Hubble universe a_n, represented by the finite circle lying on the horizontal Euclidean plane E, is infinitely curved into a limiting point b enveloping the entire finite universe. This envelopment transforms the finite Hubble universe a_n according to extension into an infinite universe b—an infinite macrocosmos—represented by the infinite sphere whose outer limiting boundary is the limiting point b (cosmological singularity) of infinite curvature. Relative to the Euclidean line of sight E, the finite magnitude a_n, is less than the maximum or infinite magnitude b such as $a_n = (a_n < b)$; relative to the non-Euclidean line of sight E', which is at right angles to the Euclidean line of sight E and at the limiting point b, the finite magnitude a_n is equal to the infinite magnitude b such as $b = (a_n = b)$. On the whole, i.e., at the limiting boundary b, we have $\lim a_n = b$, where $a_n = (a_n < b)$, $b = (a_n = b)$, $a_n = 10^{26}$ m and $b = 10^{\infty}$

If it is false that the analytic principle of the excluded third governs the universe at the cosmological distance $\approx 10^{10}$ LY $\approx 10^{26}$ m, then its negation, namely the synthetic principle of the included third is the true principle that governs the universe at this ultra-great cosmological distance.

On the basis of the formula 3, which we consider at the same time as the synthetic principle of the included third, we assert that the universe (everything) is neither (not-either) a nor a', neither limited nor unlimited. If we develop the formula 3, we obtain:

$$\left(a+a'\right)' = a'\left(a'\right)' = a'a, \tag{4}$$

in function of which the universe (everything) is both a' and a, both unlimited and limited. In other words, in the universe, the opposite determinations a and a' are both false such as $a'(a')'$ and both true such as $a'a$.

Opposites that are simultaneously false and true are said to be coexisting opposites. Thus, the negation of the analytic principle of the excluded third is equivalent to postulating the coexistence of opposites in function of which the universe (everything) is neither limited nor unlimited and both unlimited and limited. A universe, which is neither limited nor unlimited is an indeterminate, neutral universe transcending the opposites, whereas the universe which is both limited and unlimited is a complex whole comprising the opposites.

The infinite universe b is a balanced and permanent universe free of asymmetric time and coextensive with the metaphysicians' whole of infinite space or infinite whole containing everything there is at once. It constitutes the complete and real structure of the indefinitely expanding finite universe a_n and is actually known uniquely *a priori* through our mental faculty of infinite synthetic reason belonging to our infinite mind (nous, νοῦς). In fact, through our synthetic reason we intuit intellectually and immediately, with infinite and zero speeds, the whole of infinite space. In this sense, we may think of our infinite mind as the infinite universe subjectified (interiorized), and the infinite universe as the infinite mind objectified (exteriorized).

Now, if it is false that the infinite universe is governed by the analytic principle of the excluded third, then in virtue of the rule of *modus tollens*, we conclude that it is also false that the infinite universe is simple and determinate. In fact, based on our formula 3, we have demonstrated that the universe is *indeterminate* and *neutral* by virtue of being neither limited nor unlimited as well as synthetic or *complex* by virtue of being both unlimited and limited (see formula 4).

In the face of this conflict between observational data and the finite analytic paradigm of dogmatic science, there are two options: (1) to conserve the finite analytic paradigm and assign a negative meaning to the indeterminacy, neutrality, and complexity of the universe, which is synonymous to the production of ontological and epistemological absurdities such as the ontological impossibility of the universe and the epistemological impossibility of knowing the universe, or (2) to consider indeterminacy, neutrality and complexity as the very real and proper properties of the universe thought of as an infinite whole. In this second case, in which indeterminacy, neutrality and complexity have positive meanings, we are forced to replace the inadequate finite analytic paradigm by its antithesis—the infinite synthetic paradigm, considered to be the true paradigm of the indeterminate, neutral, and complex universe. Let us investigate both options and see their consequences with respect to science.

2.2 From a Negative to a Positive Meaning of the Indeterminacy and Complexity of the Universe

If anything that exists and is known is either limited or unlimited, then the negation of the analytic principle of the excluded third amounts to both the negation of the existing thing and the negation of the knowledge of the existing thing. It follows, therefore, that if our universe is at the very great cosmological distance $\approx 10^{10}$ LY $\approx 10^{26}$ m neither limited nor unlimited and both unlimited and limited, then it does not exist; and even if we assume that the universe does exist, it is impossible to know its real and complete structure, which the finite analytic paradigm postulated to be simple and determinate and therefore either limited or unlimited.

This negative meaning of the indeterminacy and complexity of our universe leads us directly to *ontological irrealism*, which is the doctrine affirming the irreality or nonexistence of the universe, and to *epistemological skepticism*, which is the doctrine that asserts the existence of the universe but the impossibility of knowing and deciding its real physical structure, whether it is unlimited or limited, open or closed, with respect to its distant past or remote future out there at the great cosmological distance $\approx 10^{10}$ LY $\approx 10^{26}$ m and its cosmological time equivalent $\approx 10^{10}$ Y $\approx 3 \times 10^{17}$ s.

During the growth of science, this epistemological impossibility gave rise to a variety of well-known theories of skepticism. For example, it engendered Nicholas of Cusa's *De Docta Ignorantia* (On Learned Ignorance) (1985), which affirms the impossibility of formulating a simple and determinate representation of the universe or nature *qua* universe; Kant's transcendental idealism, which affirms the impossibility of knowing the world as a whole existing in itself, and being either finite or infinite; Turing's unpredictability problem, which affirms the impossibility of predicting or deciding whether a given program in the machine will continue running forever or will cease to run; Gödel's incompleteness theorem of truth, which affirms the impossibility of deciding whether a proposition inferred from the axioms of a formal mathematical theory of the world is true or false;[10] and Stephen Hawking's anti-metaphysical positivism, which affirms the impossibility of reaching a complete and comprehensive mathematical theory of the universe formulated in terms of a finite number of first principles that will enable us to determine its structure at any time and stop the indefinite regression of our scientific theories (Hawking, n.d.)

[10] According to Gödel, if a consistent and formal mathematical theory of the universe (of everything) contains an indeterminate or undecidable proposition, one that is neither true nor false, neither proved nor disproved, then it is an incomplete theory. Because all consistent formal mathematical theories of the universe have undecidable propositions, it follows that our knowledge of the universe is incomplete and that it is impossible to state a univocal or sensible proposition about the universe that is either true or false. Gödel discovered the ultimate indeterminacy of all consistent mathematical theories. Instead of concluding the non-formal and non-analytic nature of the consistent mathematical theories of the universe, he preferred to interpret negatively their indeterminacy and consider it a sign of their inherent incompleteness that prevents us from ever knowing the real structure of the universe, which he tacitly assumed to be simple and determinate. For an excellent philosophical presentation of Gödel's work see Palle Yourgrau's *A World Without Time*, 2005.

All of these aspects of skepticism derive from one general epistemological impossibility: that of the finite analytic paradigm of dogmatic positivist science to give a unique, definitive answer to the cosmological questions of the general analytic form: Is the physical universe or nature *qua* universe *a* or *a′*? Is our physical universe limited or unlimited? Impossibilities and paradoxes arise whenever we expose the intrinsically complex and indeterminate physical universe to our simple and determinate analytic questioning undertaken by our finite analytic understanding reflecting the analytic functioning our finite neuronal brain.

To get out of this ontological and epistemological nihilistic trap in which nothing (no physical universe) exists and even if everything (the physical universe) exists, we know nothing about anything, we choose the option of assigning a positive meaning to the indeterminacy, neutrality, and complexity of the physical universe and consider them as real and immanent properties of the real physical universe itself. We are then forced to change via a 180° rotation the inadequate finite analytic paradigm into its antithesis. We obtain, therefore, the diametrically opposite infinite synthetic paradigm of non-dogmatic science.

According to the infinite synthetic paradigm, the universe (everything) is complex or is made up of complex parts—the universes—arranged in coexistence and having immediate contact among them that allows continuous communication. We can represent these complex universes by continuous lines. The term mathematicians use for the universe of universes—represented geometrically by the line of lines—is the *geometric continuum*, in which unity and infinite multiplicity coexist. In this sense, the geometric continuum is the universe and, conversely, the universe is the geometric continuum.

An etymological analysis of the word *universe* indicates that it derives from two Latin words: *uni*, "one," and *versus* (past participle of the verb *vetere*, "to turn"), meaning the inverse of one. The universe is simultaneously the *one* and its inverse—the *infinitely many*—thereby confirming, in etymological terms, the complex character of the universe.

In his *Metaphysics*, Aristotle clearly distinguishes the "universe" from the "individual." Anything admitting contraries (coexisting opposites) is a universe belonging to the class of universes; anything that admits contradictories and hence opposites in succession is an individual belonging to the class of individuals.[11]

In the line of lines, the number and nature of lines (universes) are indeterminate; this line of lines verifies synthetic principles of organization regarded by non-dogmatic science to be the real and true principles of the infinite physical universe existing in itself (τό καθ'αὐτό) independently of the individual observer. In this sense, synthetic principles are necessary, universal principles of the infinite universe; analytic principles, on the other hand, are contingent perceptual principles expressing, according to Aristotle, the finite analytic behavior of our sense organs—of our nerve cells—principles he erroneously took as the necessary, universal principles of nature.[12]

[11] See *La Métaphysique,* I, 10, 1058b, 30–35.

[12] For example, according to Aristotle, "No sense applied to the same object tells us that this object is simultaneously so and not so" (*La Métaphysique*, Γ, 51010b, 20, 1986). Converting this observation into a principle of thought, it becomes the analytic principle of contradiction, which stipulates

We have the case of *epistemological error* (from which the infinite synthetic paradigm liberates us) when Aristotelian dogmatic science projects these accidental analytic principles into the physical infinite universe and considers them as if they were necessary physical principles that govern its structure. In reality, however, they are accidental perceptual principles that govern the finite analytic behavior of our neuronal brain at rest.

2.3 The Synthetic Principles of the Real Physical Universe

An infinite synthetic principle is the principle of the included third, which holds that the universe (everything) is neither (not-either) a nor a': $(a + a')'$. We note here that the principle of the included third is nothing but the negation of the analytic principle of the excluded third presented in a positive manner—that is, as a necessary and universal principle that postulates the neutrality, impartiality, justice, and balance of the physical infinite universe. Indeed, the infinite universe, is free from the partiality of the individual observer who reduces the infinite universe possessing neither part a and nor part a' into a finite universe arbitrarily selecting through the finite speed of light $c = 1$ one of the alternative parts—namely either a or a'.

The principle of the included third can be also presented in the form of the synthetic principle of equivalence. Indeed, when the universe (everything) is neither a nor a', then these opposites are both false and true (see formula 4). If opposites are both false and true, then they are coexisting opposites—that is, equal opposites. From the synthetic principle of the included third, we derive the following equality of opposites:

$$(a+a')' = a'(a')' = a'a \rightarrow a = a'. \qquad (5)$$

According to the analytic principle of contradiction, this equation is an antinomy. However, if we assign a positive meaning to this antinomy, we transform it into the synthetic principle of equivalence, which stipulates the equality, unity, or coincidence of opposites (*coincidentia oppositorum*) that governs anything conceived of as a complex universe. As a matter of fact, according to the Greek Ionian philosopher Heraclitus (sixth to fifth century BCE), the synthetic principle of the unity of opposites, which he identified with the divine synthetic logos, governs the physical universe, or nature *qua* universe. The infinite physical universe is not only the unity of opposites but also, ultimately, the unity of all unities connecting everything with everything else. For Heraclitus, this natural unity of opposites—this physical and

that nothing is simultaneously a and a' (not-a). However, this analytic principle of contradiction is not a immanent and necessary principle of nature *qua* universe—or what is the same of the real physical universe itself—but a contingent principle of the analytic (digital) mode with which our finite particular senses operate when they observe the physical universe.

divine synthetic logos—is hidden (φύσις κρύπτεσθαι φιλεῖ, nature loves to hide) by lying beneath (or beyond) the surface of the observable world. He considered, therefore, the comprehensive and synthetic view as divine and cast it in opposition to the individual view, which, insofar as it is dependent on our finite analytic sense organs, is destructively chaotic, contradictory, one-sided, and unbalanced.

Another way of presenting the synthetic principle of equivalence, which is the synthetic logos or in modern terms the synthetic reason, is in the form of the principle of reflexive order (self-order). Let us define the unity or equality of opposites as the product of opposite but coexisting inequalities:

$$(a = a') = (a < a')(a' < a). \tag{6}$$

If in the right side of the equation we replace a' by its equal a, we obtain the principle of reflexive order (self-order), or spontaneous order defining the state of freedom (autonomia):

$$(a = a') = (a < a)(a < a). \tag{7}$$

On the basis of this, we deduce that the universe (everything) governed by the synthetic principle of equivalence is simultaneously governed by the synthetic principle of reflexive order (free order) and vice versa. Equivalence and freedom are correlative, mutually dependent principles.

According to the principle of reflexive order, the universe (everything) a is both cause and effect of itself and thus is free of a constraining external cause:

$$a < a, \tag{8}$$

where $<$ is the sign of inequality (before/after) and causal implication (cause/effect).

This self-causality or immanent causality determines the complete and first being—called the Divine Being—in continuous self-creation at every moment, what the ancient thinkers of the East (the Greek Ionian and Indo-Iranian philosophers of the sixth to the fifth centuries BCE) identified with the infinite universe.

Finally, the synthetic principle of reflexive order is also a principle of continuous free life according to which the living being a not only has the power of self-order and self-motion (i.e. to be cause and effect of its proper motion) but also has the power to continue to be before itself and after itself. The divine infinite universe is therefore a continuously free living being a producing life from life, order from order in conformity with the principle of self-order equally called a principle of self-similarity.

3 The Theoretical Solution of the Cosmological Problem: The Synthetic Finite–Infinite Equivalence Principle

The universe is an infinite sphere of radius 1, which under the action of contrary forces, is both infinitely extended as to be infinite and infinitely contracted as to be zero.—Ion Soteropoulos

Having exposed the primitive assumptions and principles of the infinite synthetic paradigm, we now employ them to comprehend the complete and real structure of our expanding universe. Based on the principle of the included third, the complete and real structure of our universe is neither limited nor unlimited because it is in exact balance between its opposite energies, the expansion energy generating the unlimited and the gravitational energy of its contained matter generating the limited. This indetermination of our universe, far from expressing our inability to decide its real structure (which the finite analytic paradigm assumes to be determinate and simple), manifests our power of comprehending it once we assume the opposite—namely, that its real structure is indeterminate, neutral and complex.

When Kant concluded in his *Critique of Pure Reason* the impossibility of determining whether the universe is limited or unlimited, he interpreted this cosmological indetermination (undecidability) negatively as a sign of our inability to know the universe or its structure as a whole existing in itself (τό καθ 'αὐτό) and which we call the real physical universe or the real structure of the universe.[13]Accordingly, if the universe is a whole existing in itself—that is to say a real physical universe, then, grounded in the finite analytic paradigm, Kant predicted that the universe must be either limited or unlimited, and therefore determinate and simple verifying analytic principles of organization. However, it has been shown (by synthetic reason or by observation) that the universe as a whole, that is, at the cosmological scale $\approx 10^{10}$ LY, is neither limited nor unlimited and hence is indeterminate and complex. We then conclude together with Kant that the universe—the content of all phenomena—is not a whole existing in itself—and therefore is not a real physical universe. It follows that the content of all phenomena, which we call phenomenal universe, is just a mere representation of the real physical universe.[14]

If we denote the incomplete phenomenal universe by a_n and the complete real universe by b, we express the contradiction (analytic opposition) between the phenomenal universe and the real universe by the following inequality and temporal order: $a_n < b$, where insofar the phenomenal universe a_n is just a mere representation of the real physical universe, it must have less reality, truth and information content

[13] The real physical object is that which exists in itself and independently of a particular observer. Its properties are therefore immanent, absolute and necessary. On the other hand the phenomenal object or simply the phenomenon is that which is dependent on or in relation to a particular observer. Its properties are extrinsic, relative and accidental depending on the mode of being observed. The real physical object is free of observation (phenomenon) by virtue of comprising the totality of observations (phenomena).

[14] See Kant's *Critique of Pure Reason. Chapter II. The Antinomy of Reason. Section VII. Critical Solution of the Cosmological Problem,* 304.

then the real physical universe b. Thus, the finite analytic paradigm divides our universe into two contradictory, isolated parts ordered in succession, (1) the incomplete, but empirically known phenomenal universe a_n whose indeterminate and complex structure is interpreted negatively, that is, as a source of incompleteness and ignorance and the complete but empirically unknown real physical universe b whose determinate and simple structure is interpreted positively, that is, as a source of completeness and knowledge.

Now this fundamental conflict and temporal order disappears once we replace the finite analytic paradigm employed by Kant by its negation the infinite synthetic paradigm and assume that the real physical universe or the real structure of the universe is neither simple nor determinate but complex and indeterminate. In this case, the Kantian argument runs in the opposite direction leading to the dissolution of the analytic contradiction between phenomena and things in themselves, between the phenomenal universe a_n and the real physical universe b.

On the basis of the above considerations, we develop the following counter-argument: If the universe is a whole existing in itself—called the real physical universe—then grounded in the infinite synthetic paradigm we predict that the universe must be neither limited nor unlimited, and therefore must be indeterminate and complex verifying synthetic principles of organization. And this is precisely what we discovered by means of our synthetic reason and observation. For instance, we discovered through observational data that the universe as a whole, that is, at the cosmological scale $\approx 10^{10}$ LY, is neither limited nor unlimited and hence is indeterminate and complex. Therefore the universe—the content of all phenomena—which we called phenomenal universe is a whole existing in itself, and therefore is a real physical universe. We postulate therefore the following equality: $a_n = b$, where insofar the phenomenal universe a_n and the real physical universe b are ultimately, i.e., on the whole, equivalent, they have the same reality, truth and information content. It follows that every phenomenon is a real thing and that the real physical universe or the real structure of the universe is the sum total of all phenomena. Accordingly, if the opposites a and a', namely the limited and the unlimited, are phenomena, their total sum defines the real physical universe or the real structure of the universe, which is a neutral whole:

$$a + a' = 0. \tag{9}$$

At the same time the above formula written as $(a + a')'$, defines the synthetic principle of the included third, which insofar it is an indetermination principle, it is the very negation of the analytic principle of the excluded third—known as the principle of the absolute determination of anything existing as an individual thing.

3.1 The Finite-Infinite Equivalence Principle Governs the Limiting Boundary of the Universe

We have shown in the formula 5 how the principle of the included third can be converted into the synthetic principle of equivalence stipulating the equality and simultaneity of opposites. If the opposites a and a' are the limited and the unlimited, then the synthetic equivalence principle becomes the synthetic finite-infinite equivalence principle, which is another correlative principle that governs the real physical universe or the real structure of the universe. The synthetic finite-infinite equivalence principle stipulates that everything limited is unlimited and conversely, that everything unlimited is limited. This principle holds true on the cosmological scale $\approx 10^{26}$ m according to extension, and by symmetry on the quantum scale $\approx 10^{-34}$ m according to division. Both scales are ±30 orders of magnitude further away from our familiar cellular scale $\approx 10^{-4}$ m. At these maximum scales we meet the limiting boundary of the infinite physical universe—its outer limiting point b or cosmological singularity geometrically unifying all oppositions through its infinite curvature.

In virtue of the finite–infinite equivalence principle governing the limiting boundary of the infinite physical universe the finite magnitude a_n and the infinite or maximum magnitude b are equal:

$$\lim a_n = b, \tag{10}$$

where $a_n = (10^{26}, 10^{-34})$ and $b = (10^{\infty}, 10^{-\infty})$.

The equality or proportion between the above finite and infinite magnitudes tells us at what scale of magnitudes (according to extension and division) the finite and the infinite are equal or united such that the finite becomes infinite and the infinite becomes finite. Thus, it is sufficient to reach the finite magnitudes 10^{26} m and 10^{-34} m in order to attain respectively their equivalent infinite magnitudes $10^{\infty} = \infty$ and $10^{-\infty} = 0$; and conversely, the infinite magnitudes $10^{\infty} = \infty$ and $10^{-\infty} = 0$ are actualized respectively through their equivalent finite magnitudes 10^{26} m and 10^{-34} m. The term we use for these finite magnitudes is *gateways to infinity*. They capture and enclose infinity in order to assign it to finite beings (see Fig. 3).[15]

At the very great cosmological and quantum distances $\approx 10^{26}$ m $\approx 10^{-34}$ m and at right angles to our Euclidean line of sight, which is the non-Euclidean line of sight, we claim that maximum contrary energies or forces F, F' act on the universe. Under the maximum action of the expansion energy or repulsive force F, the universe is maximally expanded into an unlimited Euclidean plane of infinite magnitude.[16] On

[15] As a matter of fact, this was the original idea of the Pythagorians who considered the infinite as being grasped and enclosed by the finite in order to assign infinity to beings (see Aristotle's *Physics* III(4)203a_{10}, 1996).

[16] In its static state of zero energy or force, the Euclidean space-time has a zero curvature, whereas in its dynamic state of energy or force—for example of expansion energy or repulsive force—the Euclidean space-time has a negative curvature, which is a false curvature. Because space-time with a negative curvature is a hyperbolic space-time, it follows that the Euclidean space-time is, in its dynamic state a hyperbolic space-time—a space-time with a negative or false curvature, which ultimately is equivalent to no curvature.

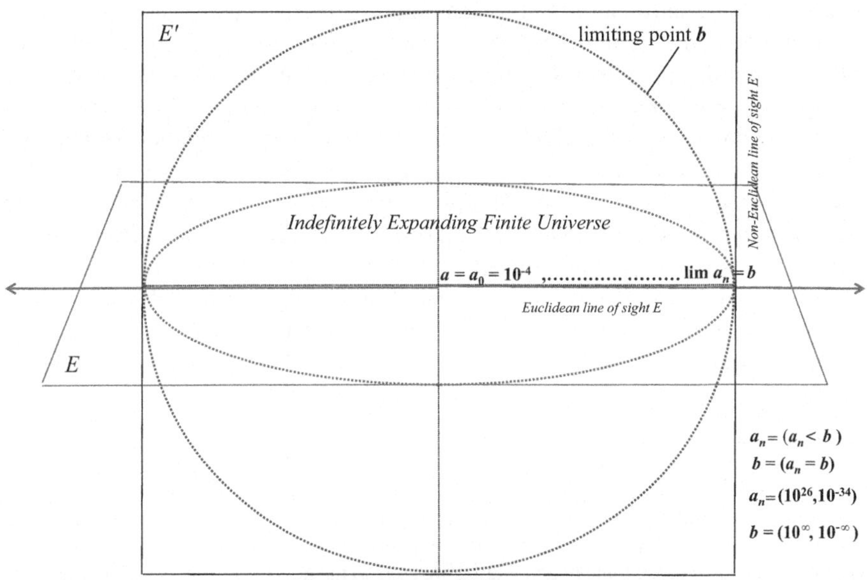

Fig. 3 At the ultra-great cosmological and quantum distances $\approx 10^{26}$ m, 10^{-34} m a part of the indefinitely expanding universe a_n, represented by the finite circle lying on the horizontal Euclidean plane E, is infinitely curved into a limiting point b enveloping the entire finite universe. This envelopment transforms the finite universe a_n according to extension and division into an infinite universe b—an infinite macrocosmos and microcosmos—represented by the infinite sphere whose outer limiting boundary is the limiting point b (cosmological singularity) of infinite curvature. Relative to the Euclidean line of sight E the indefinitely varying finite magnitude a_n, is less than the maximum or infinite magnitude b such as $a_n = (a_n < b)$; relative to the non-Euclidean line of sight E', which is at right angles to the Euclidean line of sight E and at the limiting point b, the finite magnitude a_n is equal to the infinite magnitude b such as $b = (a_n = b)$. On the whole, i.e., at the limiting boundary b, we have lim $a_n = b$, where $a_n = (a_n < b)$, $b = (a_n = b)$, and $a_n = (10^{26}$ m, 10^{-34} m), $b = (10^{\infty}, 10^{-\infty})$

the other hand under the maximum action of the gravitational energy or attractive force F' of its contained matter, the universe is maximally contracted into a limiting point b of zero volume. This limiting point b—this cosmological singularity of infinite curvature and infinite matter-energy density—embraces the unlimited Euclidean plane, which now becomes the inner part of an infinite sphere b embedded in the highest dimension—the dimension zero. Because this infinite sphere—this body of the infinite universe—is embedded in the dimension zero, that is, in nowhere, it is necessarily self-embedded and self-contained. The infinite sphere, is then, a *maximum* that cannot be further increased (relative to the highest non-Euclidean line of sight) no matter how much we increase it relative to the low Euclidean line of sight.[17]

[17] For example, the finite magnitude 10^{26} has, in function of the equation $\lim 10^{26} = 10^{\infty} = \infty$ the property of the maximum or infinite magnitude $10^{\infty} = \infty$, which no matter how much we increase this magnitude relative to the Euclidean plane or line of sight E, it remains the same and constant relative to the non-Euclidean line of sight E'at right angles to the Euclidean line E:

The outer limiting point b of the infinite sphere envelops the whole of the Euclidean finite universe a_n and assigns limit, constancy, unity, and energy to it without destroying its unlimited expansion according to extension and division. Thus, the limiting point b (cosmological singularity) does not constitute an absolute *chronological origin* in the past or an absolute *chronological end* in the future of our expanding universe a_n (the standard thesis of evolutionary singularity cosmology); it constitutes instead an absolute *ontological origin*, *principle* (ἀρχή) or *telos* that geometrically unifies at any time all oppositions a and a', and all isolated parts of the expanding universe (non-standard thesis of the steady-state singularity cosmology). As a universal unifying principle, the limiting point b transforms the unbalanced indefinitely expanding finite universe a_n into a balanced unified infinite whole b containing the entire infinite series a_n of observable parts (phenomena, observable universes) at once.[18] In this way, continuous and immediate communication among

---

$$\left(10^{\infty} \times 10^{\infty} > 10^{\infty}\right)_E \left(10^{\infty} \times 10^{\infty} = 10^{\infty}\right)_{E'}.$$

Thus, if we multiply 10^{26} an infinite number of times, we obtain an infinite magnitude $10^{26} \times 10^{\infty}$, which is infinitely greater than the finite magnitude 10^{26} relative to the Euclidean line of sight E but equal to the finite magnitude 10^{26} relative to the non-Euclidean line of sight E':

$$\left(10^{26} \times 10^{\infty} > 10^{26}\right)_E \left(10^{26} \times 10^{\infty} = 10^{26}\right)_{E'}.$$

What holds true of the finite magnitude 10^{26}, which is stipulated as equivalent to the maximum or infinite magnitude according to extension $10^{\infty} = \infty$, also holds true for its diametrically opposite finite magnitude 10^{-34}, which is stipulated as equivalent to the maximum or infinite magnitude according to division $10^{-\infty} = 0$ such as: $\lim 10^{-34} = 10^{-\infty} = 0$. Accordingly, if we multiply 10^{-34} an infinite number of times, we obtain an infinite magnitude $10^{-34} \times 10^{-\infty}$, which is infinitely smaller than the finite magnitude 10^{-34} relative to the Euclidean line of sight E but equal to the finite magnitude 10^{-34} relative to the non-Euclidean line of sight E':

$$\left(10^{-34} \times 10^{-\infty} < 10^{-34}\right)_E \left(10^{-34} \times 10^{-\infty} = 10^{-34}\right)_{E'}.$$

[18] Here it is important to note that we have two meanings of the limiting point or boundary b of a body as a unified whole: (1) the limiting point b as the absolute beginning or end of the series of parts of the body, and (2) the limiting point b as the absolute totality of the series of parts of the body. These meanings were already mentioned by Aristotle in his *La Métaphysique* Δ, 17, 1022a when he defined the limiting point or boundary of a thing as (1) "the first point beyond which it is not possible to find any part of the thing, and (2) the first point within which all points are."

Other meanings of the limiting point b mentioned by Aristotle are the limiting point as: (1) the primary being or as the residence of the primary being; (2) the form of that which has magnitude; (3) the what (*quiddité*) or ultimate constituent of everything, which is the limit of knowledge but also the limit of the known thing; and (4) the common origin (ἀρχή) and unifying principle of all things.

Because the limiting point has infinite energy density, which animates inert, soulless and lifeless matter, metaphysicians identify the outer limiting point of the infinite universe with the infinite Soul, Mind, or Life of the universe as well as the first source or cause of its infinite motion. These different aspects of the limiting point are regarded as different properties of the primary being—the Divine Being—assimilated to the infinite universe, but also to an infinite individual having the properties of the infinite universe.

To these different meanings we will add two more: (1) the *bending or breaking point b* at which something becomes something else, and (2) the *complex and indeterminate point-boundary b* at which something a is at the same time something else a'(not-a).

the distant parts of the universe is maintained as constant and undisrupted despite the indefinite expansion of their distance.

Another fundamental characteristic of the infinite universe is that it remains eternally constant—for example, it has the same matter-energy density and the same ground temperature—no matter how much it varies (that is, expands or contracts) as a variable part. This happens because its variation is symmetric and balanced and occurs simultaneously according to extension and division and under equal and opposite energies or cosmic forces such as expansion and gravitational energies, or repulsive and attractive forces.

Insofar the infinite sphere $b = (a, ab)$ equal to itself and 1 is (under the action of contrary energies or forces) both maximally expanded into an unlimited plane and maximally contracted into a limiting point embracing the unlimited plane, its radius $ab = 1$ is the complex product of infinite and zero magnitudes:

$$\lim ab = 1 = 10^{\infty} \times 1/10^{\infty} = \infty \times 0. \tag{11}$$

The above *Zenonian unit*, which consists of infinite and zero magnitudes, is the real magnitude of the infinite sphere's radius $ab = 1$.[19] The infinite sphere of radius 1 is therefore the composition and mutual neutralization of two symmetric opposite infinite spheres: The sphere of infinite radius, which is the inner enveloped part projected on the horizontal Euclidean plane E, and the sphere of zero radius, which is the outer enveloping whole assimilated to the limiting point b, which is the outer limiting boundary of the infinite sphere. If we wish to visualize the dimensionless limiting point b, we project it on the plane E', which is at right angles to the Euclidean plane E (Fig. 3).

As we notice, the above Zenonian quantity contradicts the Aristotelian analytic postulation that all quantities are simple finite things measured at any time by a unique magnitude and that no quantity of a given kind has contrary magnitudes.[20] In fact, what it shows is that the unit radius of the universe (everything) is a complex infinite unit—a Zenonian unit—measured by contrary magnitudes.

---

[19] Zeno of Elea (fifth century BCE), the Greek philosopher and brilliant dialectician, is famous for asserting propositions that defy Euclidean analytic logic: For example, he affirmed that "if there are many things, it is necessary that they are both small and large; so small as not to have magnitude, so large as to be unlimited" (Kirk, Raven, & Schofield, 1988 267). In other words, he affirmed that the multiplicity of things implies that the same things have contrary magnitudes, namely zero and infinite magnitudes, at the same time. It is not clear whether Zeno considered the above inference as an antinomy against the hypothesis that there are many things or as a manifestation of the synthetic, indeterminate nature of the multiplicity—and, as we have shown here, of the universe (everything) itself.

[20] See Aristotle's Organon, Catégories, 6, 5*b*10–15, 1984.

3.2 The Reduction of the Infinite Sphere into an Indefinitely Expanding Euclidean Plane

A finite observer—for example our finite brain or mind located at the fixed center $a = a_0$ of the infinite sphere and looking along the Euclidean line of sight E would see, the infinite sphere $b = (a, ab)$ in a reduced and incomplete form—namely, as if it were an Euclidean finite circle $a_n = (a_0, a_0a_n)$ lying on an Euclidean finite plane E indefinitely expanding under the unique action of an expansion energy or repulsive force (Fig. 1). Indeed, we claim that our finite brain at rest and operating at ground (room) temperature captures the finite part of light's real speed, which is the observable speed c taken as a constant finite unity and defined as the product of finite frequency f and finite wavelength λ: $c = f \times \lambda = 1$. Because of the finiteness of the speed of light, our finite brain perceives only the inner Euclidean finite part a_n of the spherical infinite whole b. Determined uniquely by an expansion energy or repulsive force this finite universe a_n expands with respect to extension and division without limit—without a limiting point b to partially halt and stabilize its unlimited expansion and embrace and unify its indefinitely separating parts.

But this unbalanced and time-conditioned universe a_n, whether presented as a single universe or as a sequence of universes generated and destroyed by a sequence of limiting points (cosmological singularities), is not the true and real universe, as it can neither host life nor communicate among its isolated parts. In fact, any life and communication within such an unbalanced, one-way indefinitely expanding finite universe a_n would have already been either infinitely diluted and frozen to death with respect to extension or infinitely compressed and burned to death with respect to division.

Thus, the finiteness of the speed of light produces in our finite brain or mind an erroneous perception of the infinite universe b, which, reduced into the indefinitely expanding finite universe a_n, is less than the infinite universe b in conformity with the inequality and temporal order $a_n < b$, and has less information content and structure than the complete infinite universe itself, whose outer limiting point b envelops and closes its unlimited expansion according to extension and division without destroying it.

On the other hand, the positive difference $b - a_n > 0$ measures the magnitude of the erroneous perception of the infinite universe b, which can be infinitely small or infinitely great depending on perspective. In turn, the finite brain's erroneous perception of the infinite universe b generates the indefinite approximations of observational science, which we also denote by the finite variable a_n.

These indefinite approximations a_n employ the finite speed of light and finite analytic tools to measure and understand the infinite physical universe b, which as a maximum magnitude is neither finite, nor analytic, nor measurable because it is the end of measure. In fact, what we experience, measure, and analytically understand is the inner Euclidean finite part a_n of the infinite physical universe b, which is the observable, sensible or phenomenal universe a_n, whereas the real physical universe b—that is, the non-Euclidean infinite universe itself—is an unobservable (hyper-sensible) and unintelligible physical object of our mental faculties of finite particular sensibility and analytic understanding but an intelligible (noumenal) object of our infinite mind (nous, νοῦς) and its faculty of infinite synthetic reason.

3.3 Do We Have an Infinite Brain that Physically Corresponds to Our Mental Faculty of Infinite Synthetic Reason?

Thus, while our finite brain or mind cannot perceive the real physical object spherically, nevertheless we do have the mental faculty of thinking it synthetically and spherically—that is, as a complex infinite whole comprising the totality of alternatives a and a' and verifying synthetic principles of organization.[21] Does this mean that besides our finite brain or mind, we also have an infinite brain that physically corresponds to our infinite mind's mental faculty of thinking or imagining the physical object synthetically—and which Heraclitus called divine Logos, Anaxagoras called infinite Nous, and we actually call infinite synthetic reason?

If the basis of our finite brain or mind perceiving the universe analytically as an indefinitely expanding Euclidean part a_n, is neurobiological, can we then claim that the basis of our infinite brain or mind that thinks the universe synthetically as a spherical infinite whole b, is quantum mechanical? And if this is so, can we then claim, that, when we think or imagine synthetically through our faculty of infinite synthetic reason, we think through our brain's ultimate constituents—namely our mind-atoms, or mind-singularities of zero volume, and infinite curvature?[22]

Indeed, can we conjecture that when we have infinite synthetic thoughts, we somehow access through our mind-singularities to the real speed of light $c = 1$ (or the speed of real light), which reflecting the real magnitude of the universe, is the

---

[21] We assume the brain and the mind as equal and coexisting opposites. Thus the brain is the mind when seen by an internal observer, whereas the mind is the brain when seen by an external observer. This leads us to define the mind as the totality of internal parts (phenomena), and the brain as the totality of external parts (phenomena). In this sense thought or feeling (internal mental phenomenon) generates electromagnetic energy (external brain phenomenon) and electromagnetic energy (external brain phenomenon) generates thought or feeling. However, we do not know how this reversible conversion between these two entities, mind and brain, or what is the same between mental and electromagnetic events takes place.

[22] A mind-singularity is the limiting point b that constitutes the limiting boundary of our subjective infinite mind defined as the sum total of an infinite number of internal parts, which are both mental and electromagnetic phenomena. On the other hand the cosmological singularity is the limiting point b that constitutes the limiting boundary b of our objective infinite universe defined as the sum total of an infinite number of external parts, which are both material and electromagnetic phenomena. Regardless of whether the limiting point b is mental or cosmological, the smallest part contained in everything or the greatest whole containing everything, we conjecture that it radiates extreme lights that propagate at infinite and zero light speeds, at infinite and zero wavelengths at the ground frequency f and the ground Kelvin temperature T, which are the frequency and Kelvin temperature of infrared light. These extreme lights combined together constitute the real speed of light. This will be explained in my upcoming book on the real speed of light.

product of infinite and zero magnitudes or, the product of the infinite light speed c_∞, and the zero light speed c_0?

$$c = 1 = \infty \times 0, \quad c = c_\infty \times c_0 = 1. \tag{12}$$

Finally, if by an unknown yet manner we do access through our mind-singularities the real speed of light, when we think or imagine synthetically and instantaneously the infinite whole *b* at a maximum distance from us, why then don't we sense synthetically and instantaneously this infinite whole? Why don't we have an *infinite, synthetic and spherical perception* of the infinite universe that will allow us to effectively sense what we theoretically conceive and expect through our mental faculty of infinite synthetic reason?

Only if we resolve the above metaphysical question of acquiring an infinite synthetic perception of the universe, we will have then reached a complete solution to the cosmological problem of knowing intellectually and sensuously, or theoretically and empirically the universe's real structure. Some lines of solution will come by connecting our finite brain at rest, composed of nerve cells vibrating infrared light at ground (room) Kelvin temperature and frequency, with our mind-singularities vibrating light at infinite and zero wavelengths, at infinite and zero lights speeds.[23] How can our finite brain access these extreme lights and through them their infinite and zero light speeds composing the real speed of light $c = c_\infty \times c_0 = 1$, while maintaining constant its ground (room) Kelvin temperature and frequency? In other words, how can our finite brain become an infinite brain, which is nothing but our finite brain elevated to the level of its mind-singularities at ground Kelvin temperature?

All these open questions are some of the challenges the new generation of pioneering researchers will confront with the irruption of infinity in the universe and science. If they continue to adhere to the old finite analytic paradigm, they will fail to meet them. However, if they adopt the "new" infinite synthetic paradigm, which by the way, was the original intuitive paradigm of human thought at the dawn of philosophy about 2600 years ago, humankind will have made its first majestic leap toward reaching the level of spherical completeness the infinite universe has. As a matter of fact *anthropos* (ἄνθρωπος → ἀναθρών α ὄπωπε)—the Greek word for the human being—means the being who contemplates upwards (ἀναθρεί) what it has seen (α ὄπωπε).[24] This transcendental contemplation beyond our incomplete familiar

[23] If we maintain constant the ground frequency *f*, then the speed of light *c*, varies by means of $c = f\lambda$, in function of its wavelength λ. It follows that an infinite wavelength leads to an infinite light speed, whereas a zero wavelength leads to a zero light speed, such as:

$$c_\infty = f\lambda_\infty, \quad c_0 = f\lambda_0,$$

where c_∞, designates the infinite light speed, c_0, designates the zero light speed, λ_∞, designates the infinite wavelength, and λ_0, designates the zero wavelength.

The composition of infinite and zero light speeds gives the real speed of light $c = c_\infty \times c_0 = 1$. Mutually neutralized, the infinite and zero light speeds have no bearing on the finite light speed *c*, whose ground frequency *f*, and ground temperature *T*, remain constant.

[24] Plato. *Cratylus* 399c3–6, 1989.

world constrained by the analytic principles of our neuronal brain is accomplished by the divine infinite nous of *anthropos* and aims at replicating within anthropos and its familiar world the spherical completeness of the distant physical universe in agreement with the principle of self-similarity. In doing so, the human being can at last unfold akin to a cosmic flower in its infinite potential.

References

Abell, G. (1980). *Realm of the universe*. Philadelphia: Saunders College.

Aristotle. (1984). *Organon. Catégories* (Vol. 6, 5*b*10–15) (J. Tricot, Trans.). Paris: Librairie Philosophique J. Vrin.

Aristotle. (1986). *La Metaphysique* (J. Tricot, Trans.). Paris: Librairie Philosophique J. Vrin.

Aristotle. (1996). *Physics* (R. Waterfield, Trans.). Oxford: Oxford University Press.

Barrow, J. (1999). *Impossibility: The limits of science and the science of limits*. London: Vintage.

Eddington, A. (1987). *The expanding universe: Astronomy's "great debate"*. Cambridge: Cambridge University Press.

Hawking, S. (n.d.) *Godel and the end of the universe*. Retrieved March 20, 2019, from www.hawking.org.uk/godel-and-the-end-of-physics.

Heidmann, J. (1982). *Extragalactic adventure*. Cambridge: Cambridge University Press.

Kant, I. (1991). *Critique of pure reason*. London: Everyman's Library.

Kirk, G. S., Raven, J. E., & Schofield, M. (1988). *The presocratic philosophers*. Cambridge: Cambridge University Press.

Nicholas of Cusa. (1985). *On learned ignorance* (J. Hopkins, Trans.). Minneapolis: The Arthur J. Banning Press.

Plato. (1989). *Cratylus* (H. North Fowler, Trans.). Cambridge: Harvard University Press.

Poincaré, H. (1893). Le Continu Mathématique. *Revue de Métaphysique et de Morale, 1*, 26–34.

Yourgrau, P. (2005). *A world without time: The forgotten legacy of Gödel and Einstein*. Cambridge: Basic Books.

Man as the *Ambassador of the Cosmos*: Henryk Skolimowski's Concept

Anna Małecka and Katarzyna Stark

1 Cosmism and Eco-Humanism

Henryk Skolimowski can undoubtedly be considered the founder and one of the leading thinkers in the field of ecophilosophy. Born in Poland, he taught at Oxford, University of Southern California, and University of Michigan, among others. In the 1990s, at the Lodz University of Technology (Poland), he established and was the head of the Department of Ecophilosophy, the first unit of this type to emerge in the world.

As a philosopher, Skolimowski shares several general notions with Anna-Teresa Tymieniecka's eco-phenomenology, such as the concept of the fundamental unity of things, rationality/spirituality as a result of the evolutionary development of the universe, and the human spiritual accomplishment as having roots in a metaphysical tendency of life itself. The Polish thinker understands ecophilosophy primarily in terms of "cosmism." Thus, the foundation of his overall ontological views can be found in cosmology. As Zbigniew Wróblewski pertinently remarks: "Cosmology constitutes a central domain in the system of ecophilosophy, for—as the creator of the ecological philosophy claims—the acceptance of a particular standpoint in cosmology (either explicit or implicit) determines the development of a given philosophical system" (Wróblewski, 2001, 169). Skolimowski's fundamental cosmological idea is that of the living cosmos in its entirety, as opposed to the mechanistic and atomistic vision of the world composed of separated entities. Consistently, the role of its particular parts is to be considered in relation to the holistically perceived system, as it is only the wholeness that determines the shape of its parts and their interrelations within the system.

A. Małecka (✉) · K. Stark
AGH University of Science and Technology, Krakow, Poland
e-mail: amm@agh.edu.pl; stark@agh.edu.pl

C. A. Hornbuckle et al. (eds.), *Phenomenology of the Object and Human Positioning*, Analecta Husserliana 122, https://doi.org/10.1007/978-3-030-66437-4_20

Skolimowski's "living philosophy" is basically oriented toward life in its universal dimensionality. All of existence, human beings, plants, animals and elements constitute the mutually dependent manifestations of life in its evolutionary state of creative becoming. The author of *A Vision for the New Millennium* refers first of all to Teilhard de Chardin, but also shares a general approach with Henri Bergson, Hegel and even Heraclitus, i.e. the thinkers who pointed at the processual character of reality, with its higher developmental levels representing the actualised spirituality or logos. He conceives of evolution as "a creative force, which ceaselessly articulates life in ever new forms of consciousness" (Skolimowski, 1992, 18). The holistic standpoint imposes a new vision of man—the ecological man, i.e. the one who is aware of his unity with everything that exists and approaches universal life with esteem, and in whom the cosmos itself attains self-cognition and actualization of the ingrained potential spirituality and holiness.

Accordingly, considering the status and role of man in the cosmic plane, the Polish eco-philosopher rejects a traditional "humanism," which he associates with anthropocentrism. "Humanism," he says, "exalts man at the expense of other beings, while cosmism elevates all beings, as all beings contain a divine element" (Baczulis, 2001, 385). In his opinion, the only justified solution would be an "eco-humanism," which he understands as a return to "the human authenticity, (…) the discovery of our cosmic roots, i.e. the spiritual roots, the acceptance of the fact that we are part of the cosmos and that we have no other choice" (Baczulis, 2001, 383).

The author applies the term of "the cosmic solidarity" to denote close relation and community with the whole living and existing world. As he states: "There is no radical opposition: the world and us. One element co-creates another" (Skolimowski, 1991, 102). The necessity of the "cosmic solidarity" is justified, in Skolimowski's vision, by the sacred character of the cosmos and its evolution. Possessing the divine attributes, the universe should be treated as a sanctuary, with due reverence, and its holy potential needs to be creatively actualized in cooperation with, and for the sake of all of existence. Such is the role of wise humans, who fulfil the basic "commandments" of ecophilosophy. In the view of the presented "cosmism," the position of man is to be defined primarily in terms of his reverential and creative standpoint in relation to the cosmos.

In one of the interviews, the author of *Technology and Human Destiny* states: "I dare say that I feel to be a cosmic person. I believe that everything that is and that exists has grown out of the cosmos. I also have a courage to think in terms of the cosmos, and create—within this process—new horizons of thought. It is my conviction that I have reached the state of harmony in contact with the cosmos, feeling to be part of this great cosmic matter. Consequently, I can confidently say that I am an ambassador of the cosmos" (Baczulis, 2001, 375). Further, the author grants the metaphorical "ambassadorship" status to the entire humankind, even though not all humans are yet aware of their destiny.

The question arises of how to interpret the figurative term of "cosmic ambassadorship" in terms of human essential characteristics.

Firstly, man is considered as an outcome of the cosmic evolution, its crowning point, and as such is supposed to "represent" the entirety of being on the cosmic "forum." To employ another metaphor, he is supposed to be an "advocate" of the vital "interests" of the evolutionary process. He should well understand his place in the cosmic plan and be able to unite with the cosmos in harmony, living in symbiosis with it and promoting its vital developmental tendencies. He is also expected to accept the active and participatory position of his mind in creating reality: "We have to learn a new truth that the mind is not a passive and isolated organ, but constitutes an integral part of reality to such an extent that without its participation there would be no reality" (Skolimowski, 1985, 147). The human participatory mind constantly generates knowledge out of the cosmic amorphous mass of initial data, thus contributing to the evolutionary order and meaning-directed progress.

As a *homo creator*, man co-creates the cosmos in the unceasing process of transcending. The acts of transcendence embrace the creative process, which itself constitutes the heart of the cosmic evolution and is best expressed in the conscious endeavours of man (Baczulis, 2001, 383). Skolimowski considers transcendence to be a "driving force of the whole cosmos development" (Skolimowski, 2007, 75). Its essence (inherent in the very nature of the cosmos) lies in permanently aspiring to "go beyond," to transcend the limits, to build ever more complex and sensitive structures. Human life results from transcendence, and we ourselves impose meaning upon the cosmos and life in the acts of transcendence, which transform the world into the higher supra-physical being. Transcendence means "(…) the process of going beyond, and beyond, and beyond until we reach pure spirituality" (Skolimowski, 1992, 110). Owing to transcendence, we create a dynamic image of the universe, exceeding the limits of our biological existence, and aiming towards spirituality.

Skolimowski considers spirituality to be the highest form of the so-far evolutionary process having its source in the cultivation of various forms of sensitivity, ultimately aiming at sanctity-oriented perfection. Owing to it, the most intensive transformation of external reality is possible. The sensitizing man is a creation of the sensitizing universe (Skolimowski, 1991, 80). Such is the essence of the cosmic evolution: the humanisation and continuous sensitivization of matter (Skolimowski, 1985, 120). The spiritualization of the world is accomplished by meaningful human activity.

Man develops spiritually in every field of his activity, also in the domain of art. His ingenious artistic endeavours constitute the sublimated, spiritually transformed manifestations of the cosmic intrinsic transcendental tendencies: "Great art, being an expression of transcendental yearnings of man, is obviously an expression of the transcendental aspirations of the cosmos" (Skolimowski, 2007, 79). Maria Gołaszewska comments on Skolimowski's aesthetical pro-ecological attitude stating that it is manifested by the act of communing with works of art which render the natural beauty as well as in intensive immediate contact with the beauty of nature itself (Gołaszewska, 2001, 330). Peter Harris-Jones indicates the metaphysical or even theological foundation of beauty according to the Polish ecophilosopher: "In Skolimowski's interpretation, beauty is both a vehicle of the sacred and an inherent aspect of it. It is also undergoing an evolutionary ascent; its evolutionary radiance is such a way that a more beautiful object exists more intensively than a less beautiful one because it contains more life" (Harries-Jones, 1998, 436). For the author of *The Living Philosophy*, the aspiration towards the increasingly coherent and capable structures, typical for evolution, are envisaged and actualized in the very process of beautification of the world, which belongs to "the evolutionary story of life." Spirituality is forged out of the formlessness of biology in the act of opposing forces of disintegration, and summoning God to existence by attaining divinity within oneself (Skolimowski, 1985, 104). "Holiness is not a phantasmagoria," the thinker writes, "but constitutes a state of human existence in which light, love and altruism prevail" (Skolimowski, 2007, 203).

This poetically expressed ideal of "eco-holiness" sets the ethical rules and obligations of man in relation to the world. The ecological ethic relies on the principle of reverence for all forms of life and values created by evolution. Skolimowski says that we have to develop *reverential thinking* which he defines in the following way: "To think reverentially is first of all to recognize human life as an intrinsic value; is to recognize love as an essential and indispensable modality of human existence; is to recognize creative thinking as an inalienable part of human nature; is to recognize joy as an inherent part of daily living; is to recognize the brotherhood of all beings as the basis of our epistemological paradigm" (Skolimowski, 1984, 49). The author expands the scope of his joyful and benevolent reverential ethics to the whole universe so that it includes "the global eco-community" (soils, plants, animals, rock and other beings) (Skolimowski, 1984, 49).

Reverence involves cooperation with the cosmic forces in establishing the moral order and requires effort and sacrifice. The "cosmic ambassador" sometimes has to perform a heroic Promethean role if the values delineated by the eco-ethics are threatened. First of all, the ecological man bears in mind the burden of his *global* responsibility. His reverence and worship for any single creation is tantamount to the liability not only for his own life, but also for future generations, and the future of the whole planet (Skolimowski, 1993, 30). As the author of *The Third Millenium* states, it is this "cosmic responsibility" that accompanies the evolutionally crystallized moral sense: "Responsibility perceived in the vaster cosmic plane is a late achievement of evolution. It does not emerge until self-awareness is born alongside the moral sense, the feeling that we are responsible not only for our own ego, but also for the whole world. The so-understood responsibility is a form of altruism" (Skolimowski, 1985, 100).

As an altruistic moral agent responsible for the rest of the world, man performs the role of a guardian in relation to all manifestations of life. The philosopher recalls the concept of a "reliable protector" authored by Tadeusz Kotarbiński (one of his philosophical masters). In Skolimowski's version, a reliable protector or guardian assists others, provides all beings with an unfailing custody and support, is trustworthy and sympathetic. The Polish eco-philosopher uses yet another poetical term to describe the human role as a moral guardian, "a shepherd": "We are the shepherds of humankind. (…) This responsibility implies care, reverence, and constant custody over everything that exists" (Skolimowski, 1993, 135). We should also serve as "curators of this magnificent museum created by evolution whose part we are ourselves" (Skolimowski, 1991, 95), preserving and protecting the values deposited in the universe.

The basic values of eco-ethics neither come from god nor are purely subjective or conventional—their source lies in the evolutionary process itself, in the process of our becoming the species. If the Earth is treated as a temple, the ethical principle of moderation and eco-justice is to be followed. Eco-justice applies to all and everything and implies the ecological liability for all entities mutually linked in the cosmic dimension. Ignacy S. Fiut stresses the validity of the Polish eco-philosopher's idea of universal justice for the well-being of the planet: "Skolimowski's postulate of general justice as a universal regulatory idea possesses character of categorical imperative available to man. It gives him the opportunity of deciding about his being and the being of his environment, without which, in turn, his being and the being of other animated creatures becomes problematic" (Fiut, 2013, 119). Skolimowski's global ethics is life-oriented, like his whole philosophy, and is even considered as "life-giving." He criticizes both the traditional absolutist and situational ethics, which in the view of his holistic life worship and solidary ideas lose a justifying ground (Tyburski, 2001, 133).

Following the basic precepts of Skolimowski's eco-ethic and his "evolutionary imperative," it can be concluded that the ambassador of the cosmos, the man fully conscious of his unity with the world and his responsibility for it, behaves in such a way as

> to preserve and enhance the unfolding of evolution and all its riches, (…) to preserve life and enhance life which is a necessary condition for carrying on evolution; (…) to preserve and enhance the ecosystem, which is a necessary condition for further enhancement of life and consciousness; to preserve and enhance the capacities which are the highest developed form of the evolved universe: consciousness, creativeness, compassion; (…) to preserve and enhance human life which is the vessel in which the most precious achievements of evolution are contained. (Skolimowski, 1984, 49)

This imperative is concisely expressed in Skolimowski's "eleventh commandment": "The Earth is Lord's and His is everything. Thou shall not rob the Earth and destroy its life" (Skolimowski, 1993. 130).

However, Skolimowski's final words in this field seem to pass beyond the limits of ethics itself. He writes: "When man really matures and becomes perfect, no ethics will be necessary. Because ethics are only to help imperfect man on his way to perfection" (Skolimowski, 2007: 88). Thus, the Polish philosopher introduces an ideal of the spiritual superhuman, whose perfect consciousness of the oneness with the cosmos will make all moral imperatives redundant. It can be said that the ambassador of the cosmos aspires to become the cosmically divine person.

2 Evolution Towards Divinity

The author of *Technology and Human Destiny* employs the term "theosynthesis" to render the idea of sacralisation of the cosmos out of its own inherent creative resources and emphasizes the role of man as "the focus and lens of the great creative energies of the cosmos" (Skolimowski, 2007, 53). The creativity of the cosmos leads to its sacralisation through several stages of the evolution of the primordial form of being—the light: the phases of photosynthesis, logosynthesis (where life is introduced into the sphere of reflective thinking in ethics, religion, art, and philosophy), and theosynthesis, in which light is transformed into a sacred being. Through man, light achieves self-cognition, and aspires at the actualization of the "potency to further intensification of divinity" (Skolimowski, 2007, 24). An important stage of the cosmos sacralisation (theosynthesis) includes the establishment of institutional religions, but the very concept of divinity is ingrained in the natural structure of human psyche. It is this intrinsic conviction that enables us to perceive the cosmos as holy: "Only when we approach the universe with a reverential attitude and behold it with a mind that is sacred, do we find the universe sacred" (Skolimowski, 1992, 234). As Skolimowski writes:

> We are fragments of emerging divinity. Spirituality and divinity appear at the end of the process of spiritualization of matter, rather than at the beginning. We are the actualizing God and we are bringing Him to being, so to speak, by actualizing the sensitivity-sacredness-divinity latent in us. God is at the end of the road. We are its awkward, dim, unpolished fragments for the time being. (Skolimowski, 1992, 110)

In the phase of theosynthesis, the cosmos itself summons the so-called "coryphaeuses of the sacred sphere"—monks, saints, pilgrims, yogis, divinised persons, who achieve enlightenment, and accordingly actualize the cosmic divine potential in the fullest way. By perfectly transforming themselves, their holistic reverential attitude and creative potential, by total submission to their nature aspiring at divinity, such sanctified persons gain the power to transform the cosmos into the truly holy being. The final words of ecophilosophy become the message of eco-theology, and the "ambassador" assumes a position of the divine.

References

Baczulis, S. (2001). Jestem ambasadorem kosmosu. Wywiad z Profesorem Henrykiem Skolimowskim. In A. Papuziński & Z. Hull (Eds.), *Wokół ekofilozofii. Księga jubileuszowa ofiarowana Profesorowi Henrykowi Skolimowskiemu dla uczczenia siedemdziesięciolecia urodzin* (pp. 375–385). Bydgoszcz: Wydawnictwo Akademii Bydgoskiej.

Fiut, I. S. (2013). Henryk Skolimowski's idea of eco-development. *Dialogue and Uniersalism, 33*(4), 105–126.

Gołaszewska, M. (2001). Doświadczenie ekologiczne. Szkic z pogranicza estetyki, antropologii filozoficznej i ekologii. In A. Papuziński & H. Zbigniew (Eds.), *Wokół eko-filozofii* (pp. 325–335). Bydgoszcz: Wydawnictwo Akademii Bydgoskiej.

Harries-Jones, P. (1998). Aesthetics and ecology: A nonfictional view. In P. D. Murphy (Ed.), *Literature of nature: An international sourcebook* (pp. 434–439). Chicago: Fitzroy Dearborn.

Skolimowski, H. (1984). Eco-ethics as the foundation of conservation. *The Environmentalist, 4*(Suppl. 7), 45–51.

Skolimowski, H. (1985). *Technika a przeznaczenie człowieka*. Warszawa: Wydawnictwo ETHOS.

Skolimowski, H. (1991). *Medytacje o prawdziwych wartościach człowieka, który poszukuje sensu życia*. Wrocław: Wrocławska Oficyna Wydawnicza "Astrum".

Skolimowski, H. (1992). *Living philosophy: Eco-philosophy as a tree of life*. London: Penguin/ Arkana.

Skolimowski, H. (1993). *Nadzieja matką mądrych*. Łódź: Akapit Press.

Skolimowski, H. (2007). *Geniusz Światła a Świętość Życia* (Seria wydawnicza: Biblioteka Filozoficzna). Warszawa: Oficyna Wydawnicza Vega.

Tyburski, W. (2001). Idea etyki globalnej Henryka Skolimowskiego. In A. Papuziński & H. Zbigniew (Eds.), *Wokół eko-filozofii* (pp. 125–135). Bydgoszcz: Wydawnictwo Akademii Bydgoskiej.

Wróblewski, Z. (2001). Zarys metodologicznej charakterystyki Eko-filozofii Henryka Skolimowskiego. In A. Papuziński & H. Zbigniew (Eds.), *Wokół eko-filozofii* (pp. 169–192). Bydgoszcz: Wydawnictwo Akademii Bydgoskiej.

The manufacturer's authorised representative in the EU is Springer Nature Customer Service Centre GmbH, Europaplatz 3, 69115 Heidelberg, Germany. If you have any concerns regarding our products, please contact ProductSafety@springernature.com

Printed and bound by CPI Group (UK) Ltd, Croydon, CR0 4YY

30/09/2026

02225497-0016